Resampling-Based Multiple Testing

Resampling-Based Multiple Testing

Examples and Methods for *p*-Value Adjustment

PETER H. WESTFALL
Texas Tech University
Lubbock, Texas

S. STANLEY YOUNG
Glaxo Inc.
Research Triangle Park, North Carolina

A Wiley Interscience Publication
JOHN WILEY & SONS, INC.
New York • Chichester • Brisbane • Toronto • Singapore

This text is printed on acid-free paper.

Library of Congress Cataloging in Publication Data.

Westfall, Peter H., 1957–
Resampling-based multiple testing: examples and methods for p-value adjustment / Peter H. Westfall, S. Stanley Young.
p. cm. — (Wiley series in probability and mathematical statistics. Applied probability and statistics)
"A Wiley-Interscience publication."
Includes bibliographical references and index.
ISBN 0-471-55761-7
1. Resampling (Statistics) I. Young, S. Stanley, 1943– II. Title. III. Series.
QA278.8.W47 1992
519.5′4—dc20 92-22194

To Marilyn, Amy, and Dr. & Mrs. R. D. Westfall
—P.H.W.

To Pat, David, Nicki, and Dr. & Mrs. D. A. Young
—S.S.Y.

Contents

Preface

The accurate interpretation of statistical data is a concern of physicians, politicians, sociologists, engineers, and scientists everywhere. A problem that recurs in research studies, on which these professionals depend, is the extensive analysis of data. Modern computing equipment makes extensive analysis quite inexpensive, relative to the cost of obtaining the data. Once the data is available and on the computer, researchers question and analyze it from every possible angle, to miss no information.

The result of such extensive data analysis, or "data mining" as it is sometimes called, is the increased chance of inaccurately interpreted data. In particular, spurious results may be claimed to be real; even the actual directions of associations may be misinterpreted, so the opposite directions are claimed as actual. Such false conclusions mean that results are not repeatable. Lack of repeatability is well-known to scientists, but they usually attribute the problem to faulty experimental procedures or to uncontrolled variables. To curtail these problems, scientists focus attention on the refinement of experimental technique. However, incorrect conclusions as easily arise from extensive data analysis and manipulation.

Our purpose in writing this book is to provide theory, examples, and software to address the multiple testing dilemma while highlighting the pitfalls of excessive data analysis. We hope to impress upon the users of statistics the following points: (1) The effects of multiple data manipulation and tests are real and important. (2) It is difficult to know whether bad data or data manipulation caused false conclusions. (3) Though problems due to faulty data are uncontrollable, the multiplicity problem can be controlled. (4) Finally, the method we propose for controlling multiplicity is intelligible, convenient, easy to display and interpret, and often readily accessible through statistical software.

Much of this book will be considered controversial, particularly by researchers in the academic arena, where "publish or perish" is an axiom. There is a common criteria for publication in scientific journals: a probability level, or p-value, of less than .05 is required to establish significance. Therefore researchers seek significance from their data sets to the point of "mining" the data to obtain a "statistically significant" result. It is not difficult to find such a result when extensive data analysis is performed: if 20 aspects of a data set are considered, it is *expected* that one result will be significant, even if no effects are real. Coupled with the "publish or perish" axiom that drives researchers, the multiplicity problem, in our view, has led to the dissemination of many dubious results in scientific journals. We support our contention by discussing several articles that have appeared in scientific and public media, studies whose claimed associations may easily be spurious.

Even though there are many questionable results in recent publications, there is currently a trend in the statistical and scientific community to discount the multiplicity phenomenon and its effects. Taken to its extreme, this trend may cause scientists and experimenters to ignore completely the multiplicity problem. Appropriate use of multiple testing is a difficult and controversial subject; however, ignoring the problem will make it much worse. In this book we recommend that the multiplicity problem always be recognized, and we offer strategies for multiplicity adjustment.

This book has another controversial aspect. Some readers may consider the book as promoting "p-value culture," or the distillation of all statistical information into p-values, which simply summarize tests of significance. We intend no such endorsement. Rather, to expedite understanding, we believe that graphical and other tabular analyses should support the summaries of statistical tests. Such displays are regularly provided in this book. There are inherent difficulties in significance testing methodology, but it is widely used. The dilemma of multiple testing is partly a consequence of the p-value culture which is entrenched in the mainstream of scientific research.

Our general technical solution for the multiplicity dilemma is the use of resampling methods. These methods solve many problems in multiple testing that either have been ignored or handled very crudely because adequate methodology and software were not available. Resampling techniques can incorporate relevant correlation and distributional characteristics. They often yield less conservative (more powerful) and/or more robust tests than the commonly used procedures. The class of models to which the technology applies is very broad: most types of ANOVA models for continuous data (including MANOVA and heteroscedastic models) can be analyzed; models for multivariate binary data sampled in the ANOVA framework also can be analyzed. Additionally, the methods are easily tailored to more complex methods.

While resampling is a powerful and simply applied tool, it is not without pitfalls. We provide guidelines for the appropriate construction of tests using resampling methods. A practical benefit of the resampling approach is that results are defined naturally and displayed as "adjusted p-values." Such results provide a convenient, unified method to showcase the results of multiple testing. An additional benefit of using resampling technology is that the analyst can actually *see* the false significances occurring in the simulated data sets. Visualization provides an empirical understanding of how false significances happen in actual studies. Faced with a long list of p-values from a statistical summary, the analyst's natural inclination is to scan the list for the smallest p-value. Resampling replicates this process: new lists of p-values are produced with resampling, and the lists are repeatedly scanned to seek the smallest p-value. The data analyst's method thus is replicated, and the operating characteristics of the scanning and selection process are discovered.

The "adjusted p-value" method is superior to the usual methods for multiple testing and multiple comparisons. These employ complex calculations followed by a "table look-up." Consumers of statistical data are uncomfortable with such methods: they are difficult to use and the results are not easily displayed; often, relevant tables either are approximations or are non-existent. By contrast, adjusted p-values are as easy to display as ordinary (unadjusted) p-values. Furthermore, resampling methods incorporate dependence structures of the data and tests. While the two approaches—table look-ups and adjusted p-values—are sometimes equivalent, the resampling approach extends analysis to a broader class of problems for which tabled values do not exist. A feature distinguishing this book from virtually all applied statistics books is *there are no tables of distributions*! On the other hand, numerous real, complex data sets are supplied and analyzed.

The techniques are easy to implement. Software currently exists, and the methods are fairly simple to program, using many types of mathematical and statistical software. The SAS/STAT MULTTEST procedure performs many of the analyses in this book. Algorithms are conveniently implemented using high-level programming languages: we used SAS/IML, GAUSS of Aptech Systems, Inc. (1992), MATRIX (Alexander, 1989), and RESAMPLING STATS (Simon and Bruce, 1991a). Many of the analyses were programmed using a 486 PC-compatible machine. Even simulation analyses of resampling methods (which can be extremely time-consuming) were performed using this machine. Processing times were often less than 12 hours—overnight. For routine resampling analyses, results ordinarily were obtained in minutes or even seconds.

PROC MULTTEST is the culmination of a considerable development effort. Several pharmaceutical companies funded the effort through the Pharmaceutical Manufacturers Association. In addition, statisticians from these companies worked on the project. Two predecessor programs, PROC MBIN and PROC MTEST were programmed by Youling Lin of Texas Tech University, who also contributed fast algorithms and suggestions on theory. PROC MULTTEST was donated to SAS Institute Inc. to make it available to the public. The software and a supporting theory paper, "P-Value Adjustments for Multiple Tests in Multivariate Binomial Models," have received two awards. A paper describing methodology of PROC MBIN received the "Best Statistics Paper" award at the 1988 SAS Users' Group International (SUGI) conference. "P-Value Adjustments for Multiple Tests in Multivariate Binomial Models" appeared in the *Journal of the American Statistical Association* and won the "Best Outstanding Applications Award" from the American Statistical Association in 1991. Although originally designed to address a narrow problem of the pharmaceutical industry, the techniques and their generalizations address a broad range of practical problems.

In this book we discuss many applications, including animal carcinogenicity testing, epidemiological and other surveys, human side effects analysis, and drug efficacy studies. To demonstrate that the ideas are useful beyond the Biostatistics arena, we include applications from finance and baseball! Our intent is to illustrate that resampling technology provides a generally acceptable solution to multiplicity problems arising in a wide range of applications.

For theoreticians, we provide justifications for the use of resampling-based multiple testing, showing that the methods are powerful and robust. General guidelines for the construction of resampling-based multiple testing rules in some unusual situations are also provided.

This book is meant to *supplement* rather than *supplant* the book *Multiple Comparison Procedures*, by Hochberg and Tamhane (1987). The foundations for many of the ideas in this book are given in *Multiple Comparison Procedures*. Many alternative methods are discussed in that book as well.

This book may be used as a reference book for self-instruction, or as a textbook for a seminar course at the upper division or graduate level. Students who have taken elementary statistics using the resampling approach will find the methods of this book very natural. Students with computer skills will find it fun to program the methods and algorithms of this book, while others can perform analyses using PROC MULTTEST. We recommend using a companion book with many data sets, such as *Applied Statistics: Principles and Examples*, by Cox and Snell (1981), to compare the resampling-based multiplicity-adjusted analyses with standard analyses.

While later chapters in the book build sequentially from material in earlier chapters, we invite readers to read ahead, particularly to the examples in Chapter 7. There are many examples, tables, and figures throughout the book, and readers may refer to earlier mathematical developments as needed. We assume all readers are familiar with basic statistical concepts such as t-test, p-value, and ANOVA. Readers less familiar with concepts of mathematical statistics might skim the more mathematical sections. Resampling methodology is somewhat schizophrenic: the basic idea is very simple, yet the mathematical foundations can be complex. Recent developments of these methods in journals such as *The Annals of Statistics* show that they can offer improvements over classical methods. We demonstrate these improvements in multiple testing applications.

In Chapter 1, we discuss fundamental concepts of multiple testing and resampling. In Chapter 2, we integrate the two concepts, producing the general resampling-based multiple testing methodology. Applications of this general method appear throughout the rest of the book. In Chapter 3, the application of the method to ANOVA is given, and the resampling-based approach is compared to many existing methods. In Chapter 4, the method is described for multivariate analysis applications, including MANOVA. Resampling-based analysis is quite useful in these applications since multiple testing methods are less available for MANOVA applications than they are for ANOVA applications. In Chapter 5, the method is applied to univariate and multivariate binary data. The resampling method incorporates discreteness of binary data, which can substantially reduce the amount of adjustment needed. Chapter 6 discusses additional methods and topics, including weighted p-values, power, asymptotics, and nonstandard applications. We finish the book with Chapter 7, titled "Practical Applications," a list and discussion of applications for which multiple testing can and should be applied. The chapter is broken into two parts. The first part lists examples, and discusses why multiple testing is a major concern. In the second part of Chapter 7, further examples are given with real data, and the multiplicity adjustment techniques developed in earlier chapters are applied.

We hope our readers will find this methodology a useful tool for the analysis of multiple tests in their particular areas. In addition we hope that the availability of software will make multiplicity adjustment convenient. The wide applicability of these methods, as well as the availability of the software, should eliminate objections that appropriate multiplicity adjustment methods are not available for particular applications. Readers of this book will pay more attention to the possible effects of multiple testing upon the interpretation of their data, and will learn useful methods for controlling the problem.

PETER H. WESTFALL
S. STANLEY YOUNG

Acknowledgments

We are indebted to a great many individuals and corporations for help in preparing this book. Support for the production of the book was provided by SAS Institute Inc., Glaxo Inc., Merck Sharp & Dohme Co., The Upjohn Company, Burroughs Wellcome Corporation, Pfizer Inc. and Wyeth-Ayerst Inc.

The development of the SAS software PROC MBIN and PROC MTEST (predecessors to the current PROC MULTTEST) was funded by the Pharmaceutical Manufacturers Association through contributions of the following companies: Eli Lilly and Co., Merck Sharp & Dohme Co., Upjohn Corporation, G. D. Searle and Co., Smith Kline & French Co. (now Smith Kline & Beacham), Merrell Dow Co. (now Marion Merrell Dow Co.), and Glaxo Inc.

Individuals whose time and contributions were invaluable to the development of the PROC MBIN and PROC MTEST software include Roy Tamura and Mike Murphy of Eli Lilly and Co., Keith Soper and Joseph Heyse of Merck Sharp & Dohme Co., Dror Rom of Rorer Central Research, Jeff Rostami of Upjohn Co., Dick Bittman and Art Roth of G. D. Searle Co., Lilly Sanathanan of Smith Kline & Beacham Co., Steve Ruberg and Bill Louv of Marion Merrell Dow Co., Bill Fairweather of the U.S. Food and Drug Administration, David DeLong, Rick Langston, and Russ Wolfinger of SAS Institute Inc., and John Ryan of Texas Tech University.

We owe special thanks to Lin Youling, of Texas Tech University, who originally coded PROC MBIN and PROC MTEST. He also devised clever algorithms for efficient calculations, some of which are presented in this book.

Special thanks also go to Charu Krishnamoorthy of Glaxo Inc., for her close reading and criticism of early drafts. Keith Soper of Merck Sharp & Dohme Co. also provided helpful comments and suggestions. Providing help in proofing and calculations were Wei-Wei Kao and Jiang-Ming Wu. Marilyn Westfall contributed stylistic and organizational improvements.

We are greatly indebted to all who provided assistance and financial support. Of course, we assume responsibility for any errors that remain in the book. Most importantly, we thank our families for their understanding and support.

PETER H. WESTFALL
S. STANLEY YOUNG

Resampling-Based Multiple Testing

CHAPTER 1

Introduction

Scientists and analysts commonly face the vexing problem of multiple tests. Simply stated, "multiple testing" refers to any situation in which a collection of statistical tests is formally or informally evaluated. This situation applies to virtually all data analyses where hypotheses are considered—it is most unusual that only one test will be performed on a data set. There is an increasing tendency among researchers to analyze complex data sets from every possible angle, formulating and testing myriads of hypotheses. Investigators or their sponsors, however, would like the claims to be real—highly likely to be repeatable.

Why perform such extensive data examination? A major factor contributing to the extensive use of multiple testing is the cost of obtaining additional variables compared to the cost of additional experimental units. In surveys, the cost of asking an additional question is minor, relative to the cost of interviewing an additional individual. The same applies to laboratory studies: for example, in long-term animal studies, the cost of measuring an additional characteristic of an animal is very minor, relative to the cost of maintaining the animal over the study term. As a result, data sets often contain hundreds of items that can be tested.

A second reason for performing multiple tests is to detect interesting relationships not suspected *a priori*. The availability and user-friendliness of modern statistical software has made it quite simple to perform multiple analyses. It usually takes very little time and expense to formulate and test hypotheses *once the data set is available*. On the other hand, the *acquisition* and *preprocessing* of the data is usually time consuming and expensive. Thus, once the data set is on the computer, there is a tendency to analyze every single aspect. It is not uncommon that several articles are published, all analyzing different aspects of *the same data set*. Nor is it unusual for there to be more statistical tests of hypotheses than there are experimental subjects or survey units.

Third, using alternative statistical tests can provide confirmatory evidence regarding hypotheses tested: if there is some doubt regarding the validity of a particular testing procedure, it is considered prudent to perform the test using alternative methods. When the results of the tests are essentially in agreement, the analyst feels more comfortable with the conclusion. (A problem with such "confirmatory" analysis is that the tests are highly correlated. Thus, the tests will tend to agree even when the conclusion is a false positive result.)

Multiplicity can arise from "data splitting." Suppose there is a tendency in the direction of a favorable effect in an expensive trial, but the effect is not "statistically significant." It is sometimes argued that the treatment might be more effective for some subpopulation, so a search is conducted over various subdivisions of the experiment. A single test of control versus treated becomes many tests: control vs. treated for males, for females, for young, for old, for young females, for old males, and so on. A humorous account of how such subdivisions can be used to guarantee statistical significance is given by Graham Martin, in a letter to the editor in the December 22/29, 1984 issue of *Lancet*. Titled "Munchausen's Statistical Grid, which Makes all Trials Significant," Martin reports "The great virtue of this device is that it can resurrect a significant result from a foundering therapeutic trial." The letter goes on to explain that one simply subdivides the data according to sex, age and other variables, until the "desired result" is found. If the results are still not satisfactory, Munchausen's grid allows subdividing classes "along some other line," which "provides a second chance at winning." The letter appears in its entirety as Appendix I of this book, courtesy of Dr. Martin.

There are good reasons for performing multiple tests; however, as suggested by "Munchausen's Statistical Grid," a major drawback of multiple testing is the greatly increased probability of declaring "false significances," or statistically significant associations where none exist in reality. A related negative feature is that it is very easy to *overstate* the evidence for a particular association, if the statistical test that best supports a given hypothesis is chosen. These problems are particularly pernicious when one considers that published statistical analyses often give the results of just a few of the many tests considered or computed. There is usually no indication of how much data investigation preceded the final report. The use of excessive statistical hypothesis testing is often called "data snooping" or "data dredging" in an attempt to shame investigators. The results of such data manipulation are commonly artifacts of the techniques employed, and should be dismissed as "spurious results," but often are not.

What are the consequences of data snooping? Practically every day one finds an article in a newspaper or magazine indicting some food, chemical, drug, or human practice as dangerous to health. The article usually refers to a scientific journal where statistical evidence is given of an association between a common human activity and some untoward event. It is ironic that never before have people lived so long and productively, yet many common substances or activities are considered hazardous. Even scientifically literate individuals can have difficulty making sense of the conflicting information.

How can general health be so good when so many things are unhealthy? It has been charged by Feinstein (1988) that much of the reporting is fear-mongering based on faulty methods. Among the methods Feinstein indicts is a misuse of statistics and computers, "data dredging": hundreds of possible effects are tested, and the few results that are "statistically significant" are reported as dire. With larger studies there is more opportunity for such false significances to appear, as richer data sets allow for exploration of more research questions. Researchers should get as much information as possible for their research dollars; however, at some point they and their sponsors should worry about the quality of the information and the validity of their conclusions. In other words, there ought to be some assurance that the "statistically significant" effects reported are not just data-dredging artifacts.

Funding of extensive surveys and experiments makes large data sets available, and modern computers make multiple analyses possible. Often the scientific articles where the results are published do not explicitly state how many questions were asked; nor do they indicate whether the effects tested were pre-planned or generated by the study itself. To cover these methodological flaws, the journal article usually contains a warning such as this: "These results are preliminary and should be independently verified." Such caveats are extremely important, but they are frequently overlooked or underemphasized by newspapers and magazines, leading to confusion and doubt in the general public.

Researchers long have been able to shrug off the multiplicity problem because of the unavailability of statistical tools for its resolution. We now have computer-intensive methods to address the multiplicity issues, which produce p-values that are adjusted for the particular set of questions asked. Thus, readers and investigators can judge the veracity of particular findings against the standard p-values, which are not corrected for multiplicity, and also against the adjusted p-values, which do contain a multiplicity correction.

For example, a particular survey may identify a small p-value, say $p = .005$, and claim that the associated effect is "statistically significant." This p-value is interpretable as follows: when there is no causal basis for the effect, there is only a 0.5% probability of observing a result as extreme as the observed result. On the other hand, it is possible that the multiplicity adjusted p-value is .15 (Adj_p = .15), which is not statistically significant. This adjusted p-value incorporates the multiple tests performed and can be interpreted as follows: when there is no causal basis for *any effect* tested, there is a 15% chance that *somewhere in the experiment* a result as extreme as the observed result of .005 will appear.

What can be done to address the multiplicity problem inherent in complex investigations? Investigators should not deceive themselves. At the simplest level, investigators should keep track of the number of questions asked and of whether the questions are preplanned or after-the-fact. Any after-the-fact "significances" which do not remain after multiplicity adjustment should not be considered causal in the absence of *a priori* theory. Investigators also should review their results against the ten criteria of causation given by Sir A. B. Hill (1966), Table 1.1. In particular, the issue of whether a result is *reproducible*, covered by the *Consistency* and *Reasoning by Analogy* rules (2. and 9.) of Hill, should be given high priority. While Hill's criteria were directed towards a medical audience, the ideas apply equally well to all statistical applications, with appropriate changes in nomenclature.

Editors and referees of scientific journals have responsibilities with regard to the multiple testing dilemma. They should attempt to reduce the number of spurious and/or exaggerated claims that may result from excessive data dredging. They also should attempt to ascertain how much statistical manipulation was required to obtain

Table 1.1 A. B. Hill's Rules for Judging Causality*

Evidence	Description
1. Strength of association	What is the absolute size of the effect?
2. Consistency	Have the effects been observed in other studies?
3. Specificity	Is the cause associated with one or a few specific effects?
4. Relationship to time	Do the results occur after the cause?
5. Biological gradient	Is there a dose response?
6. Biological plausibility	Is there a known mechanism supporting cause?
7. Coherence of evidence	Does all the evidence fit together? Does the evidence fit with a natural history?
8. Experiment	Is the evidence the result of a planned experiment?
9. Reasoning by analogy	Is the observed result similar to other disease or drug effects?
10. Common sense	Do the results satisfy elementary rules of logic, data collection, etc.?

* Condensed from *Principles of Medical Statistics, 8th Edition* (1966) by A. B. Hill, with permission of Dr. I. D. Hill.

results, since statistical significances can be manufactured easily from virtually any set of data. Finally, they should require that all data used to derive published results be made available in electronic form (Fienberg et al., 1985), so re-analysis is possible.

In this chapter we review concepts of multiple testing and resampling methods. There have been several recent developments in both fields, and each field represents an area of vigorous ongoing research. A general benefit of resampling methods is that larger and more realistic problems can be analyzed. One of our main goals is to increase the range and scope of multiple testing methodology using resampling methods, so that more complex (and typical) applications can be addressed.

In Section 1.1, the multiple testing problem is introduced using a fictitious example and simulated data. Use of simulated data demonstrates the connection to simulation-based procedures used in this book. We prefer to use real data, and there are many worked examples using real data in later chapters. In Section 1.2, the *Familywise Error Rate*, or FWE, is discussed. The methods of this book all are oriented toward protecting the FWE. Section 1.3 describes the adjusted p-value, and its connection to the FWE. Resampling methods are introduced in general terms in Section 1.4. In Section 1.5 we discuss controversies associated with the multiple testing paradigm; and in Section 1.6, we give a systematic structure for the planning and analysis of studies involving multiple tests.

1.1 THE MULTIPLE TESTING PROBLEM

We begin by illustrating the multiple testing problem in a way that naturally suggests resampling methods. The research question is as follows: Does eating oat bran reduce cholesterol level? This appears to be a simple, single question, but when the question is asked of various age/sex/geographic/ethnic subdivisions of a data set, a single question can become multiple questions, resulting in a large collection of tests. In this situation, multiple testing is particularly appealing, as there is little prior information that the effect is real. If the effect is real, there is little prior information to suggest that it is more prevalent in any subdivision of the population. Nevertheless, the question of whether oat bran reduces cholesterol is a question of great interest, and many studies have been conducted to answer it.

Anything that might lower blood cholesterol would be of interest given the proven relation between high cholesterol levels and heart disease. So it might happen that numerous investigators all test essentially the same hypothesis, unknown to one another. If there is really no effect, it is likely that most of these studies will end up in the "file drawer" and never see the light of day. But some few of the studies will have a statistically significant effect (and if the hypothesis is tested in several ways in each study more than a few studies will be reported as statistically significant). Journal editors will see only those studies with significant effects.

The following "oat bran study" is presented as a multicenter trial. But the general question, Does roughage in diet lower blood cholesterol levels? might become very popular among nutritionists and food processors so that a large number of uncoordinated trials are run addressing essentially the same question. If a few particular tests are selected from an entire collection of tests, it is very probable that evidence concerning association between oat bran and cholesterol will be overstated. The following example demonstrates how this can happen.

Example 1.1 A Simulation of a Multicenter Study

At each of ten study centers, a simple control (C) versus treated (T) experiment is performed with 20 subjects per group. It is common to analyze the data for each center separately, as well as to combine them over centers. The data from such a hypothetical study are given in Table 1.2.

For this problem, the summary t-statistics are computed for each center as

$$t = \frac{\bar{y}_T - \bar{y}_C}{\sqrt{(s_T^2 + s_C^2)/20}}$$

where $\bar{y}_j$ and s_j refer to the sample mean and standard deviation of group j, respectively. The p-values are the lower tail probabilities from the t-distribution with 38 degrees of freedom.

Notice that centers 2 and 3 show significant ($p < .05$) reductions in blood cholesterol for the treatment group. (Center 5 happens to show a significant *increase*, but a food producer would consider only the lower tail, choosing to consider any increase the result of chance.) Are the significances real? We happen to know that they are not real, because the data were randomly generated using the same mean for treatment and control groups within each center, although the mean values were allowed to differ from center to center.

Producers of oat bran food products are extremely interested in demonstrating reductions. If these ten studies were run as uncoordinated trials, then it is quite likely that the two significant studies would be reported, and perhaps published in reputable journals. Conversely, the oat bran researchers are very unlikely to report

Table 1.2 First Multicenter Oat Bran Study Using Simulated Data

Center	Group	$\bar{y}$	s	t-Statistic	p-Value (Lower-Tailed)
1	Treated	219.1	7.0	.30	.616
	Control	218.3	9.8		
2	Treated	212.6	11.3	−1.76	.043*
	Control	218.5	9.8		
3	Treated	207.5	11.6	−1.79	.041*
	Control	213.6	9.9		
4	Treated	212.5	10.4	.76	.774
	Control	209.6	13.5		
5	Treated	211.9	8.5	1.90	.968
	Control	206.6	9.1		
6	Treated	222.3	13.4	.06	.523
	Control	222.1	7.5		
7	Treated	212.0	7.4	.04	.515
	Control	211.9	8.9		
8	Treated	217.4	8.6	.82	.792
	Control	215.0	9.8		
9	Treated	220.7	10.7	1.28	.895
	Control	217.2	6.0		
10	Treated	222.9	9.1	−.45	.326
	Control	224.4	11.6		

* p-value less than .05.

the nonsignificance of oat bran as a cholesterol-reducing agent, particularly after a *competitor* has reported that its product reduces cholesterol levels! The eight nonsignificant studies would go to the file drawer and a "true, confirmed" effect would be established for the two sites where significance is found.

The centers with insignificant results may decide to collect fresh data, and new researchers may jump in and study the issue. This may give rise to data such as in Table 1.3, which we generated using simulation. Again, there is a small p-value, .026, for Center 7. Also note an opposite effect for Center 1. □

Although this example appears (and is) artificial, it highlights the problem of why multiple testing is an important practical issue. With simulated data, it is easy to point to multiplicity as the *cause* of the small p-values. In real studies, it is not easy to make this determination. Because time and human commitment have been devoted to the study, experimenters rationalize the significant results. At a particular center showing "significance," scientists are inclined to believe that the effect is real; after all, why should the mere *existence* of other centers in the study affect the outcome at the given center? Even good investigators will attempt to explain *any* significant result in terms of underlying causal mechanisms. Exacerbating this problem is the fact that better investigators tend to offer more likely-sounding explanations.

Table 1.3 Second Hypothetical Oat Bran Study

Center	Group	$\bar{y}$	s	t-Statistic	p-Value (Lower-Tailed)
1	Treated	214.6	9.2	1.90	.968
	Control	209.3	8.4		
2	Treated	213.9	8.7	1.21	.884
	Control	210.2	10.5		
3	Treated	217.6	7.6	.59	.720
	Control	216.0	9.5		
4	Treated	215.5	6.2	1.59	.940
	Control	211.7	8.7		
5	Treated	211.6	9.6	1.24	.889
	Control	208.1	8.2		
6	Treated	220.1	8.7	.069	.527
	Control	219.9	9.6		
7	Treated	210.3	5.9	−2.00	.026*
	Control	215.0	8.7		
8	Treated	212.2	9.8	−1.55	.065
	Control	217.7	12.5		
9	Treated	217.3	8.8	.79	.784
	Control	215.0	9.5		
10	Treated	212.2	11.2	.53	.700
	Control	210.5	9.0		

* p-value less than .05.

This problem is particularly troublesome in real studies, since the analyst often has no way to verify whether the unusual event is real or artificial. In the oat bran "study" of Table 1.2, the centers might be examined for differences to explain why oat bran was effective at Centers 2 and 3. If enough explanatory variables are examined, a subpopulation specific to these centers might be identified to "explain" these significances. However, in many cases there is no such evidence to confirm or contradict the result. As Example 1.1 demonstrates, the possibility of false positive results is very real, and can lead to serious misinterpretation by analysts: it is human nature to rationalize any dramatic—statistically significant—change that is noted.

False positive (i.e., erroneously significant) results can and do occur in scientific endeavors. The following is a list of "Rules" about such events:

Young's False Positive Rules

- With enough testing, false positives *will* occur.
- Internal evidence will not contradict a false positive result.
- Good investigators will come up with a possible explanation.
- It only happens to the other person.

The third and fourth "Rules" are particularly interesting. A researcher is not inclined to believe that his/her data are somehow leading to erroneous conclusions—false positives only happen to other people! Such a researcher is inclined to develop an elaborate, perhaps completely erroneous, explanation of the result.

A probabilistic analysis of the multiplicity problem seen in the previous example is as follows: under the assumptions of normality, independence, and within-center homoscedasticity, the null distribution of each *individual* p-value is uniform on [0, 1], or $\mathcal{U}[0, 1]$ as shown in Figure 1.1. For example, if the interval from zero to one is divided equally into 20 parts, 0.00 to 0.05, 0.05 to 0.10, ..., the probability that a p-value falls within any particular interval is 0.05. Thus, the probability of declaring a *particular* test significant at the .05 level is exactly .05. On the other hand, the probability of declaring *at least one* of the ten results significant is

$$\Pr\left(\min_i p_i \leqslant .05\right) = 1 - \Pr(\text{all } p_i > .05) = 1 - (1 - .05)^{10} = .401$$

If significance is determined at the $\alpha = .10$ level, the probability jumps to .651. In general, the probability of declaring at least one result significant at level α is $F(\alpha) = 1 - (1 - \alpha)^{10}$; differentiating with respect to α yields the density $F'(\alpha) = f(\alpha) = 10(1 - \alpha)^9$, a special case of the class of *beta* family of distributions ($\mathcal{B}[1, 10]$), also plotted in Figure 1.1.

The density functions in Figure 1.1 clearly show the high probability concentration near zero for the minimum p-value distribution; this concentration becomes much more exaggerated when the number of tests increases. It is likely that a true null hypothesis is erroneously declared false if ten independent .05-level tests are computed, which corroborates the simulations presented in Tables 1.2 and 1.3.

Since the usual "$\alpha = .05$" protection level does not apply when a large collection of tests is considered, how can one tell whether a result is truly significant or due

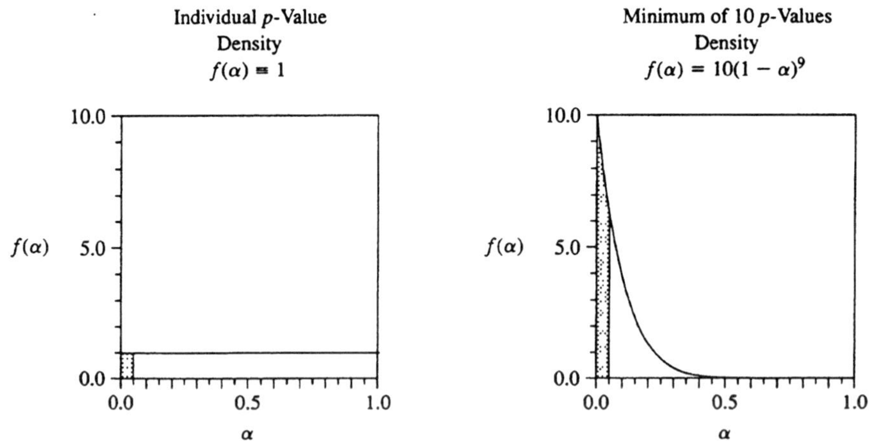

FIGURE 1.1. Probability density of an individual p-value compared to the density of the minimum of ten independent p-values.

to chance? One method is to use the *minimum* p-value distribution instead of the individual p-value distribution. If instead of just two simulations, thousands were performed, then the shaded probability in the minimum p-value density plot would be well estimated as the proportion of simulations for which the minimum p-value is less than .05. In this case, we do not have to perform such simulations, since the true distribution is known and the probability can be computed analytically. However, in more complicated situations the probabilities must be estimated. Often, the relevant distributions can be reliably estimated by resampling the data at hand, and the required probabilities can be estimated using thousands of simulations.

The idea that resampling can be used to estimate the distribution of p-values is central to resampling-based testing. In multiple testing, we are interested in the minimum p-value. Throughout this book we describe methods for estimating distributions of statistics related to the minimum p-value. Such distributions can be evaluated analytically in isolated situations (such as the preceding "oat bran" example). However, these distributions may not be evaluated in most situations because of failure of assumptions or lack of available methodology. In these cases, resampling provides a good, easily computed approximation.

1.2 FAMILYWISE ERROR RATES

In the usual hypothesis testing vocabulary, H refers to a particular hypothesis (the *null* hypothesis), and H' refers to the alternative hypothesis. (The prime denotes "complement," e.g., the complement of $H\colon \theta \leqslant 0$ is $H'\colon \theta > 0$.) In multiple testing applications, there are many such null/alternative pairs to be evaluated, say H_1 vs. H_1', H_2 vs. H_2', ..., H_k vs. H_k'. A specific claim is made when one *rejects* a hypothesis H_i; otherwise the unspecific conclusion *accept* H_i, or more properly, *fail to reject* H_i, is made.

One solution for solving the multiplicity dilemma is to make the individual tests more conservative, or more difficult to arrive at a *reject* H_i conclusion. Such a procedure is often called a *Simultaneous Test Procedure* or STP, (e.g., by Hochberg and Tamhane, 1987), which means any procedure by which *each* of the k hypotheses $H_i, i = 1, \ldots, k$, is determined to be *rejected* or *accepted* at a particular level α. Using the terminology of Hochberg and Tamhane, STP's are commonly devised to control the *Familywise Error Rate* (hereafter abbreviated FWE).

We now define two kinds of FWE's—FWEC, calculated under the *complete null hypothesis* (meaning that all H_i are true); and FWEP, computed under the *partial null hypothesis* (meaning that some subcollection of nulls, say $H_{j_1}, \ldots, H_{j_t}$, is true). Defined loosely, an FWE is the probability of rejecting *any* true null hypothesis. Formally,

$$\text{FWEC} = \Pr\big(\text{Reject at least one } H_i \mid \textit{all } H_i \text{ are true}\big), \tag{1.1}$$

and

$$\text{FWEP} = \Pr\big(\text{Reject at least one } H_i,\ i = j_1, \ldots, j_t \mid H_{j_1}, \ldots, H_{j_t} \text{ are true}\big). \tag{1.2}$$

Clearly, FWEP depends upon *which particular subset* $\{j_1, \ldots, j_t\}$ of null hypotheses is true. A simultaneous test procedure is said to control the FWE in the *weak* sense if FWEC $\leqslant \alpha$; it controls the FWE in the *strong* sense if FWEP $\leqslant \alpha$, regardless of which subset $\{j_1, \ldots, j_t\}$ of hypotheses happens to be true (Hochberg and Tamhane, 1987, p. 3). It is clearly more desirable that an STP *strongly* control the FWE.

The FWE may also called Type I FWE: in the standard univariate hypothesis testing jargon, a Type I error refers to the rejection of a hypothesis that is true. Extended to multiple testing, the Type I FWE refers to the rejection of any true hypothesis. Type II errors refer to the acceptance of a false hypothesis.

Kaiser (1960) defined "Type III" error, the probability of misclassifying the *sign* of the effect. For example, if the hypothesis $H\colon \mu = 0$ is rejected in favor of the alternative $H'\colon \mu \neq 0$ using an $\alpha = .05$ level test, is it then permissible to claim that the *sign* of the true mean μ is the same as the sign of the estimated mean $\bar{y}$? For typical t-statistic and z-statistic procedures, the answer is "yes" because of the close correspondence between hypothesis tests and confidence intervals. Specifically, suppose a test is based on the t-statistic $t = \bar{y}/(s/\sqrt{n})$, with rejection of H if $|t| > t_{.975}$, where $t_{.975}$ is the .975 quantile of the t-distribution with $n - 1$ degrees of freedom. The associated confidence interval for μ is $\bar{y} \pm t_{.975}\, s/\sqrt{n}$. In this case, a two-tailed rejection of H at the $\alpha = .05$ significance level with a sample mean $\bar{y} > 0$ is equivalent to having the entire 95% confidence interval lie above the hypothesized value 0:

$$t > t_{.975} \iff \bar{y} - t_{.975}\, s/\sqrt{n} > 0$$

Conversely, two-tailed rejection of H at the $\alpha = .05$ significance level with a sample mean $\bar{y} < 0$ is equivalent to having the entire 95% confidence interval lie below the hypothesized value 0:

$$t < -t_{.975} \iff \bar{y} + t_{.975}\, s/\sqrt{n} < 0$$

Thus, a rejection of H in favor of the two-sided alternative at the level of significance α allows one to claim that the *sign* of the parameter is the same as the sign of the estimated parameter, with the same margin of error α.

When a hypothesis has been rejected, it is extremely desirable that the true sign of the effect can be claimed to be in the direction of the estimated sign. In univariate testing applications, this practice is frequently allowed because of the direct correspondence between two-sided confidence intervals and two-sided tests.

In a large experiment where many questions are asked, it is possible that the *sign* of one or more effects is misclassified. For example, factor A causes a *decrease* in the response, but a given experiment shows factor A *increases* the response. This outcome is even more troublesome than declaring a real effect when no effect exists.

For multiple testing applications, the Type III FWE is the probability that the *sign* of *any* tested effect is misclassified. When an STP controls the Type III FWE, one may claim that the true signs of all significant effects are in the same directions as the estimated signs. As in the univariate testing situation, this property may be demonstrated when there is a direct correspondence between the simultaneous test procedure and the simultaneous confidence interval procedure. Specific applications for which these correspondences hold are given in Chapters 3 and 4.

1.3 ADJUSTED p-VALUES

In most cases it is possible to compute an *adjusted* p-value, $\tilde{p}_i$, $i = 1, \ldots, k$, for each test of H_i vs. H_i', so the decision to *reject* H_i at FWE $= \alpha$ is obtained merely by noting whether $\tilde{p}_i \leqslant \alpha$. The reasons for using adjusted p-values are the same as those for using ordinary p-values: (1) the statistician does not force a particular α upon the consumer of the report, and (2) cumbersome table look-ups are avoided. Table look-ups are usually difficult (if not impossible) in many STP applications, because of limited tables. Later, we will describe methods for obtaining multiplicity-adjusted p-values directly, in an extremely wide variety of applications. For some of these examples, alternative methods are simply unavailable. In addition to being more convenient to use than table look-ups, adjusted p-values are also easily determined using resampling methods, as we demonstrate in Chapter 2.

Similar to the definition of an ordinary (unadjusted) p-value, the mathematical definition of an adjusted p-value, $\tilde{p}_i$, is as follows:

$$\tilde{p}_i = \inf\{\alpha \mid H_i \text{ is rejected at FWE} = \alpha\} \tag{1.3}$$

In words, $\tilde{p}_i$ is the smallest significance level for which one still rejects H_i, given a particular simultaneous test procedure.

Various forms of multiplicity-adjusted p-values have been considered by Shafer and Olkin (1983), Farrar and Crump (1988), Heyse and Rom (1989), Westfall and Young (1989a), and Dunnett and Tamhane (1992).

An alternative to using adjusted p-values is to use the method by which multiple testing is most commonly performed—tabled critical values to reach an "accept H" or "reject H" decision. Using the "oat bran" setting of Section 1.1 for illustration, one might tabulate the percentiles of the minimum p-value distribution graphed in Figure 1.1. For example, the α quantile of the distribution of the minimum of ten p-values is the value c_α for which

$$\int_0^{c_\alpha} 10(1-p)^9\,dp = \alpha$$

Solving, we obtain

$$c_\alpha = 1 - (1-\alpha)^{1/10}$$

For general k, the quantiles are $c_{k,\alpha} = 1 - (1-\alpha)^{1/k}$. The following table presents $c_{k,\alpha}$ values for different k's and α's.

Number of Tests, k	Quantile, α .01	.05	.10
1	.0100	.0500	.1000
5	.0020	.0102	.0209
10	.0010	.0051	.0105
15	.0007	.0034	.0070
20	.0005	.0026	.0053

Having observed a p-value, say $p = .013$, when $k = 5$, the decision would be to reject H if FWE $= .10$, and accept H if FWE $= .05$. This type of reporting is obviously much more cumbersome than reporting the adjusted p-value, $\tilde{p} = .063$. By reporting the adjusted p-value, it is immediately clear which FWE levels allow a rejection of the hypothesis, and which allow acceptance of the hypothesis. Furthermore, the need for table look-ups is eliminated.

In the case where $p = 0.013$, the adjusted p-value is calculated from the equation $.013 = 1 - (1 - \tilde{p})^{1/5}$, yielding $\tilde{p} = 1 - (1 - .013)^5$. In general, we prefer that computer software be used to compute all adjusted p-values.

Even though tabled values of critical constants abound for multiple testing applications, they are always quite limited by the number of quantiles presented, as well as by experimental characteristics such as sample sizes or correlation structures. Unfortunately, tabling of this sort would have to be impossibly extensive to handle all the situations that arise in practice.

Before widespread availability of computed p-values from statistical software packages, the reporting of simple tests was similarly couched in the "Reject/Accept" mode, requiring table look-ups, and often interpolation. Now that fast and accurate algorithms for evaluating t- and other sampling distributions are available, computer software packages report p-values from statistical tests. There is little need for practitioners to consult tables to find significance levels, and hypothesis testing is much more easily accomplished.

It is time that the results of multiple tests also be reported using p-values. With this method of reporting, multiple testing methodology is as easy to use and interpret as univariate testing methodology.

1.4 STATISTICAL RESAMPLING

In ordinary usage the phrase *resampling methods* refers to methods in which the observed data are used repeatedly, in a computer-intensive simulation analysis, to provide inferences. In simple terms, resampling does with a computer what the experimenter would do in practice, if it were possible: he or she would repeat the experiment. In resampling, the observed variable's values are randomly re-assigned to treatment groups, and the test statistics are re-computed. These reassignments and recomputations are done thousands of times. The original test statistic is considered unusual if it is unusual compared to the resampling distribution.

For example, instead of comparing the means of two groups using a p-value obtained from the t-distribution, one might compute the p-value via resampling. The original data values are randomly resampled, and pseudo-data is created. Since the data are randomly assigned to the groups, there should be no difference in the groups beyond that which can occur by chance. The t-statistic is then re-computed using the pseudo-data, and is compared with the original t-statistic to see whether the original t-statistic is typical or unusual. The resampling-based p-value is then the proportion of resampled data sets yielding a t-statistic as extreme as the original t-statistic.

A definition of the resampling method given by Simon and Bruce (1991b) suffices for our purposes.

> Using the entire set of data you have in hand, or using the given data-generating mechanism (such as a die) that is a model of the process you wish to understand, to produce new samples of simulated data, and to examine the results of those samples.... In some cases it may also be appropriate to amplify this procedure with additional assumptions.*

This definition encompasses parametric and nonparametric *bootstrap* methods (e.g., Efron and Gong, 1983), as well as various types of "exact" permutation analysis (e.g., Fisher, 1935; Edgington, 1987). In some situations resampling is not necessary, since the answers can be computed analytically; nevertheless, we still refer to the methods as *resampling methods* to emphasize that the analytic results can be directly verified through simulation.

Resampling methods considered in this book include the bootstrap method, permutation (or rerandomization) analysis, and parametric simulation methods (e.g., Edwards and Berry, 1987). Other sample re-use methods such as jackknifing and data-splitting are not considered since they do not easily lend themselves to the estimation of probabilities and distributions that are required in multiple testing applications.

A general benefit of using resampling methods is that larger classes of statistical problems may be analyzed. Diaconis and Efron (1983) write:

> ...the new [resampling] methods free the statistician to attack more complicated problems, exploiting a wider array of statistical tools.†

This is certainly true for multiple testing applications: the methods encompass many existing parametric multiple testing methods as special cases, and provide multiple testing solutions in situations where alternative methods are unavailable.

An additional benefit of resampling methods is the ability to incorporate distributional characteristics, making the tests more robust. In cases involving count data (which can be extremely nonnormal in sparse applications), resampling makes the

*Reprinted from *Chance* 4(1) 22–32 (1991), with permission of Springer-Verlag and J. L. Simon and P. Bruce.

†Reprinted from "Computer-Intensive Methods in Statistics," by Diaconis and Efron, *Scientific American* (1983), with permission of *Scientific American*.

procedures much more powerful as well. Requisite dependence structures inherent in multiple testing applications are naturally incorporated in the resampling-based estimates, further improving the power of the tests.

While unknown dependence structures and nonnormal distributional characteristics are incorporated *approximately* via resampling, they are often not captured *exactly*. However, neither are they captured *exactly* using typical competing methods. Resampling provides a convenient and, in many cases, asymptotically valid method for incorporating these characteristics. Even without the requisite large sample sizes, use of resampling with alternative models (e.g., heteroscedastic variances vs. homoscedastic variances) allows an individual to assess the effects of deviations from an assumed model *using the available data*.

The fact that resampling-based methods produce adjusted p-values as the natural form of output further enhances the attractiveness of these methods. Classical multiple testing methods require critical values (usually table look-ups), whereas adjusted p-values appeal to the weight-of-evidence judgmental nature of experimenters, rather than to the dichotomous accept/reject jargon of classical hypothesis testing methods.

Critical values are also less convenient than adjusted p-values from a computational standpoint. To compute a critical values, the test statistic is computed for each simulated data set. These values are saved and sorted, and the desired quantile is obtained. Saving and sorting is computationally expensive, and adds complexity to the code. On the other hand, adjusted p-values are easily computed using simple count variables, and no sorting is required. Also, calculation of the resampling standard error of a critical value is complicated, involving the unknown density of the resampling distribution. The resampling standard error of an adjusted p-value can be computed from the familiar and simple binomial standard error formula.

The main disadvantage of resampling-based methods is the heavy computation involved and availability of computer software. Many of the methods described are currently available in the SAS/STAT MULTTEST procedure; however, other applications are not currently available and require individually crafted programs. In addition, once software is in place, the thousands of simulations required can be uncomfortably time-consuming, particularly for those with limited computing facilities.

Some statisticians (e.g., Sen's comments in Kempthorne et al., 1991) remain uncomfortable with simulation-based analyses, fearing (1) students will become too reliant on computers and use less analytical reasoning, and (2) different analysts who use the same set of data may come to different conclusions. Regarding the first criticism, Simon and Bruce (1991b) comment that analytical reasoning can (and should) be taught side-by-side with simulation methods in statistics. The second criticism is no longer in issue with the computing machinery we enjoy: Analyses with negligible simulation error are obtained in minutes, or even seconds, depending upon the capability of the computing equipment.

An example of using resampling to estimate quantities related to the sampling distribution of statistics is instructive. For example, if $Y_1, \ldots, Y_n$ are a random sample from a large population with mean μ and standard deviation σ, it is of interest to know the value of

$$\alpha_n = \Pr\left(\frac{\bar{Y} - \mu}{s/\sqrt{n}} \geqslant 1.96\right)$$

where $\bar{Y}$ and s denote the sample mean and standard deviation of the n observations. If the data are sampled from a normal population, then this probability is computed exactly from the familiar t-distribution with $n - 1$ degrees of freedom. If the underlying distribution is unknown, this probability cannot be determined. As n gets large, α_n tends to .025 (by the central limit theorem), but how can we estimate its value with a fixed, finite sample size?

Resampling methods, particularly the bootstrap (Efron, 1979), provide a convenient method for approximating this probability: If G is the distribution function of the Y_i, then the probability α_n *depends on* G. This dependence is explicitly displayed as follows:

$$\alpha_n = \Pr\left(\frac{\bar{Y} - \mu(G)}{s/\sqrt{n}} \geqslant 1.96 \mid G\right)$$

Since μ is a function of G, this dependence is also displayed explicitly.

One way to estimate this probability is to substitute an estimate $\hat{G}$, the empirical distribution function, for G. This function is defined as

$$\hat{G}(y) = \frac{1}{n}[\text{\# of } Y_i\text{'s} \leqslant y]$$

Substituting this estimate, we obtain

$$\hat{\alpha}_n = \Pr\left(\frac{\bar{Y}^* - \mu(\hat{G})}{s^*/\sqrt{n}} \geqslant 1.96 \mid \hat{G}\right) \tag{1.4}$$

The superscript asterisk is used to highlight the fact that the distribution of this random quantity is induced by $\hat{G}$, unlike $\bar{Y}$, whose distribution is induced by G. The value $\mu(\hat{G})$ is the familiar

$$\mu(\hat{G}) = \bar{Y}$$

which is the maximum likelihood estimate of μ under normal distributions. In fact, since the empirical distribution function $\hat{G}$ is the *nonparametric maximum likelihood* estimate of G, the estimate $\mu(\hat{G})$ may be called the *nonparametric maximum likelihood* estimate of μ (Efron, 1979).

In bootstrap resampling, one creates random variables having distribution $\hat{G}$. These random variables are then used in a simulation analysis to approximate the probability (1.4). Let $Y_1^*, \ldots, Y_n^*$ each be generated randomly and independently as follows:

$$Y_i^* = \begin{cases} Y_1, & \text{with probability } \frac{1}{n} \\ Y_2, & \text{with probability } \frac{1}{n} \\ \vdots & \\ Y_n, & \text{with probability } \frac{1}{n} \end{cases}$$

Then the distribution of each Y_i^* is the empirical distribution function $\hat{G}$, since $\Pr(Y_i^* \leq y) = \{\#\text{ of } Y_i\text{'s} \leq y\}/n$. A typical random sample from this distribution may be obtained by randomly sampling the data $Y_1, \ldots, Y_n$ *with replacement*: the value of Y_1^* is a random selection from the observed data $Y_1, Y_2, \ldots, Y_n$, say Y_j. Next Y_j is *returned* to the collection, and Y_2^* is sampled from the same collection from which Y_1^* was sampled. This process continues until all values through Y_n^* have been selected. At each step, the value selected is *returned* to the pool of data. The term *with replacement sample* refers to samples constructed in this way. Note that some of the original Y_j data values are likely to appear more than once among the sampled values Y_j^*; others will not appear at all.

A related sampling method is *without replacement* sampling, in which the values are *not* replaced after having been selected. In this case, each original value Y_j appears once and only once among the sampled values Y_j^*. In other words, the Y_j^* are a random shuffle, or permutation, of the original Y_j. This method is not useful for estimating the probability that the t-statistic exceeds 1.96, since the values $\bar{Y}^*$ and s^* are constant regardless of the order of the data values. However, the *without replacement* sampling method is quite useful to replicate the process of assigning experimental units to treatment groups in experimental studies, and for determining exact significance levels. These applications are considered in succeeding chapters.

Note that the observed data Y_j are sampled again, or *resampled*, in this procedure. Thus the procedure is called a *resampling* method.

Generation of *with replacement* samples and *without replacement* samples is convenient using various mathematical and statistical software packages. Suppose the data vector (or matrix) is $\mathbf{Y}$, with n rows, one for each observation, and v columns, one for each response variable. To create a resampled version $\mathbf{Y}^*$, generate an $(n \times 1)$ vector $\mathbf{U}$ of independent and identically distributed uniform random variables. For *with replacement* sampling, generate the index vector denoting the units to be sampled as $\mathbf{i} = \text{trunc}(n\mathbf{U} + \mathbf{1})$. Here, "trunc" denotes the truncation operator, which returns the integer portion of the argument, and "**1**" refers to a column vector of ones. Note that the index vector $\mathbf{i}$ is now a *with replacement* sample of the integers $1, \ldots, n$. The *with replacement* sample $\mathbf{Y}^*$ is then obtained by "evaluating" the original $\mathbf{Y}$ at the index vector $\mathbf{i}$.

For example, suppose there are ten rows for the data matrix $\mathbf{Y}$. For a particular sample, the index vector $\mathbf{i}$ might be $(2, 3, 5, 1, 3, 7, 10, 1, 7, 7)'$. Note that some indexes 1–10 appear more than once, and some do not appear at all. Then the first row of the bootstrap matrix $\mathbf{Y}^*$ is row 2 of $\mathbf{Y}$, the second row of $\mathbf{Y}^*$ is row 3 of $\mathbf{Y}$, the third row of $\mathbf{Y}^*$ is row 5 of $\mathbf{Y}$, and so on.

Table 1.4 shows commands used by four different software packages to construct resampled data sets $\mathbf{Y}^*$. For all of the indicated software, the ability to create many such resampled data sets, operating upon them and storing the results, is easily accomplished by putting the resampling statements within a loop or conditional execution group. Appendix IV contains complete sample programs that perform resampling within a loop.

To obtain a *without replacement* sample, the same general method may be used, but the index vector $\mathbf{i}$ must be a randomly created *without replacement* sample from the integers $1, \ldots, n$. A simple method for doing this is to let $\mathbf{i}$ be the *vector of ranks* of

Table 1.4 Generation of Resampled Data Sets

Software	With Replacement	Without Replacement
SAS PROC IML	`n=nrow(Y);` `U=ranuni(J(n,1,seed`[1]`));` `I=int(n*U+J(n,1,1));` `Ystar=Y(|I,|);`	`n=nrow(Y);` `U=ranuni(J(n,1,seed));` `I=rank(U);` `Ystar=Y(|I,|);`
GAUSS	`n=rows(Y);` `U=rnds(n,1,seed);` `I=trunc(n*U+ones(n,1,1));` `Ystar=Y[I,.];`	`n=rows(Y);` `U=rnds(n,1,seed);` `I=rankindx(U,1);` `Ystar=Y[I,.];`
MATRIX	`n=rows(Y)` `U=wn(seed,n,1,2)` `I=int(n*U+fill(n,1,1))` `Ystar=Y[I;]`	`n=rows(Y)` `U=wn(seed,n,1,2)` `Ystar=Y` `Aux=sort(U,Ystar)`
RESAMPLING STATS	`Generate n`[2]` 1,n I` `Take Y I Ystar`[3]	`Shuffle Y Ystar`[3]

[1] "seed" must be a specified integer.
[2] Manually input n.
[3] For single column Y.

the uniform random number vector **U**. Equivalently, one might append the vector **U** to **Y**, sort the resulting matrix by the first column, then reclaim the vector $\mathbf{Y}^*$ from the sorted matrix. This method is shown for the "MATRIX" software (Alexander, 1989).

Again, methods for doing these tricks using four different software packages are given in Table 1.4.

Note that the "RESAMPLING STATS" software has particularly convenient commands for resampling. This simplicity makes the software ideal for statistical novices. However, this software has limited range of applications, and therefore will not appeal to those interested in programming more complex resampling applications.

Table 1.4 is by no means a complete listing of available software to obtain resampled data sets, rather, it is just a list of packages with which the authors happen to be familiar. Resampling certainly may be performed using other software: for example, Noreen (1989) provides BASIC, FORTRAN, and Pascal code to perform various types of resampling. Nor does Table 1.4 provide algorithms which are optimal in terms of processing speed. The point of presenting the given algorithms is to demonstrate that resampled data sets are simple to generate, using available software. More computationally efficient methods exist, but are more difficult to program.

Using *with replacement* sampling, the probability $\hat{\alpha}_n$ of Equation (1.4) may be estimated using the following algorithm:

Algorithm 1.1 A Bootstrap Probability Estimate

0. Initialize a counting variable: $COUNT = 0$.
1. Generate resampled data $Y_1^*, \ldots, Y_n^*$, a *with replacement* sample from the original data $Y_1, \ldots, Y_n$.

2. If

$$\frac{\bar{Y}^* - \bar{Y}}{s^*/\sqrt{n}} \geqslant 1.96$$

then increment the count variable: $COUNT \leftarrow COUNT + 1$.

3. Repeat steps 1–2 N times, creating N resampled data sets. The estimated value of $\hat{\alpha}_n$ is $\hat{\alpha}_n^{(N)} = COUNT/N$.

There are two sources of error here, the simulation error (a function of N), and the error inherent in using $\hat{G}$ in place of G. The simulation error is from the *binomial* distribution in this case: note that the final value of the total $COUNT$ is the number of "successes" in N trials. In each trial, a "success" refers to the event $(\bar{Y}^* - \bar{Y})/(s^*/\sqrt{n}) \geqslant 1.96$. The probability of "success" is constant from trial to trial: it is simply $\hat{\alpha}_n$. Furthermore, all trials are independent, assuming that the software generates independent random variables.

Letting "SE" denote the simulation standard error, the binomial formula gives the following estimate for the accuracy of the simulation:

$$\mathrm{SE}(\hat{\alpha}_n^{(N)}) = \left(\frac{\hat{\alpha}_n(1 - \hat{\alpha}_n)}{N}\right)^{1/2}$$

This error can be brought arbitrarily close to zero with enough simulations—with sufficient computing power the simulation error can be ignored! However, there are two non-simulation sources of error that must not be ignored. First, the large-sample validity of the estimates depends on the ability of the software's random number generator to produce ideal random variables. No software generates random variables that are perfect from a probabilistic standpoint. Nevertheless, it is our experience that existing random generators produce very accurate estimates; and that, even with extremely large simulation sizes, simulation error far exceeds any error that may result from imperfect random number generators.

A second, and more important source of non-simulation error is the error incurred using $\hat{G}$ in place of the unknown G. What has made the bootstrap appealing is that this error (the error incurred by replacing G with $\hat{G}$) approaches zero for increasing sample sizes, since $\hat{G}$ generally approaches G. This convergence is often extremely rapid, faster than the convergence rates for classical methods. The bootstrap procedure has even been called a "small sample asymptotic" procedure because of its utility in moderately small sample sizes. Relevant references on this property are Singh (1981), Bickel and Freedman (1981), Beran (1982), Babu and Singh (1983), to name a few.

Most of the resampling-based methods in this book are discussed within a theoretical framework similar to that of bootstrap theory. In particular, permutation resampling is similar to bootstrap resampling, but the observations are sampled *without replacement* instead of *with replacement*. (Other distinctions between bootstrap and permutation resampling are given in Chapters 4 and 5.) Much of the classical parametric statistical theory naturally falls into a resampling framework very similar to that of bootstrap resampling, and we treat it as a special case of the general bootstrap method. The sampling distributions that interest us usually involve distributions of p-values, as suggested by the "oat bran" example in Section 1.1.

1.5 MULTIPLE TESTING CONTROVERSIES

While the methods described in this monograph expand the set of applications to which multiple testing methods can be applied, a central controversy is *whether* multiple testing methods should be applied at all. Even if it is agreed that multiple testing should be applied in a given circumstance, there remains the question of how many tests (k) should be included in the family. Miller (1981) and Hochberg and Tamhane (1987) are inclined to let the family be defined by *all tests performed within a single experiment.* This definition is reasonable; we certainly do not want the family to include all experiments performed in a statistician's lifetime. However, this definition remains somewhat ambiguous: what if a large experiment consists of several sub-experiments; e.g., evaluation of a compound's tumorigenicity in each sex of rats and mice? Should the "family" consist of *all sex/species*? Or should we have a separate family, one for each sex/species combination? In addition, what if new information becomes available *while analyzing the data*, suggesting additional hypotheses to be tested? Should these new hypotheses be added to the previous collection to create a larger family, or should they be treated separately?

1.5.1 Opposing Views

There are no definitive answers to these questions because there can be no universal agreement on this controversy. Statisticians have argued back and forth (sometimes vehemently) over this issue, often arriving at dramatically different conclusions. On the one hand, Mantel and Haenszel (1959) write:

> Repeated confirmation is an essential element. Otherwise, a very specific association may be a reflection of the multiple comparison problem; If enough contrasts are created by fractionation of a single data set, some apparently significant result is likely to appear.

Diaconis (1985) writes:

> ...with no tradition, or with multiplicity far in excess of tradition, adjustment seems mandatory...

On the other hand, Saville (1990) recommends $k = 1$, no adjustment at all! Instead, he recommends that analyses be unadjusted for multiplicity, but that the context be made clear. For example, if 500 tests are examined and the most extreme five tests are culled from this study, then these five tests should be unadjusted for multiplicity; however, it should be made clear to the consumer of the statistical report that these tests are in fact selected from 500 tests. While not formally addressing the multiplicity problem, Bailar (1991) essentially concurs:

> State, with maximum clarity and precision, which hypotheses were developed entirely independently of the data and those which were not, so readers will know how to interpret the results. *

While we agree that all information from experiments should be divulged, it is not the case that "readers will know how to interpret the results" when a selection of, for instance, five tests is made from a collection of 500. The amount of "surprise" one should expect at observing a small p-value, say .004, from such a study should depend not only upon the number of tests scanned, but also upon distributional characteristics of the data. For example, if the data are binary and sparse, then it may be impossible or highly unlikely that a p-value as low as .004 could have been generated for most of the 500 tests. Thus the operational number of tests, or family size, from which the five suspects were culled is substantially less than 500. Formal quantification of the level of "surprise" one should have in observing the value .004 should incorporate distributional characteristics; without such quantification it is doubtful that even those familiar with the subject matter will "know how to interpret the results," while those unfamiliar with the subject matter will be largely at the mercy of the author.

We argue that experimenters should use adjusted p-values to quantify the level of "surprise." Rather than require that the reader guess what interpretation should be given to small p-values culled from large studies, why not provide such a specific quantification? Of course, the *family* of tests, as well as the testing methods, should still be stated with "maximum clarity and precision."

Multiple testing is ubiquitous, and what is needed is a recognition of and a formal approach to the problem. Originally, univariate p-values and concomitant experimental design and analysis were devised to make routine what had been difficult practice. Much of human endeavor is art where long experience and talent are needed for sound judgments. As good techniques are developed and passed on through education, that part of the world that was art becomes science. The difficult job of judging multiple p-values from large experiments or surveys requires subject matter knowledge and much experience if sound judgments are to be made on the questions "What is a repeatable result?" and "What is a meaningful result?" Meaningfulness depends upon costs and alternatives, and is likely to remain more art than science. Sir R. A. Fisher and those who followed largely removed repeatability of results from art and placed it into the realm of science for univariate experiments. Multiplicity adjusted p-values are a first step in moving from art to science in multiple outcome experiments. Our position is that subject matter experts need to have both raw and multiplicity adjusted p-values to help them make more informed judgments on the repeatability of their multiple outcome experiments.

Many practitioners as well as statisticians share the view that multiplicity adjustment leads to valuable insights. Not only have several books been written on the subject, including Hochberg and Tamhane (1987) and Miller (1981), but the problem has been noted in applications journals such as *Science*, vol. 251, p. 1566, and in an article by Feinstein (1988) in the same journal. In animal carcinogenicity studies, the

*Reprinted from *Chance* 4(2) 27–38 (1991), with permission of Springer-Verlag and J. C. Bailar.

occurrence of "false positive" indicants of tumorigenic effect is known to result from the multiple testing problem (Haseman, Winbush, and O'Donnell, 1986).

By considering several examples in diverse applications, we hope to show readers how great the effects of sampling variation can be when multiple outcomes are considered simultaneously. The adjusted p-values give *specific quantifications* of the effects of sampling variability, and often the disparity between the raw p-values and the adjusted p-values is remarkable. In considering both measures of evidence, unadjusted and adjusted p-values, we hope that users of statistics will become better aware of the difficulty inherent in drawing firm conclusions from studies where many questions are addressed, and of the specific dangers involved in "data-snooping."

Resampling methodology does not *solve* all multiple testing controversies, since some problems are over and above calculation, but it does provide an excellent, easily usable methodology for multiplicity adjustment. Ultimately, the choice of the *family* of tests is subjective, and different choices of a *family* can yield dramatically different adjustments. There is as much opportunity to "cheat" when multiplicity adjustment is used as when it is not used: the analyst might try various choices of a "family" of tests until a particular family shows the result that he or she desires. Therefore, proper reporting of the results of a multiplicity-adjusted analysis should include (1) a clear description of the *family* of tests considered, as well as some justification for why that particular family was chosen (ideally, it is defined before the experiment is run or data are collected), and (2) *unadjusted* as well as adjusted analyses should be given.

When should an analyst be concerned about the inflated error rates? We suggest the following situations (not necessarily mutually exclusive) where multiple testing concepts are important.

- It is considered plausible that all null hypotheses tested may in fact be true (e.g., in parapsychology studies, and in the "oat bran" example of Section 1.1).
- The analyst plans to make a serious claim whenever *any* result $p < .05$ is found.
- The analyst is prepared to perform much data manipulation to find a "significant" result. (For example, a doctoral candidate may require a significant result for a defensible thesis.)
- The analysis is *planned* to be exploratory in nature, yet the analyst still wishes to claim that any significant result is in fact real.
- The experiment or survey is expensive and is unlikely to be repeated before serious actions are taken.

We return to the issue of selecting a family of tests in Chapter 7, where we discuss how multiple testing concepts apply to real scientific studies.

1.5.2 The Bayesian Approach

Belief of the plausibility of the null hypothesis (the first "•" in the preceding list) suggests that *a priori* reasoning has a place in whether and how multiple testing should be performed. Indeed, some claim that the multiple testing controversy disappears if one adopts a Bayesian viewpoint (Berry, 1988, and Lindley, 1990). As an alternative to multiplicity adjustment, one may incorporate prior knowledge to adjust the significance levels of multiple tests using Bayesian statistical methods. For example, if prior theory is sufficiently clear that a particular effect is real, then the Bayesian method will allow weak statistical evidence to confirm that the effect is real. On the other hand, if prior theory suggests that there is likely to be no effect, then it will take very strong statistical evidence to conclude an effect exists.

While there are many strong supporters of the Bayesian paradigm, others, such as Tukey (1977), are not convinced:

> I have yet to see a Bayesian account in which there is explicit recognition that the numbers we are looking at are the most favorable of k. Until I do, I doubt that I will accept a Bayesian notion of this sort as satisfactory.*

The Bayesian method is quite appealing; however, it is also difficult to implement, as researchers are often unwilling or unable to state their prior beliefs in the required mathematical form. Instead of using *a priori* analyses, significances are often defended *post hoc*: given a small p-value, the investigator develops a plausible explanation for the observed significance. As we have said before, this problem is compounded by the fact that the cleverest investigators are able to offer the most likely-sounding explanations.

There are many valuable insights that can be gained from the Bayesian viewpoint, but it is not a panacea. On the contrary, it reinforces the commonly held notion that firm conclusions are difficult to reach in studies where many questions are asked. In multiple testing applications, one is forced to use multivariate priors. The difficulty inherent in selecting univariate priors increases dramatically when one moves to the multivariate case, since prior correlations and joint prior probabilities must be selected as well. Often point masses must be considered (Berry, 1988), further compounding the problem of multivariate prior determination. Many statisticians feel that any prior (whether univariate or multivariate) cannot be uniquely determined; rather, there is a *class* of reasonable priors that one might consider (this is the "robust Bayesian" view, discussed by Berger and Wolpert, 1984, for example). In the multivariate case, the class of such priors can be extremely wide, leading to drastically different inferential measures for different members within the class. Thus, the Bayesian viewpoint provides neither a perfect nor an easily implemented solution to the multiplicity problem; however, it does support the notion that there are difficulties interpreting complex data sets, especially when many questions are addressed simultaneously.

Rather than abandon the Bayesian viewpoint, we suggest that practitioners study its utility with regard to the multiple testing situation. Many commonly used frequency-based multiple testing protocols are in fact based loosely on Bayesian considerations. It has been suggested that if a particular hypothesis is suspected based on *a priori* grounds, then that hypothesis should *not* be adjusted for multiplicity (Smouse and

*From J. W. Tukey, "Some Thoughts on Clinical Trials, Especially Multiplicity," *Science* **198** 679–684, with permission of *Science*.

Williams, 1982, Pocock, Geller, and Tsiatis, 1987). For example, suppose prior research and/or subject matter knowledge suggest that, among all of factors A, B, C, D, ..., factor A is likely to affect a response variable. Then the statistical test for factor A's effect should not be adjusted for multiplicity, even though factors B, C, D, ... are also measured and tested. This view reflects differentiation of prior likelihoods of hypotheses, a Bayesian notion.

1.5.3 Meta-Analysis

Another alternative to multiple testing is called *meta-analysis* (Hedges and Olkin, 1985). A fundamental tenet of meta-analysis is that reinforcing tests should be *combined*, rather than examined individually, when obtaining evidence for or against particular hypotheses and collection of hypotheses. Rosenthal (1990) summarizes this position as follows:

> Because meta-analytic procedures faithfully record p levels and effect sizes, there is no need for the continued indefensible practice of claiming a .05 result as evidence against a null and a .06 result as evidence for a null. Two .06 results are much stronger evidence against the null than one .05; and 10 p's of .10 are stronger evidence against the null than 5 p's of .05.*

A major assumption for the appropriateness of such meta-analysis procedures is that all tests considered in a study or collection of studies are essentially testing the same hypothesis. If so, then one would expect that a combined test would be more appropriate, and often more powerful, than the examination of many tests individually, particularly when such individual tests are followed by multiplicity adjustment.

The ability to claim *site- or location-specific* significances differentiates the adjusted p-value methodology of this book from the meta-analysis methodology. If one is simply not interested in specific hypotheses, and if there is prior reason to suspect that all hypotheses are related, then the meta-analytic point of view is very attractive. However, once meta-analysis is adopted, one completely loses the ability to say *which particular* results are significant. If a researcher does not wish to sacrifice the ability to make location-specific conclusions, then he/she may prefer to look at individual tests, with specific conclusions allowed only if results are significant after multiplicity adjustment.

As when suspected tests are isolated, and not included in the family over which multiplicity adjustment is performed, prior considerations also can help to decide whether all tests ought to be considered individually, or whether they ought to be combined. The multicenter "oat bran" study of Example 1.1 is ideally suited for meta-analysis, since all hypotheses are essentially the same for all centers. Any differences in the hypotheses are due to center × treatment interaction, which in this case should not be large, and it should be very reasonable to combine the data for all centers. A meta-analysis of the "oat bran" data of Example 1.1 would then involve constructing a single test that combines data for all centers, rather than constructing individual tests for each center. Although it loses information on possible site-specific effects, this is a useful method for avoiding the multiplicity problem. The statistically significant results in Example 1.1 are commonly called "selection effects," and would be avoided using most meta-analytic procedures. In addition to avoiding such selection effects, meta-analytic procedures are often more powerful when the effect is real.

Combination of tests via meta-analysis is most appropriate when the hypotheses tested are very similar, as they are in Example 1.1. The determination of whether the various hypotheses are asking essentially the same question requires *a priori* subject matter knowledge. From a Bayesian standpoint, one might specify prior distributions that are highly correlated in such an example: if oat bran is effective at a particular site, then it is quite likely to be effective at all sites. When more *a priori* correlation is modelled in prior distributions, the posterior measures of evidence concerning location-specific effectiveness of oat bran become more dependent on data from *all* locations. Non-Bayesian meta-analytic tests combine data for all locations in a related fashion.

1.5.4 Concluding Remarks

Specific quantification and calibration of prior distributions for routine analysis of multiple tests, whether site-specific or meta-analytic, would be very useful. Crude initial attempts have been made by Meng and Dempster (1987), Berry (1988), and Breslow (1990); however, much work remains to create a general Bayesian methodology acceptable for general use by the scientific community.

The methods of this book are blatantly frequentist with no Bayesian connection. Resampling methods provide an excellent frequentist solution to the multiple testing problem, but do not account for *a priori* knowledge. As a crude compromise between frequentist and Bayesian methods, we suggest that *a priori* knowledge be incorporated wherever it is sensible to do so, and that the specific method for incorporating such knowledge be made very clear in the presentation of the statistical results. For example, if particular null hypotheses (among several) are suspected to be false based on *a priori* considerations, then it is sensible either to leave them out of the "family" over which multiplicity adjustment is to be performed, or to allocate a large portion of the FWE to these suspected hypotheses. Also, if every test in a collection concerns essentially the same hypothesis, with little or no differentiation, then statistical intuition as well as formal Bayesian analysis suggest that combination of the results via meta-analysis is appropriate.

1.6 A STRUCTURE FOR MULTIPLE TESTING DATA ANALYSIS

In most studies, some degree of multiple testing is required. Of course, the more tests one considers, the greater the chance of incorrectly claiming (or overstating evidence about) significant results. The probability of incorrectly claiming a significance can be controlled using multiple testing methods if one includes all hypotheses into a family of comparisons, then performs multiplicity adjustment over that family. However, larger families yield greater adjustments, which in turn make the detection of significant results more difficult. There is a trade-off between size of the family and ability to detect significance: on one hand, larger families allow more questions to be investigated, but on the other hand, it becomes more difficult to detect associations when multiplicity adjustment is performed using these larger families. Conversely, when large families are considered without multiplicity adjustment, it is too easy to detect associations. As the number of questions becomes very large, some adjustment seems mandatory; anything less borders on charlatanism.

By carefully planning a study in advance, it is possible to reduce number of hypotheses included in the family. In large complex data sets one must have a clear conceptual picture of the data and of how the questions are to be asked. We now give a step-by-step description of data layout and analysis which is very flexible, and conforms with the available PROC MULTTEST software for resampling-based multiple testing analysis. This analysis is similar to the methods described by Wishart (1938), Box (1950), and Grizzle and Allen (1969).

Figure 1.2 gives a data and analysis diagram, with each box denoting a collection of numbers or characters, arranged in rows and columns. The two starting matrices are a treatment/block/covariate identification matrix **X** and a response matrix $\mathbf{Y}_1$. In the **X** and $\mathbf{Y}_1$ matrices, each row corresponds to one experimental unit. Each column of the **X** matrix denotes a different treatment, block, or covariate variable. Each column of the $\mathbf{Y}_1$ matrix denotes a particular response variable of interest.

A treatment variable in the **X** matrix could be as simple as a column of 0's and 1's, each denoting one of two treatments. Or the treatment column could be more complex, encoding multiple treatment groups. A blocking variable in **X** is optional. An example of a blocking variable is differing locations (or centers) where a treatment or procedure is tested. All treatments are nested within each block. The **X** matrix also may contain *covariate* columns, or more generally, the matrix may contain any collection of predictor variables. In regression analysis, the matrix **X** may contain several columns of predictors.

For each experimental unit, there are usually multiple response variables in the $\mathbf{Y}_1$ matrix. These might be disparate, such as height, weight, and blood cholesterol; or they might be similar, such as body weights collected at weekly intervals over a fixed period of time. The columns of the $\mathbf{Y}_1$ matrix denote the multiple measurements taken on each experimental unit. The measurements may be continuous, such as blood cholesterol, or qualitative, such as "with/without liver tumor," often coded as a binary $(0, 1)$ response; they could also be multinomial or categorical, ordered or not ordered. Different testing methods are required for handling continuous and binary response. PROC MULTTEST of SAS allows both types of data, and it allows mixed types within the same analysis. These measurements may or may not be the items of direct interest.

The $\mathbf{Y}_2$ box should contain variables, condensed or otherwise derived from $\mathbf{Y}_1$, which *are* of direct interest, and which are to be used in the primary analysis. These might be selected original variables from $\mathbf{Y}_1$, transformations of original variables

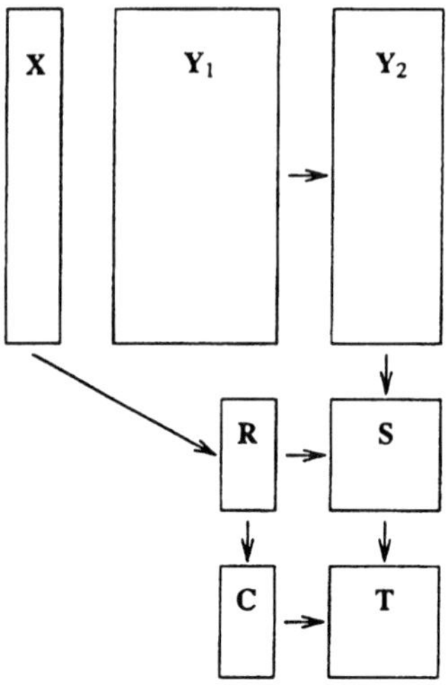

X contains the identification variables; e.g., treatment codes and blocking variables.

$\mathbf{Y}_1$ contains the multiple response variables.

$\mathbf{Y}_2$ contains variables derived from $\mathbf{Y}_1$; e.g., linear contrasts, selected variables.

R contains "reduced form" of **X**; e.g., group indicators.

S contains summary statistics for each treatment/block combination. An element within a row/column combination of **S** can be a set of summary statistics, e.g., mean, std. dev., *n*.

Rows of **C** are contrasts of treatment groups to be tested.

T contains summary tests computed for **C** contrasts of **S** statistics.

Each row of $\mathbf{X} \mid \mathbf{Y}_1 \mid \mathbf{Y}_2$ corresponds to one experimental unit.

Each row of $\mathbf{R} \mid \mathbf{S}$ corresponds to a summary, e.g., a treatment group and block combination.

Each row of $\mathbf{C} \mid \mathbf{T}$ corresponds to a linear contrast of the treatment groups (pooled over blocks if applicable.)

FIGURE 1.2. Diagram of data and analysis.

(such as $\log_{10}$ or square root), or more complex functions of several original variables, such as linear combinations measuring rate of growth in a repeated-measures study, or perhaps even principal components scores computed from the columns of $\mathbf{Y}_1$. With many $\mathbf{Y}_1$ variables, it is good practice to reduce the number of variables while retaining the information contained in $\mathbf{Y}_1$ that is appropriate for the experimental questions. Careful construction of the $\mathbf{Y}_2$ variable set allows such condensation. A trivial, but often useful condensation is simply to omit many of the $\mathbf{Y}_1$ variables from the primary analysis. As an example of when this might make sense, consider a toxicology experiment where *death* of an animal is the primary outcome (or *endpoint*), but where various disabilities (possible causes of death) are recorded. These disabilities, tumors for example, might be secondarily analyzed to determine the cause of a response on the primary variable, death, but omitted for the purposes of primary analysis. The point of this discussion is to indicate a general method by which many experiments can be analyzed, noting that subject matter expertise is essential when creating $\mathbf{Y}_2$ from $\mathbf{Y}_1$ for any particular application.

The **R** box is the "reduced form" of **X**: in many applications there are multiple observations for each treatment/block combination, and **R** lists all such combinations, but without repeats. In general, **R** denotes the summary statistics that are considered for each variable. For example, in multivariate regression applications **R** may denote a collection of regression parameters to be estimated for each variable.

The **S** data box contains summary statistics computed for each variable and each element of **R**. These statistics may include any combination of means, standard deviations, standard errors or sample sizes. Each "entry" in the **S** data matrix is therefore a *collection* of numbers representing summary statistics. Usually, it is the inspection of these statistics that gives the scientist the overall impression of "what happened" in the experiment or survey.

The **C** box contains contrasts of treatment groups that are of interest. For example, if there are two treatment groups and the measurements are continuous, there might be only one contrast of interest,

$$\text{Contrast} = (\,+1 \quad -1\,)$$

If there are multiple treatment groups, then a set of linear contrasts comparing the various groups might be appropriate. For example, in a (2×2) factorial experiment there are four groups, and the linear contrasts might measure main effects and interactions:

$$\begin{aligned}\text{Contrast 1} &= (\,+1 \quad +1 \quad -1 \quad -1\,)\\ \text{Contrast 2} &= (\,+1 \quad -1 \quad +1 \quad -1\,)\\ \text{Contrast 3} &= (\,+1 \quad -1 \quad -1 \quad +1\,)\end{aligned}$$

The **T** box contains test statistics, often *t*-distributed with continuous measurements, for evaluating the significance of the various contrasts (rows) indicated by the **C** box, for each variable in the **S** box (columns).

It is in the selection of the $\mathbf{Y}_2$ variables and **C** contrasts among the treated groups that scientists translate their questions into statistical tests of hypotheses. The construction of the $\mathbf{Y}_2$ and **C** boxes should be done very carefully so that all relevant questions are answered somewhere in the final matrix **T** of test statistics; however, to reduce problems of multiplicity, only questions that are of *genuine interest* should be addressed in **T**.

The final "family" of tests to be considered is the collection of tests appearing in the final matrix **T**. Along with the test statistics, *p*-values should be provided. In order to protect against the possibility of incorrectly declaring a particular contrast significant, adjusted *p*-values should be provided in addition to the raw *p*-values. The multiplicity-adjusted *p*-values in **T** give the probability that an unadjusted *p*-value as small as the observed value could occur *somewhere*, resulting from random fluctuations. These *p*-values are based on the particular collection of tests in **T** and the correlation structures. Here, correlations may come from two sources: (1) from correlations among the columns of the $\mathbf{Y}_2$ matrix, and (2) from correlations induced by multiple comparisons involving the same treatment group. To protect against false significances, these adjusted p-values should be of primary importance in interpreting the results.

One must use the data analysis paradigm rather rigidly to account for multiplicity. Strictly speaking, the $\mathbf{Y}_2$ and $\mathbf{C}$ matrices must be *preplanned* for multiplicity adjustments to be valid. With modern statistical software and electronic data sets, it is both fun and easy to explore many questions; however, simply jumping into the analysis of a large, complex data set without theory to guide the exploration invites misinterpretation. Even skilled data analysts and subject matter experts need a *plan* to keep track of the *number* of questions asked and the *different ways* these questions are asked. When the "family" is identified *post-hoc* it is easy to "cheat" by selecting a particular collection of tests that proves the researcher's desired claim. Nevertheless, there is value in using multiplicity adjustments even when families are developed *post-hoc*, for example, to educate a naive researcher about the effects of multiplicity on his or her study.

This data analysis paradigm is sufficiently flexible to apply to many complex examples, and we will use it often. The method is illustrated in the following example and in examples found throughout the book.

Example 1.2 Growth Curve Data

It is instructive to work through an example using the data analysis paradigm. Rat growth data taken from Box (1950) are used for this purpose. There are three groups of animals, a control group and two treated groups, one treated with Thyroxin and the other with Thiouracil. Initial body weights, W_0, and four weekly weights, W_1–W_4, were collected to assess the growth of the rats. W_0–W_4 form $\mathbf{Y}_1$ and are given in Table 1.5. Rat growth typically follows a sigmoid curve relationship over time: first there is a period of slow growth, then a period of more rapid linear growth, and finally there is a plateau at the animal's adult weight. In this example we do not know where on the typical growth curve these animals were started. If toxic doses are employed, rats tend to grow more slowly and often their final weight reaches a lower plateau. With these thoughts in mind, two $\mathbf{Y}_2$ variables were selected. The variable *gain*, $W_4 - W_0$, will measure the growth rate of the animals. It has the advantage of taking the initial weight into account. Analysis of this variable is like a paired t-test. Next, it would be nice to have a simple variable to measure the *shape* of the growth curve. If we compute the deviation of W_2 from the average of W_0 and W_4, then we have a measure of the nonlinearity of the growth curve: if the *shape* variable is positive, then a downward concavity is suggested; if *shape* is negative, an upwardly concave (convex) shape is suggested. Depending upon where we are on the typical rat growth curve and the effects of the treatments, our shape measure will detect nonlinearity of the rat growth curve. So *Gain* and *Shape* are the two $\mathbf{Y}_2$ variables.

The reduction matrices $\mathbf{R}$ and $\mathbf{S}$ are given in Table 1.6. Each entry in $\mathbf{S}$ contains a sample mean ($\bar{y}$), a standard error (SE $= s/\sqrt{n}$), and a sample size (n). The shapes and comparative positions of the growth curves are also displayed graphically in Figure 1.3. From the summary statistics and from the graph, it is clear that Control and Thyroxin rats grow about equally rapidly, but Thiouracil treated rats grow more slowly. From the *Shape* variable for the control rats, it appears that we are in the

Table 1.5 X, Y_1, and Y_2 Matrices for Box Growth Curve Data*

X	Y_1					Y_2	
Treatment	W_0	W_1	W_2	W_3	W_4	*Gain*	*Shape*
Control	46	70	102	131	153	107	2.5
Control	49	67	90	112	140	91	−4.5
Control	49	67	100	129	164	115	−6.5
Control	51	71	94	110	141	90	−2.0
Control	52	77	111	144	185	133	−7.5
Control	56	81	104	121	151	95	0.5
Control	57	82	110	139	169	112	−3.0
Control	57	86	114	139	172	115	−0.5
Control	60	93	123	146	177	117	4.5
Control	63	91	112	130	154	91	3.5
Thyroxin	52	70	105	138	171	119	−6.5
Thyroxin	52	73	97	116	140	88	1.0
Thyroxin	54	71	90	110	138	84	−6.0
Thyroxin	56	75	108	151	189	133	−14.5
Thyroxin	57	72	97	120	144	87	−3.5
Thyroxin	59	85	116	148	177	118	−2.0
Thyroxin	59	85	121	156	191	132	−4.0
Thiouracil	46	61	78	90	107	61	1.5
Thiouracil	51	75	92	100	119	68	7.0
Thiouracil	51	75	101	123	140	89	5.5
Thiouracil	53	79	100	106	133	80	7.0
Thiouracil	53	72	89	104	122	69	1.5
Thiouracil	56	78	95	103	108	52	13.0
Thiouracil	58	69	93	114	138	80	−5.0
Thiouracil	59	80	101	111	122	63	10.5
Thiouracil	59	88	100	111	122	63	9.5
Thiouracil	61	86	109	120	129	68	14.0

* From G. E. P. Box, "Problems in the Analysis of Growth and Wear Curves," *Biometrics* **6** 362–389 (1950), with permission from The Biometrics Society.

linear portion of their growth curve. There is some indication that there is non-linear growth for the Thiouracil treated rats.

We would like to know if each treated group differs from the control, so the treatment contrasts of interest are the Dunnett (1955) style contrasts, $1, -1, 0$ and $1, 0, -1$. Applying these contrasts to the mean values of the different treatment groups, we get the statistics displayed in the **T** matrix of Table 1.6. There are four p-values displayed. Note that there were five Y_1 variables, and if we had used all possible comparisons among the three treatments there would be three treatment contrasts for a total of 15 comparisons. By selecting Y_2 variables of interest and two of the three treatment contrasts, we have greatly reduced the multiplicity problem. Not all of these comparisons are of equal interest. Changes in the *Gain* variable are much more important than changes in *Shape*. To control the FWE at .05, it might

Table 1.6 R, S, C, and T Matrices for Box Growth Curve Data

R	Statistic	Gain	Shape
Control	$\bar{y}$	106.6	−1.30
	SE	4.9	1.59
	n	10	10
Thyroxin	$\bar{y}$	108.7	−5.07
	SE	5.9	1.89
	n	7	7
Thiouracil	$\bar{y}$	69.3	6.45
	SE	4.9	1.59
	n	10	10
C	**T**		
Control vs Thyroxin (1, −1, 0)	$\bar{y}$	−2.11	3.77
	SE	7.64	2.47
	t	−0.28	1.53
	p^*	0.78	0.14
Control vs Thiouracil (1, 0, −1)	$\bar{y}$	37.30	−7.75
	SE	6.94	2.24
	t	5.38	−3.46
	p^*	0.000	0.002

* p-values are two-sided, unadjusted for multiplicity.

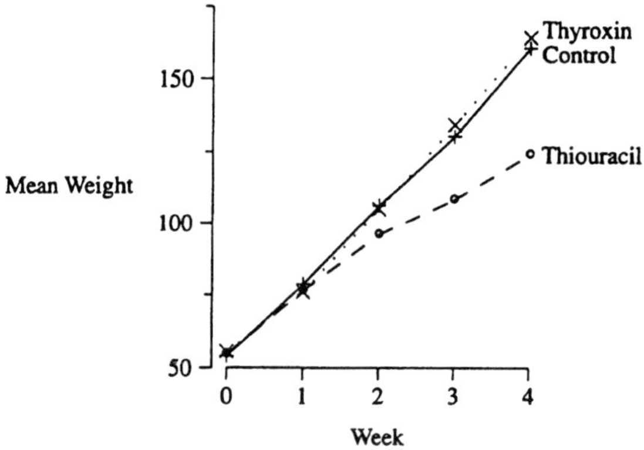

FIGURE 1.3. Sample means for Box growth curve data.

be reasonable to allocate .02 for each of the *Gain* contrasts and .005 for each of the *Shape* contrasts. We allocate higher α-values to contrasts that we feel are more important from the experimental standpoint.

Using the Bonferroni inequality

$$\begin{aligned}
&\Pr(\text{Reject at least one } H_i \mid \text{all } H_i \text{ are true}) \\
&\quad \leqslant \Pr(\text{Reject } H_1 \mid H_1 \text{ is true}) + \cdots + \Pr(\text{Reject } H_4 \mid H_4 \text{ is true}) \\
&\quad = .02 + .02 + .005 + .005 = .05
\end{aligned}$$

We can therefore conclude with a FWE of .05 the following: the growth of Thyroxin treated animals is not significantly different from the control animals, Thiouracil animals grow more slowly than control animals and the shape of their growth is more downwardly concave—they appear to be approaching a plateau whereas the control animals are still growing linearly.

Note that the $\mathbf{Y}_2$ variables are much more interpretable than an analysis of the weekly weights; they were selected to match the experimenter's knowledge of rat growth and possible treatment effects. Selection of appropriate treatment contrasts focuses the analysis on the important questions. In this case the FWE was computed using the Bonferroni inequality. This example shows the general direction taken by this book, though there are many refinements. □

CHAPTER 2

Resampling-Based Adjustments: Basic Concepts

To perform multiple testing using resampling methods, we require three things:

- Methods for testing individual hypotheses (t-tests, for example),
- A predefined collection of k hypotheses to be tested (e.g., the $\mathbf{T}$ matrix of Section 1.6) and
- A probabilistic model for the data-generating process. The model may be *parametric*, such as normal distributions with common variances; or *nonparametric*, such as a model where the data are determined by experimental randomization.

Given these items, resampling-based multiplicity adjustments are easily obtained for many situations of practical importance.

The methods presented in this chapter encompass the usual Bonferroni procedure, as well as the more complex step-down methods developed by Holm (1979), Shaffer (1987), and Holland and Copenhaver (1988). The Bonferroni-style methods use probability inequalities, avoiding correlation structures and other distributional characteristics. By contrast, the resampling methods incorporate correlation structures and distributional characteristics. Many parametric methods, for example Tukey's method, are special cases of the general resampling-based adjustments; but resampling-based methods enlarge the scope of applications to account for possible nonnormalities. For ease of presentation and comparison, we cast all methods in the framework of p-value adjustment. Incorporating correlation structures usually makes the adjusted p-values *smaller*; therefore, the power of resampling-based methods to detect significant effects is usually greater than the power of methods that fail to account for correlation structure.

More recent developments including *step-up* adjustments may be found in Simes (1986), Hochberg (1988), Hommel (1988), and Rom (1990). These methods again rely on Bonferroni-type inequalities, and can therefore be made more precise by incorporating dependence structures. Dunnett and Tamhane (1992) demonstrate this in a simple parametric application, and also note that the *step-up* analysis gives results similar to the *step-down* analysis. It would be interesting to use resampling procedures for *step-up* analyses, and we leave the development of this method as an open research question. Throughout this book, we consider only *step-down* methods.

While it is important to incorporate relevant correlation structures correctly, the effect of nonnormal distributions on maintaining the correct α-level can be much more important. Since maximal (or minimal) statistics are often considered in multiple testing, effects of slight deviations in the "tail" areas of the marginal sampling distributions are magnified. This can make standard parametric multiple testing procedures extremely liberal or extremely conservative—even when the marginal (univariate) testing procedures are approximately valid. Resampling methods offer protection against such troublesome distributional characteristics by utilizing the distributional form of the sampled data.

The ideas presented in this chapter are the basis for all applications and methods discussed in the remainder of the book. As such, the concepts are presented in fairly general terms, and the mathematical level is somewhat higher than that of these other chapters. For practical purposes, we suggest that the reader at least read the examples, definitions, and algorithms (particularly Algorithms 2.5 and 2.8), and skim the remaining material for basic ideas.

Requisite notation is developed in Section 2.1. The concept of *pivotality* is extremely important in univariate and multiple testing, and its role is discussed in Section 2.2. Single-step multiplicity adjustments are defined in Section 2.3, including the classic Bonferroni method as well as a method designed to account for explicit correlation structures. Control of the FWE for single-step methods is also given in Section 2.3. The general resampling method for multiple testing applications is given in Section 2.4. In Section 2.5, the need for resampling is indicated using an example with multiple skewed populations, a situation in which classical methods fail miserably. Sections 2.6–2.7 extend the ideas to step-down methods, still incorporating correlation structures and other distributional characteristics. Strong control of the FWE is demonstrated for step-down resampling-based methods in Section 2.8.

2.1 DEFINITIONS AND PRELIMINARY MATERIAL

Assume that the observable data is a vector $\mathbf{Y}$, generated from a multivariate probability distribution function G. All relevant sample sizes, correlation structures, and other distributional characteristics are captured in G. Usually, G is unknown, but is assumed to belong to a *set* (or *family*) of distribution functions $\mathcal{G}$. Each hypothesis H_i imposes restrictions on $\mathcal{G}$: $\{H_i \text{ is true}\} \Rightarrow G \in \mathcal{G}_i \subseteq \mathcal{G}$.

For example, if the data $Y_i, i = 1, \ldots, n$ are assumed to be generated independently from the normal distribution with mean θ and variance σ^2, then the data vector $\mathbf{Y}$ belongs to the family of multivariate normal distributions

$$\mathcal{G} = \{\mathcal{N}_n(\theta\mathbf{1}, \sigma^2\mathbf{I}) \mid -\infty < \theta < \infty, 0 < \sigma^2 < \infty\}$$

Here, "**1**" is the column vector of ones, and "**I**" refers to the identity matrix.

Consider the hypotheses $H_1: \theta \geqslant 0$ and $H_2: \sigma^2 \geqslant 10$. Each hypothesis places restrictions on the allowable family of distributions $\mathcal{G}$:

$$\mathcal{G}_1 = \{\mathcal{N}_n(\theta\mathbf{1}, \sigma^2\mathbf{I}) \mid 0 \leqslant \theta < \infty, 0 < \sigma^2 < \infty\}$$

and

$$\mathcal{G}_2 = \{\mathcal{N}_n(\theta\mathbf{1}, \sigma^2\mathbf{I}) \mid -\infty < \theta < \infty, 10 \leqslant \sigma^2 < \infty\}$$

The complete null hypothesis is $H^C = \{\text{All } H_i \text{ are true}\}$, abbreviated as $H^C = \cap_{i=1}^k H_i$. Under the complete null hypothesis it follows that $G \in \cap_{i=1}^k \mathcal{G}_i = \mathcal{G}^C$, which denotes the space of allowable distributions for $\mathbf{Y}$ under the complete null hypothesis. In the example above,

$$\mathcal{G}^C = \{\mathcal{N}_n(\theta\mathbf{1}, \sigma^2\mathbf{I}) \mid 0 \leqslant \theta < \infty, 10 \leqslant \sigma^2 < \infty\}$$

Assume that the evidence against hypothesis H_i has been summarized using a p-value p_i, $i = 1, \ldots, k$, each of which is a function of the observable data: $p_i = p_i(\mathbf{Y})$. It is important that the marginal test procedures which yield the individual p-values be chosen to have good power and robustness properties. For example, when data are heavy-tailed, it may be appropriate to log- or rank-transform the data before testing. In theory, the p-values themselves may result from a resampling analysis (e.g., Noreen, 1989; Romano, 1989); however, since we plan to resample the distribution of these p-values, this analysis can easily become too computer-intensive. In most cases we assume that the p-values have been computed from standard normal, χ^2, t, or F approximations or perhaps analytically derived from a permutation distribution.

Distinguish the random p-value from the experimentally observed p-value using capital and lower case letters, respectively. Then a random vector of p-values is $\mathbf{P} = (P_1, \ldots, P_k)$, and the observed vector is $\mathbf{p} = (p_1, \ldots, p_k)$. The multivariate distribution function for $\mathbf{P}$ depends upon which $G \in \mathcal{G}$ generated the data.

Individual p-values for each hypothesis are usually computed under a particular setting $H_{0i} \subseteq H_i$. For example, when testing $H_1: \theta_1 \leqslant 0$ vs. $H_1': \theta_1 > 0$, the p-value is computed under $H_{01}: \theta_1 = 0$. Corresponding to H_{0i} is the restricted class of distributions $\mathcal{G}_{0i}$, as well as the corresponding complete null class $\cap_{i=1}^k \mathcal{G}_{0i} = \mathcal{G}_0^C$. Under the complete null specification, $H_0^C = \cap_{i=1}^k \{H_{0i} \text{ is true}\}$, we assume there is a particular, unknown $G_0 \in \mathcal{G}_0^C$ which must have generated the data. Using the H_1 and H_2 above, there is only one G_0 in the class $\mathcal{G}_0^C$, namely the normal distribution $\mathcal{N}_n(\mathbf{0}, (10)\mathbf{I})$. In general, there are many members of the class $\mathcal{G}_0^C$, with particular members indexed by "nuisance parameters" or other unknown distributional characteristics.

If G_0 is known (a rare case), we may simulate data vectors $\mathbf{Y}$ having this distribution, and thereby evaluate the distribution of the minimum p-value under G_0. In other cases, the vectors may be simulated from an approximation of G_0, say $\hat{G}_0$. Resampling methods implicitly estimate such a $\hat{G}_0$. Often, the estimation of G_0 and the simulation of data vectors are performed simultaneously. For example, *with replacement* sampling procedures implicitly estimate G using the empirical distribution function, as discussed in Section 1.4.

2.2 GUIDELINES FOR RESAMPLING-BASED TESTING

Hall and Wilson (1991, p. 757) give the following guidelines for bootstrap testing in univariate situations.

> The first guideline says that care should be taken to ensure that even if the data might be drawn from a population that fails to satisfy H_0, resampling should be done in a way that reflects H_0. The second guideline argues that bootstrap hypothesis testing should use methods that are already recognized as having good features in the closely related problem of confidence interval construction. This often amounts to pivoting, which usually means correcting for scale.*

While these guidelines were given for *univariate* testing, they are equally important in *multiple* testing. Before moving to multiple testing applications, it is instructive to consider these two guidelines in more detail for a simple application.

2.2.1 Resampling under the Null Hypothesis—Centering

Noreen (1989, pp. 74–75) presents a simulation study to evaluate the effectiveness of bootstrap testing when sampling from a skewed population. The data values are generated from a lognormal population, $Y = e^Z$, where $Z \sim \mathcal{N}(0, 1)$. Let the distribution of Y be denoted by G_0. The mean of G_0 is $\mathrm{E}(Y) = \sqrt{e}$, and the distribution is strongly skewed in the positive direction, with a skewness coefficient of $+6.2$. Data $Y_1, \ldots, Y_n$ are sampled independently from this lognormal population, and the hypothesis $H\colon \mathrm{E}(Y) = \sqrt{e}$ is tested. Alternatives may be upper- or lower-tailed. The test will be based upon the sample mean, $\bar{Y} = \sum Y_i/n$.

The first guideline of Hall and Wilson requires that resampling be done to reflect the null hypothesis. It is quite easy to violate this guideline. Since the observed deviation $\bar{y} - \sqrt{e}$ measures deviation from the null, it seems logical to evaluate the probability of observing a deviation as extreme as $\bar{y} - \sqrt{e}$ using bootstrap resampling. Naively, one might simply resample the y_i with replacement, obtaining y_i^*, and count the proportion of samples for which $\bar{y}^* - \sqrt{e}$ exceeds $\bar{y} - \sqrt{e}$. The following algorithm presents this incorrect and inefficient method.

Algorithm 2.1 An Incorrect Bootstrap Testing Method (Poor)

0. Initialize a counting variable: $COUNT = 0$.
1. Generate resampled data $y_1^*, \ldots, y_n^*$, a *with replacement* sample from the original data $y_1, \ldots, y_n$.

*From P. Hall and S. R. Wilson, "Two Guidelines for Bootstrap Hypothesis Testing," *Biometrics* **47** 757–762, (1991), with permission from The Biometric Society.

2. If

$$\bar{y}^* - \sqrt{e} \geqslant \bar{y} - \sqrt{e}$$

then increment the count variable: $COUNT \leftarrow COUNT + 1$. (For lower-tailed tests, replace "$\geqslant$" with "$\leqslant$.")

3. Repeat 1–2 N times. The estimated p-value is $p = COUNT/N$.

This method is absolutely awful, since the mean of the bootstrap statistic $\bar{Y}^*$ is in fact $\bar{y}$, the observed sample mean, implying that the bootstrap p-value is always in a neighborhood of .5. Clearly, the first guideline of Hall and Wilson is violated, since resampling is not done under the null hypothesis: if the true distribution G is far from the null distribution G_0, the data y_i^* are sampled from $\hat{G}$, which approximates G, not G_0. In more general applications, Hall and Wilson report that such a method loses power to detect alternative hypotheses.

Recognizing this defect, Noreen (1989) presents a "shift method" for testing a hypothesis about the mean. In this method there is a very slight but important distinction: instead of comparing $\bar{y}^* - \sqrt{e}$ with $\bar{y} - \sqrt{e}$, one compares $\bar{y}^* - \bar{y}$ with $\bar{y} - \sqrt{e}$, for each resampled data set. Thus, even if the alternative hypothesis is true, the resampling distribution of $\bar{Y}^* - \bar{y}$ is similar to the true null distribution of $\bar{Y} - \sqrt{e}$ in that both are centered at zero. Thus, if the observed value $\bar{y} - \sqrt{e}$ is very *far* from zero, then the probability that $\bar{Y}^* - \bar{y}$ is more extreme (the bootstrap p-value), should be very small. As a result, the "shift method" has reasonable power to detect the alternative hypothesis.

The shift method is best motivated by distributions whose *shape* is constant, but whose *location* may be shifted under the alternative hypothesis. In this case, the resampling distribution of $\bar{Y}^* - \bar{y}$ will have approximately the same shape, as well as the same center, as the true null distribution of $\bar{Y} - \sqrt{e}$.

Algorithm 2.2 The Shift Method (Better)

0. Initialize a counting variable: $COUNT = 0$.
1. Generate resampled data $y_1^*, \ldots, y_n^*$, a *with replacement* sample from the original data $y_1, \ldots, y_n$.
2. If

$$\bar{y}^* - \bar{y} \geqslant \bar{y} - \sqrt{e}$$

then increment the count variable: $COUNT \leftarrow COUNT + 1$. (For lower-tailed tests, replace "$\geqslant$" with "$\leqslant$.")

3. Repeat 1–2 N times. The estimated p-value is $p = COUNT/N$.

Algorithm 2.2 is a definite improvement over Algorithm 2.1 in terms of power, but can be improved still further by making its actual significance level closer to the nominal level. This is accomplished by incorporating the second guideline, pivotality.

2.2.2 The Role of Pivoting

The second guideline of Hall and Wilson is that the statistics should be pivotal. In other words, under the null hypothesis, their sampling distributions should not depend upon which distribution G generated the data $\mathbf{Y}$. An example of a pivotal statistic is $T = (\bar{Y} - \mu)/(s/\sqrt{n})$, where s is the ordinary sample standard deviation. If the data are sampled from the normal distribution $\mathcal{N}(\mu, \sigma^2)$, then the distribution of T is the central t-distribution with $\nu = n - 1$ degrees of freedom, independent of the values of μ and σ. If the data are randomly sampled from a *nonnormal* population with mean μ and variance σ^2, then the statistic is *asymptotically pivotal*—regardless of μ, σ, or the actual population distribution, the distribution of T approaches $\mathcal{N}(0, 1)$ for increasing n, an application of the central limit theorem.

On the other hand, the statistic $T_\sigma = n^{1/2}(\bar{Y} - \mu)$ is *not* pivotal—its asymptotic distribution is $\mathcal{N}(0, \sigma^2)$. One may apply bootstrap testing using T_σ or the pivotal statistic T: both are valid with large enough sample sizes. However, as pointed out by Babu and Singh (1983), Abramovitch and Singh, (1985), and Hall and Wilson (1991), application of bootstrap resampling to pivotal statistics has *better* asymptotic properties. They show that the *rate* of convergence of the actual significance level to the nominal significance level is more rapid when pivotal statistics are resampled.

While the use and benefit of resampling pivotal statistics is well-accepted by theoreticians, this point remains unappreciated by many practitioners. Noreen (1989) employs the shift method of Algorithm 2.2, which does not use pivotal statistics, and finds that the bootstrap fares no better than the usual t-test. He comments:

> This is good news and bad news for the bootstrap. It is good news in the sense that the bootstrap does so well compared to the conventional test, even though the t test is known to be robust. It is bad news in the sense that there is no apparent advantage to using the bootstrap test when testing the mean for these populations even though the assumptions of the t test are violated.*

It turns out that the poor approximation of the bootstrap in Noreen's simulation study is mainly due to resampling the non-pivotal statistic T_σ, rather than the pivotal T. Theoretical explanations for this phenomenon may be found in Babu and Singh (1983), who prove that the bootstrap approximation is asymptotically more accurate for the t-statistic $n^{1/2}(\bar{y} - \mu)/s$ than for the centered statistic $n^{1/2}(\bar{y} - \mu)$; and that the same bootstrap approximation is asymptotically more accurate than the standard normal approximation.

In the context of Noreen's simulation study involving the mean of the lognormal distribution, the pivotal statistic of interest is

$$t = (\bar{y} - \sqrt{e})/(s/\sqrt{n})$$

The following algorithm is a simple extension of Algorithm 2.2, incorporating pivoting.

*Reprinted from *Computer-Intensive Methods for Testing Hypotheses*, by E. Noreen, copyright 1989, with permission from John Wiley & Sons, Inc.

Algorithm 2.3 The Shift and Pivot Method (Best)

0. Initialize a counting variable: $COUNT = 0$.
1. Generate resampled data $y_1^*, \ldots, y_n^*$, a *with replacement* sample from the original data $y_1, \ldots, y_n$.
2. If

$$\frac{\bar{y}^* - \bar{y}}{s^*/\sqrt{n}} \geqslant t$$

(where t is the original t-statistic) then increment the count variable: $COUNT \leftarrow COUNT + 1$. (For lower-tailed tests, replace "$\geqslant$" with "$\leqslant$.")
3. Repeat 1–2 N times. The estimated p-value is $p = COUNT/N$.

One may evaluate these methods by simulating the probability that an observed p-value falls below a pre-specified significance level, α. This gives us an estimate of the true significance level, α_0. When data are generated from a null hypothesis population, we would like α_0 to be as close as possible to α. If $\alpha_0 \gg \alpha$, then the method rejects the null hypothesis too often, leading us to believe that H' is true when in fact H is true. If $\alpha_0 \ll \alpha$, then the method rejects the null hypothesis too rarely, leading to a procedure with low power. The method is called "level-robust" when α_0 is reasonably close to α.

Since the p-values are themselves computed as a result of simulation, these types of studies can be extremely computer-intensive. In the following algorithm, we introduce two sample sizes, *NBOOT* and *NSIM*. The bootstrap sample size is *NBOOT*: for each simulated data set, the resampling-based p-value is based on *NBOOT* simulations. This process is replicated *NSIM* times, so *NBOOT* is the *inner loop* sample size, and *NSIM* is the *outer loop* sample size.

We replicate Noreen's study using the following algorithm.

Algorithm 2.4 Evaluation of Resampling-Based p-Values

1. Initialize a simulation counter: $COUNTS = 0$.
2. Initialize a bootstrap counter: $COUNTB = 0$.
3. Generate $y_1, \ldots, y_n$ from the standard lognormal, and compute

$$t = \frac{\bar{y} - \sqrt{e}}{s/\sqrt{n}}$$

4. Center the data: $c_i = y_i - \bar{y}$, $i = 1, \ldots, n$.

   4a. Let $y_1^*, \ldots, y_n^*$ be a *bootstrap* sample of the c_i, and compute

$$t^* = \frac{\bar{y}^*}{s^*/\sqrt{n}}$$

4b. If $t^* \geqslant t$, increment the bootstrap counter: $COUNTB \leftarrow COUNTB + 1$. (Use $\leqslant$ for lower-tailed tests.)

4c. Repeat 4a–4b *NBOOT* times. The bootstrap p-value p^* is

$$p^* = \frac{COUNTB}{NBOOT}$$

5. If $p^* \leqslant \alpha$, increment the simulation counter: $COUNTS \leftarrow COUNTS + 1$.
6. Repeat 2–5 *NSIM* times. The estimated actual significance level is

$$\hat{\alpha} = \frac{COUNTS}{NSIM}$$

Note that there are $NSIM \times NBOOT$ total simulated data sets of size n. In our study we used $NSIM = 4000$ and $NBOOT = 1000$; Noreen's study used $NSIM = 1000$ and $NBOOT = 99$. The larger sample sizes remove some simulation "noise" from the results, but do not alter the general conclusions.

The results of the simulation studies are presented in Table 2.1. We find that the centered and pivoted bootstrap test approximates the nominal levels *substantially* better than both the t-test and the simple shift method. In sharp contrast with Noreen's analysis, our results plainly show that the bootstrap test, when properly constructed, outperforms the t-test, when the underlying distribution is nonnormal.

We conclude this section with some general comments regarding choice of *NSIM* and *NBOOT*. If an analysis is to be performed on real data, then one should choose *NBOOT* to be as large as computing facilities and time requirements comfortably allow. In this way, there will be very little variation in the reported results: two people analyzing the same set of data will essentially come to the same conclusion. In most cases, this means that *NBOOT* (or *NRESAMPLE*, for non-bootstrap resampling) should be in the thousands, say 10,000 or more.

However, if the performance of resampling-based p-values is evaluated via simulation, it is much more critical that the *NSIM* sample size (the outer loop) be large. To further analyze this problem, we follow the mathematical development found in Hope (1968).

Using Algorithm 2.4 and the associated experiment with lognormal data, the "true bootstrap p-value" is

$$P = \Pr(T^* \geqslant t \mid \hat{G})$$

which is a function of the original sample. This is a random quantity varying from outer simulation to outer simulation: each outer simulation results in a different sample $y_1, \ldots, y_n$, in turn implying a different empirical distribution estimate $\hat{G}$. The value P could be obtained by setting $NBOOT = \infty$ in the inner loop of Algorithm 2.4. In our simulation experiment, we are trying to estimate the quantity

$$\alpha_0 = \Pr(P \leqslant \alpha)$$

Table 2.1 Proportions of Rejections for Resampling-Based Tests Using Lognormal Data

Test	$n = 20$	$n = 40$	$n = 80$	$n = 160$
$\alpha = .05$				
Upper-Tail Test				
t-test[1]	0.008	0.010	0.015	0.022
shift method bootstrap[1]	0.014	0.014	0.018	0.027
shift and pivot bootstrap[2]	0.030	0.038	0.044	0.043
Lower-Tail Test				
t-test[1]	0.153	0.149	0.124	0.109
shift method bootstrap[1]	0.180	0.167	0.136	0.119
shift and pivot bootstrap[2]	0.106	0.087	0.076	0.066
$\alpha = .10$				
Upper-Tail Test				
t-test[1]	0.032	0.038	0.062	0.067
shift method bootstrap[1]	0.041	0.040	0.053	0.060
shift and pivot bootstrap[2]	0.082	0.085	0.092	0.094
Lower-Tail Test				
t-test[1]	0.210	0.211	0.178	0.152
shift method bootstrap[1]	0.229	0.216	0.191	0.162
shift and pivot bootstrap[2]	0.163	0.143	0.134	0.119

[1] Noreen (1989), p. 74: *NSIM* = 1000, *NBOOT* = 99, reprinted from *Computer-Intensive Methods for Testing Hypotheses*, by E. Noreen, copyright 1989, with permission from John Wiley & Sons, Inc.
[2] Our study: Algorithm 2.4, *NSIM* = 4000, *NBOOT* = 1000

When the outer simulation and associated data $y_1, \ldots, y_n$ are held fixed, the final bootstrap total *COUNTB* is a binomial random variable with parameters P (now a *fixed* quantity, since the outer simulation counter is fixed), and *NBOOT*. The (conditional) probability that the bootstrap p-value is less than or equal to α is then

$$\begin{aligned}\Pr(P^* \leqslant \alpha \mid P) &= \Pr(T^* \geqslant t \text{ in } \leqslant NBOOT \times \alpha \text{ samples} \mid P)\\ &= \sum_{i=0}^{[NBOOT\times\alpha]} \binom{NBOOT}{i} P^i(1-P)^{NBOOT-i}\end{aligned} \tag{2.1}$$

where $[NBOOT \times \alpha]$ refers to the greatest integer in $NBOOT \times \alpha$.

Suppose now that the outer loop had infinitely many simulations, or $NSIM = \infty$. Using (2.1), the resampling-based estimate of α_0 is then

$$\begin{aligned}\Pr(P^* \leqslant \alpha) &= \mathrm{E}\{\Pr(P^* \leqslant \alpha \mid P)\}\\ &= \int_0^1 \sum_{i=0}^{[NBOOT\times\alpha]} \binom{NBOOT}{i} p^i(1-p)^{NBOOT-i} g(p)\, dp\end{aligned} \tag{2.2}$$

where $g(p)$ is the density function for the random variable P. Actually, the random variable P is discrete, as there are only finitely many distinct bootstrap samples which may be selected. However, the distribution is very nearly continuous for most applications involving moderate n, since there are n^n possible bootstrap samples. In common parametric applications, the density of the p-value is continuous and uniform. For example, if the outer loop data in Algorithm 2.4 were generated from a normal rather than lognormal distribution, and if the inner loop data were also generated from a normal distribution, then the distribution of P would be exactly uniform. Using the uniform distribution to approximate $g(p)$, equation (2.2) yields

$$\int_0^1 \sum_{i=0}^{[NBOOT \times \alpha]} \binom{NBOOT}{i} p^i (1-p)^{NBOOT-i}\, dp = \frac{[NBOOT \times \alpha] + 1}{NBOOT + 1}$$

which is bounded between $\alpha_0 - [\alpha_0/(NBOOT + 1)]$ and $\alpha_0 + [(1 - \alpha_0)/(NBOOT + 1)]$. Thus, the resampling-based estimate of α_0 is within $1/(NBOOT + 1)$ of the true value under the simplifying assumption of a uniform p-value distribution, and assuming $NSIM = \infty$.

Suppose, on the other hand, that $NBOOT = \infty$, and $NSIM$ is finite. In other words, the value P is evaluated without error each time within the inner loop. In this case, the outer loop total $COUNTS$ is binomially distributed with parameters α_0 and $NSIM$. Thus, the standard error is

$$\text{SE}(COUNTS/NSIM) = \{\alpha_0(1 - \alpha_0)/NSIM\}^{1/2}$$

As a point of specific comparison, consider the errors from Algorithm 2.4 when (1) $(NBOOT, NSIM) = (500, \infty)$ and (2) $(NBOOT, NSIM) = (\infty, 500)$, assuming the true value of α_0 is .05. Using the uniform distribution to approximate $g(p)$, the probability estimate is within $1/501 = .002$ of the true value .05 with *certainty* in case (1). However, in case (2), we can say only that the estimate is within $1.96\sqrt{(.05)(.95)/500} = .019$ of the true value, and only with 95% probability! The margin of error in case (2) is approximately *ten times* the margin of error for case (1), and the margin of error occasionally could be larger.

While the assumption of a uniform distribution for $g(p)$ is definitely incorrect, it provides a useful approximation. Evaluation of the magnitude of the error is much more complex for nonuniform $g(p)$, although Hope (1968) provides some bounds and approximations. The main point of the preceding analysis is that it is much more important for the outer loop simulation size $NSIM$ to be large than it is for the inner loop simulation size to be large. Ideally, of course, we would like to have both sizes equal to infinity, but this is not practical. When performing such massive simulation studies, time constraints dictate that resources be allocated. The preceding analysis suggests that it is better to allocate more to $NSIM$ than to $NBOOT$.

For further analysis of the allocation of simulation sample sizes, see Oden (1991).

2.2.3 Multiple Testing Guidelines

Hall and Wilson's guidelines may be extended to multiple testing. In this case, we shall construct test statistics and resampling methods so that (1) the guidelines apply to the marginal distributions of all test statistics, (2) resampling reflects the *complete* null hypothesis, and (3) all test statistics are pivotal.

In addition to the univariate pivotality requirement, a multivariate pivotality condition is also desirable. We define such a concept, and call it *subset pivotality*. It represents an ideal situation regarding the joint distribution of the test statistics used to test the individual hypotheses, and is described as follows.

Condition 2.1 Subset Pivotality
The distribution of **P** has the *subset pivotality* property if the joint distribution of the subvector $\{P_i;\ i \in K\}$ is identical under the restrictions $\cap_{i \in K} H_{0i}$ and H_0^C, for all subsets $K = \{i_1, \ldots, i_j\}$ of true null hypotheses.

The subset pivotality condition is important for two reasons. First, resampling is particularly convenient under this condition: resampling is done under the complete null hypothesis H_0^C, rather than under partial hypotheses H_0^K. Second, when subset pivotality holds, it will be shown that such resampling-based methods control the FWE in the *strong* sense (at least approximately, under asymptotic subset pivotality). Without this condition, resampling under H_0^C can be assumed to control only the FWEC.

In many cases, we consider analyses which are approximate in an asymptotic sense. Suppose the data **Y** is assumed to come from a *sequence* of experiments involving increasing amounts of data. Indexing the experiments by $t = 1, 2, \ldots$, the p-value for testing H_i is indexed correspondingly as P_{it}, and the entire vector is $\mathbf{P}_t$. The following condition is the large-sample analog of subset pivotality, and is less restrictive than Condition 2.1:

Condition 2.2 Asymptotic Subset Pivotality
Again let $K = \{i_1, \ldots, i_j\}$ denote any set of *true* null hypotheses. The distribution of $\mathbf{P}_t$ has the *asymptotic subset pivotality* property if the joint distribution of the subvector $\{P_{it};\ i \in K\}$ converges to the same limit law (as $t \rightarrow \infty$) under the restrictions $\cap_{i \in K} H_{0i}$ and H_0^C.

The subset pivotality condition requires that the multivariate distribution of any subvector of p-values is unaffected by the truth or falsehood of hypotheses corresponding to p-values not included in the subvector. In typical parametric analyses, the marginal distribution of any P_i is $\mathcal{U}[0, 1]$ whenever H_{0i} is true; therefore, Condition 2.1 refers only to dependence structures in these cases.

When data are not normally distributed, a univariate test procedure has asymptotic validity if the probability of rejecting H converges to α under H_0 for large sample sizes, for any α between zero and one. Mathematically, as $t \rightarrow \infty$,

$$\Pr(\text{Reject } H \mid H_0 \text{ is true}) \rightarrow \alpha$$

for all $\alpha \in (0, 1)$. This implies that the p-value distribution converges to $\mathcal{U}[0, 1]$, asymptotically, under H_0. Thus, as long as the marginal test procedures are asymptotically valid, Condition 2.2 concerns only the asymptotic dependence structure.

Example 2.1 An Example of Subset Pivotality
Suppose the sampled data consist of 6-dimensional vectors

$$\mathbf{Y}_i' = (\, Y_{i1} \quad \ldots \quad Y_{i6}\,),\ i = 1, \ldots, n$$

Suppose the n data vectors are a random sample from a multivariate normal process with mean vector $\boldsymbol{\mu} = (\,\mu_1 \quad \ldots \quad \mu_6\,)$ and covariance matrix $\Sigma = \{\sigma_{ij}\}$. Hypotheses of interest might involve the means: H_j: $\mu_j = 0$, for example. Such tests are used when the variables are differenced values, or the result of linear contrasts. Univariate t-tests commonly are used for each individual test: T_j is used to test H_j.

Suppose a subset of hypotheses 1–6 is true, for example, hypotheses $\{3, 5, 6\}$ are true but the remaining are false. The joint distribution of the test statistics $\{T_3, T_5, T_6\}$ is a special multivariate t distribution described by Siddiqui (1967), depending only upon n and the correlations among $\{Y_3, Y_5, Y_6\}$. In particular, the distribution does not depend on the values of the remaining means $\{\mu_1, \mu_2, \mu_4\}$. The distribution of $\{T_3, T_5, T_6\}$ is the same, whether all six hypotheses are true, or whether just H_3, H_5, and H_6 are true. Similar results hold for any arbitrary subset of $\{T_1, T_2, \ldots, T_6\}$. The subset pivotality condition holds for the vector of p-values, since the p-values are constructed from these t-statistics. □

Example 2.2 An Example Where Subset Pivotality Fails
Consider the same problem setting as Example 2.1, but suppose we are interested in testing whether the correlations are zero: H_{jk}: $\rho_{jk} = 0$. In large samples, the commonly used test statistics are equivalent to $T_{jk} = n^{1/2} r_{jk}$, where r_{jk} denotes the ordinary sample correlation between variables j and k.

Suppose that hypotheses H_{12} and H_{13} are true, with all others false. Using the results found in Aitken (1969, 1971), the joint distribution of $\{T_{12}, T_{13}\}$ is approximately normal, with means zero, variances one, and correlation ρ_{23}. Since this distribution depends upon the value of ρ_{23}, the subset pivotality condition fails.

□

2.3 SINGLE-STEP METHODS

Loosely stated, *single-step* methods are simultaneous test procedures, or STP's, that perform equivalent multiplicity adjustments for all tests, regardless of the *ordering* of the observed p-values $p_1, \ldots, p_k$, and without considering any predetermined sequence of hypotheses (Hochberg and Tamhane, 1987). On the other hand, *stepwise* STP's allow different adjustment techniques for different hypotheses, depending upon how the hypotheses are ordered. It is natural to order the hypotheses according to the size of the p-values, $p_{(1)} \leqslant \cdots \leqslant p_{(k)}$. In certain situations, experimental or logical constraints might dictate that some particular predetermined sequence of hypotheses be tested.

2.3.1 Bonferroni-Type Methods

The simplest single-step method is the Bonferroni method, which rejects hypothesis H_i when the p-value p_i is less than α/k, where α is the preset FWE level and k is the number of tests. Using definition (1.3), this leads to the *Bonferroni single-step adjusted* p-value

$$\tilde{p}_i = \min(kp_i, 1) \tag{2.3}$$

Protection of FWEC motivates the Bonferroni single-step method:

$$\begin{aligned}\Pr(\text{Reject at least one } H_i \mid H_0^C) &= \Pr\Big(\min_{1\leqslant i\leqslant k} P_i \leqslant \alpha/k \mid H_0^C\Big) \\ &\leqslant \sum_{i=1}^{k} \Pr(P_i \leqslant \alpha/k \mid H_0^C)\end{aligned}$$

The notation "$\mid H_0^C$" means "given H_0^C," or "assuming all H_{0i} are true." The probability inequality is the well-known Bonferroni inequality. Assuming that all p-value distributions are $\mathcal{U}[0,1]$ under their respective null hypotheses H_{0i}, the upper bound becomes $k(\alpha/k) = \alpha$. Since the actual probability of rejecting at least one null hypothesis is *less than* the nominal FWE level α, the Bonferroni method is conservative when the marginal p-value distributions are uniform.

If the distribution of $\{\min P_i \mid H_0^C\}$ were known, the Bonferroni method could be improved in two areas: first, there would be no need for the conservative approximation via Bonferroni's inequality; second, the assumption of uniform p-value distributions would not be required. In the case of nonnormal data, the p-value distributions can be markedly *non*uniform, which affects the operating characteristics of the procedure. The Bonferroni method can be extremely conservative, more than one would expect by ignoring correlation structure alone. This can occur with sparse binary data, as well as in continuous applications where the tails of the sampling distributions are extremely light. On the other hand, for heavy-tailed sampling distributions, the Bonferroni method can be extremely *anti*conservative, implying that null hypotheses are rejected much more frequently than the stated FWE level. A specific illustration of this phenomenon is given in Section 2.5.

Very closely related to the Bonferroni method is the *Šidák method*, which rejects hypothesis H_i when the p-value p_i is less than $1-(1-\alpha)^{1/k}$, where α is the preset FWE level and k is the number of tests. This leads to the *Šidák single-step adjusted* p-value

$$\tilde{p}_i = 1-(1-p_i)^k \tag{2.4}$$

The method is exact for protecting the FWEC if all p-values are independently distributed as $\mathcal{U}[0,1]$ since

$$\begin{aligned}
&\Pr(\text{Reject at least one } H_i \mid H_0^C) \\
&= \Pr\Big(\min_{1\leqslant i\leqslant k}\{1-(1-P_i)^k\} \leqslant \alpha \mid H_0^C\Big) \\
&= 1-\Pr\Big(\min_{1\leqslant i\leqslant k}\{1-(1-P_i)^k\} > \alpha \mid H_0^C\Big) \\
&= 1-\Pr(P_i > 1-(1-\alpha)^{1/k} \text{ for } \textit{all } i \mid H_0^C) \\
&= 1-\prod_{i=1}^{k}\Pr(P_i > 1-(1-\alpha)^{1/k} \mid H_0^C) \quad \text{(assuming independence)} \qquad (2.5) \\
&= 1-\{(1-\alpha)^{1/k}\}^k \quad (\text{assuming } P_i \sim \mathcal{U}[0,1]) \\
&= \alpha
\end{aligned}$$

Without the independence assumption, the equality (2.5) becomes the inequality "$\leqslant$" for very general conditions. This implies that the probability of rejecting at least one H_i, given that all H_i are true, is *at most* α when the p-values have uniform marginal distributions. Like the Bonferroni method, the Šidák method is conservative.

There is extensive literature on this probability inequality: it is commonly called the Šidák inequality from Šidák (1967). One version of this inequality is given by

$$\Pr(|Z_1| \leqslant c_1, |Z_2| \leqslant c_2, \ldots, |Z_k| \leqslant c_k) \geqslant \prod_{i=1}^{k}\Pr(|Z_i| \leqslant c_i)$$

where the Z_i are jointly distributed random variables and the c_i are positive constants. Šidák shows that this inequality holds when the vector $\mathbf{Z}$ is multivariate normal with mean zero, and Jogdeo (1977) shows that the inequality applies to a larger collection of applications, including multivariate t- and F-distributions.

The use of the Šidák-Jogdeo inequality for multiple testing applications is as follows. In equation (2.5), the condition $P_i > 1-(1-\alpha)^{1/k}$ is often equivalent to $|Z_i| < c_{i,k,\alpha}$, where $c_{i,k,\alpha}$ is the $(1-\alpha)^{1/k}$ quantile of the distribution of $|Z_i|$. The Z_i are often distributed as standard normal or Student's t. If the Šidák-Jogdeo inequality applies for the joint distribution of such Z_i, then the Šidák multiple testing method is conservative. Holland and Copenhaver (1987) provide particular applications and additional references.

We call the adjusted p-values (2.4) the *single-step Šidák* adjustments.

While the Šidák adjustments usually are conservative, they are less conservative than the Bonferroni adjustments, meaning that the adjusted p-values are smaller, as indicated in Table 2.2. For very small p-values, there is apparently little difference between these methods. The difference depends upon p and k, as may be seen from the Taylor-series expansion

$$1-(1-p)^k = kp - \tfrac{1}{2}k(k-1)(1-\xi)^{k-2}p^2$$

Table 2.2 Comparison of Bonferroni ($\tilde{p}_B$) and Šidák ($\tilde{p}_S$) Adjusted p-Values

Unadjusted	Adjusted	k			
p-Value	p-Value	5	10	20	00
.0001	$\tilde{p}_B$	.0005	.0010	.0020	.0040
	$\tilde{p}_S$	.0005	.0010	.0020	.0040
.0005	$\tilde{p}_B$	.0025	.0050	.0100	.0200
	$\tilde{p}_S$	.0025	.0050	.0100	.0198
.0010	$\tilde{p}_B$	.0050	.0100	.0200	.0400
	$\tilde{p}_S$	.0050	.0100	.0198	.0392
.0050	$\tilde{p}_B$	.0250	.0500	.1000	.2000
	$\tilde{p}_S$	.0248	.0489	.0954	.1817
.0100	$\tilde{p}_B$	.0500	.1000	.2000	.4000
	$\tilde{p}_S$	.0490	.0956	.1821	.3310
.0500	$\tilde{p}_B$	.2500	.5000	1.0000	1.0000*
	$\tilde{p}_S$	.2262	.4013	.6415	.8715

* $\tilde{p}_B$ truncated to 1.0000.

Or, since the Šidák and Bonferroni adjustments are $\tilde{p}_S = 1 - (1-p)^k$ and $\tilde{p}_B = kp$, respectively, we have

$$\tilde{p}_S = \tilde{p}_B - \tfrac{1}{2}k(k-1)(1-\xi)^{k-2}p^2$$

where ξ is between p and 0. This expression shows (1) that the Šidák adjustments must be smaller than the Bonferroni adjustments, (2) that the difference between the two becomes smaller with smaller p, and (3) that the difference becomes larger with larger k.

2.3.2 Resampling Method

The Bonferroni-style methods fail to incorporate dependence and distributional characteristics of the P_i's. If we knew the joint distribution of the p-values, we could compute the *single-step adjusted* p-values, defined as

$$\tilde{p}_i = \Pr\left(\min_{1 \leqslant j \leqslant k} P_j \leqslant p_i \mid H_0^C\right) \qquad (2.6)$$

For example, in a resampling scheme, $\tilde{p}_i$ is the probability that the smallest p-value in the resampled data set is smaller than the observed p-value p_i in the original experiment.

Sometimes this probability can be computed analytically. Using Example 1.1 (the fictitious "oat bran" study), the p-values are distributed independently as $\mathcal{U}[0, 1]$ under the complete null hypothesis, thus

$$\begin{aligned}\tilde{p}_i &= \Pr\Big(\min_{1\leqslant j\leqslant k} P_j \leqslant p_i \mid H_0^C\Big)\\ &= 1 - \Pr(P_j > p_i \text{ for all } j \mid H_0^C)\\ &= 1 - (1 - p_i)^k\end{aligned}$$

These correspond to the Šidák adjustments in this case. If the p-values were not independent, the adjustments (2.6) ordinarily would be *smaller* (resulting in more powerful tests) than the Šidák adjustments, owing to the Šidák-Jogdeo inequality.

Using the data from Table 1.2, the single-step Šidák adjusted p-values are $\tilde{p}_1 = 1 - (1 - .616)^{10} = .9999$, $\tilde{p}_2 = 1 - (1 - .043)^{10} = .3557$, etc. To be considered FWE significant, an unadjusted p-value must be less than $1 - (1 - \alpha)^{1/k}$ so that $\tilde{p} < \alpha$; when $\alpha = .05$ and $k = 10$ this means that $p_i < .0051$. The Bonferroni critical value, $\frac{.05}{10} = .0050$, is almost identical.

In situations involving dependence among the p_i, the distribution of $(\min_{1\leqslant j\leqslant k} P_j \mid H_0^C)$ is usually intractable. However, in many cases vectors $\mathbf{P}^*$, having the same distribution as the original p-value-vector $\mathbf{P}$, may be easily simulated. When the original distribution is unknown, an estimate can be substituted via resampling or other simulation methods. The adjusted p-values, $\tilde{p}_i$, may then be approximated by repeatedly simulating the minimum p-value statistic. For example, one might randomly generate data sets such as those displayed in Tables 1.2 and 1.3, calculate the minimum p-value for each simulated data set, and note the proportion of simulations for which the minimum p-value falls below an individual observed p-value of interest.

A more formal description of this process is given in the following algorithm. The sample size N corresponds to *NBOOT* (or, more generally, to *NRESAMPLE*) of previous algorithms. There is no outer loop simulation size *NSIM* in this algorithm, so the definition of N is clear.

Algorithm 2.5 Single-Step Resampling Method

0. Initialize counting variables: $COUNT_i = 0$, $i = 1, \ldots, k$.
1. Generate a p-value-vector $\mathbf{p}^* = (p_1^*, \ldots, p_k^*)$ from the same (or at least, approximately the same) distribution as the *original* p-values under the complete null hypothesis. Usually, this step consists of two parts: first generate data $\mathbf{Y}^*$ from an estimated complete null distribution, then compute the p-values $\mathbf{p}^*$ from the simulated data $\mathbf{Y}^*$. The resampled p-values $\mathbf{p}^*$ are computed using the same calculations which produced the original p-values $\mathbf{p}$ from the original experimental data $\mathbf{Y}$.
2. If $\min_{1\leqslant j\leqslant k} p_j^* \leqslant p_i$, then $COUNT_i \leftarrow COUNT_i + 1$.
3. Repeat 1–2 N times. The estimated value of $\tilde{p}_i$ is $\tilde{p}_i^{(N)} = COUNT_i/N$.

It should be noted here that, when applying this algorithm, one must be careful how one generates data vectors $\mathbf{Y}^*$. One should consider carefully the form of the underlying distribution, which parameters should be estimated, and *how* the parameters are to be estimated. While simple *with replacement* sampling works well in many situations, the procedure must be modified in other situations. For example, in Chapters 3 and 4 we discuss the importance of resampling *residuals*, rather than raw data values, in linear model applications. The bottom line is that resampling should be *model-driven*: assume the form of the model *first*, then construct a resampling scheme to reflect the model's characteristics.

Another extremely important caveat: Corresponding to guidelines of Hall and Wilson (1991), it is *crucial* that the complete null hypothesis $H_0^C = \cap_{i=1}^k H_{0i}$ be true for the estimated distribution $\hat{G}_0$ of the $\mathbf{Y}^*$. Otherwise, p-values will be simulated from non-null distributions, potentially skewing the adjustments severely toward non-significance. For example, when testing for a difference in means of two groups, one should force the group means to be identical in the resampled data. If the group means are different for the resampled data, the resampled p-values will tend to be small (significant), in turn skewing the resampling-based p-value toward non-significance.

The simulation standard error of the estimate $\tilde{p}_i^{(N)}$ comes from the binomial distribution, and is simply

$$\mathrm{SE}(\tilde{p}_i^{(N)}) = [\{\tilde{p}_i(1 - \tilde{p}_i)\}/N]^{1/2}$$

This error can be brought arbitrarily close to zero with larger N and increasingly precise random number generators. In practice, $N \geqslant 10{,}000$ is suggested, as indicated by Table 2.3. With increasingly available and powerful personal computers, large simulations are very feasible.

Recent application of techniques such as *importance sampling* might be useful in reducing the number (N) of simulations. These ideas have been explored by Do and Hall (1991), Efron (1990), Graham, Hinkley, John, and Shi (1990), Mehta, Patel, and Senchaudhuri (1988), and Morganthaler and Tukey (1991). If only an Accept/Reject decision is required at a particular α, sequential analysis also may be used to reduce the simulation size, as reported by Besag and Clifford (1991).

Example 2.3 Homoscedastic Normal ANOVA

Let $Y_{ij} = \mu_i + \epsilon_{ij}$, where $i = 1, \ldots, g$; $j = 1, \ldots, n_i$, and with the $\{\epsilon_{ij}\}$ independent and identically distributed as $\mathcal{N}(0, \sigma^2)$. This model completely defines the class $\mathcal{G}$ of multivariate normal distributions for $\mathbf{Y}$: a particular $G \in \mathcal{G}$ is determined by the particular settings of the parameters μ_i and σ^2.

Hypotheses of interest might be linear contrasts among the means, $H_l\colon c_{l1}\mu_1 + \cdots + c_{lg}\mu_g = d_l$, or $H_l\colon \mathbf{c}_l'\boldsymbol{\mu} = d_l$, with one- or two-sided alternatives. Letting $SSE = \sum\sum(Y_{ij} - \bar{Y}_i)^2$, $s^2 = SSE/\sum(n_i - 1)$, and $\Delta = \mathrm{diag}\{n_i\}$, the usual test statistic for each H_l is

$$T_l = \frac{(\sum_i c_{li}\bar{Y}_i - d_l)}{s(\mathbf{c}_l'\Delta^{-1}\mathbf{c}_l)^{1/2}}$$

with the p-value computed from the central t-distribution having $\nu = \sum(n_i - 1)$ degrees of freedom.

Table 2.3 Standard Error of $\tilde{p}_i^{(N)}$ as a Function of N

	$\tilde{p}_i$		
N	.1000	.0500	.0100
100	.0300	.0218	.0099
1,000	.0095	.0069	.0031
10,000	.0030	.0022	.0010
100,000	.0009	.0007	.0003

In this example, the p-values have the subset pivotality property (Condition 2.1). The joint null distribution of any subvector $\{T_j; j \in K\}$ does not depend on the truth or falsehood of any hypotheses H_j, $j \notin K$. To see this, note that the test statistics are completely invariant to the means μ_i under the null hypothesis. In fact, the joint distribution of any subvector $\{T_j; j \in K\}$ is a *multivariate t* distribution (discussed by Hochberg and Tamhane, 1987, pp. 374–375, for example).

While the exact functional form of the complete null distribution of the p-value vector is known, it might be extremely complicated. The beauty of the resampling method is that the actual form of the distributions need not be specified; one needs only to be able to simulate vectors $\mathbf{P}^*$ having this joint distribution. Since the t-statistics are independent of the mean values μ_i and the error variance σ^2 under their respective null hypotheses, we may take the mean values to be identically equal to zero and the variance equal to one in the resampling model. When the data Y_{ij}^* are simulated from such a model, the distribution of the simulated test statistics $(T_1^*, \ldots, T_k^*)$ is identical to the joint null distribution (under the complete null hypothesis) of the *original* test statistics $(T_1, \ldots, T_k)$: both distributions are central multivariate t with the same correlation and degrees of freedom parameters. Thus, the joint distributions of the p-value vectors $\mathbf{P}$ and $\mathbf{P}^*$ are also identical.

The resampling method is demonstrated in the following algorithm.

Algorithm 2.6 Normal Linear Model Application

0. Initialize counting variables as before: $COUNT_l = 0, l = 1, \ldots, k$.
1. Generate data

$$y_{ij}^* = \epsilon_{ij}^* \tag{2.7}$$

   where the ϵ_{ij}^* are independent and identically distributed as $\mathcal{N}(0, 1)$.
2. Compute the statistics

$$t_l^* = \frac{(\sum_i c_{li}\bar{y}_i^*)}{s^*(\mathbf{c}_l'\Delta^{-1}\mathbf{c}_l)^{1/2}}$$

   and the associated p-values p_l^* from the central t_ν distribution.

3. If $\min_{1 \leqslant m \leqslant k} p_m^* \leqslant p_l$, where the p_l are the p-values computed from the original experiment, then increment the count variable: $COUNT_l \leftarrow COUNT_l + 1$.
4. Repeat 1–3 N times. The estimated value of $\tilde{p}_l$ is $\tilde{p}_l^{(N)} = COUNT_l/N$. □

Further applications using algorithms related to this one are given in Chapter 3.

This resampling method may be made more efficient by generating $\bar{y}_i^*, i = 1, \ldots, g$, independently from the normal distributions $\mathcal{N}(0, 1/n_i)$; and generating $(s^*)^2$, distributed independently of the $\bar{y}_i^*$, $i = 1, \ldots, g$ as a χ_ν^2/ν random variable. This will speed up the simulations. However, there are two potential benefits from using the approach in Algorithm 2.6. First, if software is readily available for computing tests based on the data structure of the one-way ANOVA model, then one need only to supply the software with data y_{ij}^* from model (2.7), and to save the necessary output from each analysis. Second, model (2.7) illustrates a connection that this approach has with the bootstrap method: entire data sets are re-created, and the same analysis that was performed on the original data is performed on each re-created data set. This type of analysis provides the experimenter with complete replications of the original experiment, and visibly demonstrates the variability of p-values as a function of the number of questions asked.

The adjusted p-values given by Algorithm 2.6 could have been made using the test-statistics T_i, further simplifying the calculations. If all alternative hypotheses are two-tailed, the following definition of the adjusted p-value is equivalent to definition (2.6):

$$\tilde{p}_i = \Pr\left(\max_{1 \leqslant j \leqslant k} |T_j| \geqslant |t_i| \mid H_0^C\right) \tag{2.8}$$

Here T_i and t_i represent the random and observed test statistics, respectively.

While equation (2.8) perhaps is more familiar to those who have studied multiple testing arguments, and while the computational burden is generally less when simulating (2.8) than it is when simulating (2.6), equation (2.8) lacks flexibility for general use, and can lead to *unbalanced* multiplicity adjustments. For example, suppose some tests were upper-tailed and others were two-tailed. The two-sided test statistics $|T_i|$ will then dominate the upper-tailed test statistics T_i: the distribution of the maximum is primarily determined by the absolute values, since they are stochastically larger. This will result in a lack of balance: not all tests contribute equally to the adjusted p-values.

As a more extreme example, suppose the overall F-test for $H: \mu_1 = \cdots = \mu_g$ were included in the collection of tests. It is clear that the maximum of both t-distributed random variables and F-distributed random variables would not be appropriate, since the F-statistic tends to be much larger than the t-statistic. If multiplicity-adjusted significance is determined using the distribution of $\max(|T_1|, \ldots, |T_k|, F)$, then it will be very difficult to achieve significance for the t-tests: they must be significant relative to the upper quantile of the F-distribution in this case. On the other hand, there will be relatively little multiplicity adjustment for the overall F test, since the random variables $|T_i|$ will contribute little to the distribution of $\max(|T_1|, \ldots, |T_k|, F)$.

In some instances, lack of balance is desirable. When certain tests are considered more important, then they should be given more "weight," and multiplicity adjustments should be less. The experimenter can input such weights directly into the multiplicity adjustment procedure. Resampling-based methods for performing such weighted analyses are discussed in Chapter 6.

In order that all tests have similar weight (as occurs with Bonferroni-style corrections, when all k tests are given equal weight), it is desirable that the null distributions of the measures of evidence against null hypotheses be comparable. Thus, when minimal (or maximal) statistics are considered, all individual test statistics will have similar weight in determining the distribution of the minimum (or maximum). The p-values are natural quantities to use in this regard since their null distributions are all uniform on [0, 1] (or at least approximately so). The property that the marginal p-value distributions be comparable (as they are when all are $\mathcal{U}[0, 1]$) is analogous to the "balance" property described by Beran (1988a), who uses bootstrap methods to construct simultaneous confidence sets.

Fast and accurate algorithms for evaluating standard cumulative probability functions such as the t-distribution make computer-intensive resampling of p-values feasible from the standpoint of processing time. However, in many cases all tests in an experiment utilize the same test statistic, for example, a two-sided t-test. In these cases, the resampling of t-statistics is equivalent to resampling the p-values derived from such t-statistics, and calculation of the p-values within each resampled data set wastes processing time.

2.3.3 Error Rate Control for Single-Step Adjustments

Among resampling-style p-values, a *single-step adjusted* p-value is the simplest to understand and explain: informally, it is the proportion of replicated (resampled) experiments for which the minimum p-value is smaller than the given experimentally observed p-value. The formal definition is given by equation (2.6). Step-down adjustments (discussed in Sections 2.6–2.8) offer improved power (smaller adjusted p-values) while maintaining strong control of the FWE.

A benefit of the resampling-based single-step adjustments is that they allow us to claim directional associations when two-sided tests are used. The methods often control the Type III error rate referred to in Section 1.2: if the *sign* of the effect is positive, and if the effect is significant according to the single-step two-tailed adjusted p-value, then, in many cases, one may claim that the effect is indeed positive. This property may be demonstrated easily when there is an equivalence between multiple testing and the simultaneous confidence intervals. Such correspondences are shown for two-sided tests and continuous data in Chapters 3 and 4. Whether resampling-based *step-down* methods offer the same level of protection against type III errors is an open question.

In this section, we present details regarding control of the Type I FWE error for *single-step* adjustments in general applications. Control of the Type III FWE occurs frequently as well. In Chapters 3 and 4, we demonstrate this property in univariate and multivariate linear model applications.

Since the single-step adjusted p-values (2.6) are all computed under the complete null hypothesis H_0^C, it should not be surprising that the FWEC is controlled. If a null hypothesis H_i is rejected when $\tilde{p}_i \leqslant \alpha$, where $\tilde{p}_i$ is computed from (2.6), then

$$\Pr(\text{Reject at least one } H_i \mid H_0^C) = \Pr(\text{at least one } \tilde{P}_i \leqslant \alpha \mid H_0^C)$$

Corresponding to the notation of Section 2.1, $\tilde{P}_i$ denotes the *random* (i.e., prior to its realization) adjusted p-value.

Using the single-step adjusted p-value definition of equation (2.6), it is seen that $\tilde{P}_i \leqslant \alpha$ if and only if $P_i \leqslant c_\alpha$, where c_α denotes the α *quantile* of the minimum p-value distribution (see Figure 2.1). Formally, letting S denote the sample space of the minimum p-value,

$$c_\alpha = \max\left\{p \in S \mid \Pr\left(\min_{1\leqslant j\leqslant k} P_j \leqslant p \mid H_0^C\right) \leqslant \alpha\right\}$$

In words, c_α is the *largest* observable value of the random variable $\min_{1\leqslant j\leqslant k} P_j$ for which the cumulative probability is *at most* α. Quantiles are not uniquely defined for discrete distributions. Some authors define quantiles to be the *smallest* observable value of the random variable for which the cumulative probability is *at least* α. This alternative definition can yield a *higher* value of c_α.

Noting the equivalence

$$\{\tilde{P}_i \leqslant \alpha\} \Longleftrightarrow \{P_i \leqslant c_\alpha\}$$

we have

$$\Pr(\min \tilde{P}_i \leqslant \alpha \mid H_0^C) = \Pr(\min P_i \leqslant c_\alpha \mid H_0^C) \leqslant \alpha$$

so that this STP controls the FWEC at or below the level α. If the distribution of the minimum p-value is continuous in a neighborhood of c_α, then the procedure controls the FWEC *exactly* at the level α.

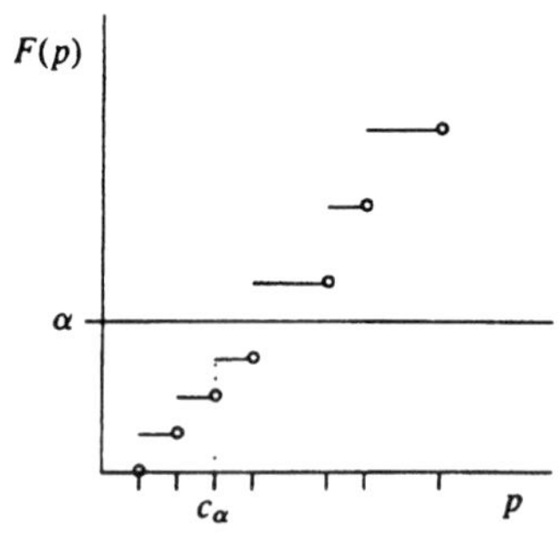

FIGURE 2.1. Cumulative probability distribution of the minimum p-value (discrete case).

An extremely important caveat: the preceding demonstration assumes that we are able to evaluate the adjusted p-values according to definition (2.6) *without error*, either from simulation or from the estimated distribution function. In some cases this assumption is valid, for example, when the adjusted p-values are evaluated using analytic distributions such as the studentized range, or when they are evaluated analytically from permutation distributions. However, in many applications, as when using bootstrap sampling to estimate possibly nonnormal distributions, this assumption is not met. In such cases, we can only claim *approximate* control of the FWE, where the closeness of the approximation depends upon the accuracy of the resampling-based estimates of the underlying distributions. Simulation analysis (as given in Section 2.5, for example) can be helpful when attempting to evaluate such approximations.

What about controlling the FWE under *partial* null hypotheses? Since all probabilities have been computed under the *complete* null hypothesis, it might seem that partial error rates are not controlled. In some cases this is true. However under the subset pivotality condition of Section 2.2.3, the FWE is controlled in the *strong* sense by the rule $\tilde{P}_i \leqslant \alpha$. To demonstrate this fact, suppose that $K = \{i_1, \ldots, i_j\}$ is the collection of hypotheses H_{0i} which are true. Then

$$
\begin{aligned}
&\Pr(\text{at least one } H_{0i} \text{ is rejected for } i \in K \mid \cap_{i \in K} H_{0i}) \\
&\quad = \Pr\Big(\min_{i \in K} \tilde{P}_i \leqslant \alpha \mid \cap_{i \in K} H_{0i}\Big) \\
&\quad = \Pr\Big(\min_{i \in K} P_i \leqslant c_\alpha \mid \cap_{i \in K} H_{0i}\Big) \\
&\quad = \Pr\Big(\min_{i \in K} P_i \leqslant c_\alpha \mid H_0^C\Big) \quad \text{(by subset pivotality)} \\
&\quad \leqslant \Pr\Big(\min_{1 \leqslant i \leqslant k} P_i \leqslant c_\alpha \mid H_0^C\Big) \\
&\quad \leqslant \alpha
\end{aligned}
$$

demonstrating strong control of the FWE. Again, it is assumed that the adjusted p-values are evaluated without error. If this is not the case then the method *approximately* controls the FWE in the strong sense.

2.4 ESTIMATING THE p-VALUE DISTRIBUTION

So far we have assumed that the distribution of the p-value vector is known under H_0^C, or at least that p-value vectors $\mathbf{P}^*$ can be simulated from this distribution. Usually this condition applies only to simple parametric problems. In general, the true distribution depends upon unknown parameters or distributional characteristics which must be estimated. The estimated distribution should be as close as possible to the true distribution; at the very least, we want it to converge in an appropriate asymptotic sense. Maximum likelihood estimates (parametric or nonparametric) are quite useful in this regard.

In some cases the p-value distribution may be estimated indirectly by using the distribution G_0 of the raw data $\mathbf{Y}$. We simulate data $\mathbf{Y}^*$ from the estimated distribution

$\hat{G}_0$, and compute p-values $\mathbf{P}^*$ from the simulated data $\mathbf{Y}^*$. Algorithm 2.5 shows how to do this.

In other cases, computational simplifications are possible. In Example 2.3, it was seen that the data y_{ij}^* may be generated independently from the $\mathcal{N}(0, 1)$ distribution, which is not the distribution of the original data values Y_{ij}. What is important is that the complete null distribution F_0 of the $\mathbf{P}$ vector be well approximated by the distribution $\hat{F}_0$ of the $\mathbf{P}^*$ vector. In Example 2.3 these distributions are in fact identical. In other situations one must estimate F_0. Types of estimation schemes that may be considered are:

- *Parametric Estimates:* Suppose $F_0 = F_0(\Theta)$, where F_0 is a known function and Θ is a vector of unknown parameters. Then $\hat{F}_0 = F_0(\hat{\Theta})$, where $\hat{\Theta}$ is a maximum likelihood (or at least a consistent) estimate of Θ.
- *Nonparametric Estimates:* Here F_0 has unknown form, depending perhaps on location shift or other parameters. In this case consistent estimates (sometimes nonparametric maximum likelihood) can be constructed using empirical distributions of the original data $\mathbf{Y}$. This method is most commonly thought of as the "bootstrap," proposed by Efron (1979).

Given an estimated distribution $\hat{F}_0$, the resampling-based adjusted p-values are estimated as described in Algorithm 2.5, where the distribution of $\mathbf{P}^*$ is $\hat{F}_0$. The validity of the resulting approximate adjustments depends upon how close $\hat{F}_0$ is to F_0. One factor affecting the accuracy of this approximation is the *experimental sample size*: if a consistent estimate is used, then larger sample sizes will yield better approximations. For a given sample size, another factor affecting the accuracy is the *complexity* of F, which is measured by number of parameters or other distinct characteristics that must be estimated. Allowing more parameters or characteristics for F_0 increases the realism of the model; however, it may decrease the efficiency of the procedure by introducing greater variability in the estimates. Therefore, in any small-sample situation, it is desirable to assess the vulnerability of the estimated adjustments to the approximation $\hat{F}_0$ by considering alternative estimates and/or adjustment strategies.

Particular choices of estimation schemes, parametric or nonparametric, depend upon the particular data and the assumed model for the data-generating process. These vary from application to application. Various types of estimates are considered in Chapters 3–7 in the context of specific applications. There is no "generic" bootstrap that applies to all possible situations. Rather, the choice of an appropriate bootstrap resampling method depends upon the type of model assumed. For example, in Chapter 3, we define *homoscedastic* and *heteroscedastic* bootstraps, both for nonparametric and for parametric (normal) models. Every model results in a different resampling scheme, and the choice of which method is most appropriate largely depends on what assumptions one is willing to make.

The following schematic diagram is adapted from Efron and Tibshirani (1986) and Efron (1986), with special application to the multiple testing problem.

Population Model and Analysis — Resampling Model and Analysis

$$\left\{\begin{matrix} G_0 & \rightarrow & \mathbf{Y} \\ & & \downarrow \\ & & \mathbf{P} \end{matrix}\right\} \Longrightarrow \left\{\begin{matrix} \hat{G}_0 & \rightarrow & \mathbf{Y}^* \\ & & \downarrow \\ & & \mathbf{P}^* \end{matrix}\right\}$$

In this diagram, the distribution G_0 is the true data-generating process under the complete null hypothesis H_0^C. The observed data is $\mathbf{Y}$, from which the p-value-vector $\mathbf{P}$ is computed. If we were fortunate enough to know G_0, we could determine all characteristics of interest concerning $\mathbf{P}$, for example the minimum p-value distribution, by repeatedly simulating data vectors $\mathbf{Y}$ from G_0. In the resampling model, the entire process of the population model is duplicated exactly, only with $\hat{G}_0$ replacing G_0. In this case, the distribution $\hat{G}_0$ *is* known, and properties of the random vector $\mathbf{P}$ in the population model may be estimated by repeatedly sampling $\mathbf{P}^*$ in the resampling model.

2.5 THE NEED FOR RESAMPLING IN MULTIPLE TESTING

Much of the multiple testing literature concerns methods for incorporating correlation structures correctly, usually in the context of homoscedastic normal models, and usually constructed in the critical value-based Accept/Reject mold. For example, when comparing all $k = \binom{g}{2}$ pairs of means in a balanced one-way ANOVA with g treatments, critical values of "Tukey's T Method" are derived from the studentized range distribution (e.g., Hochberg and Tamhane, 1987). These methods exploit the correlation structure between the test statistics, and have somewhat smaller critical values than the Bonferroni-style critical values which use α/k for determining significance. Thus it is easier to obtain a statistically significant result when correlation structures are incorporated.

In the context of adjusted p-values, the incorporation of correlation structure results in a smaller adjusted p-value (Equation 2.6) than the Bonferroni-style adjustment (Equation 2.3 or 2.4), again resulting in more powerful tests. As an extreme example, consider what would happen if all p-values were *perfectly* correlated and uniformly distributed under the complete null hypothesis: in this case Equation 2.6 would become

$$\Pr\left(\min_{1 \leq j \leq k} P_j \leq p_i \mid H_0^C\right) = \Pr(P_i \leq p_i \mid H_0^C) = p_i$$

implying no adjustment at all! This suggests that incorporation of correlation structures can be very important when the correlations are extremely large.

Moderate correlation structures occur frequently in multiple testing applications in univariate models. For example, when comparing means in a balanced homoscedastic ANOVA, the correlation between comparisons $\bar{Y}_1 - \bar{Y}_2$ and $\bar{Y}_1 - \bar{Y}_3$ is exactly .5. In multivariate applications, variables which are highly correlated are often recognized by experimenters. Often, certain variables are recognized as duplicating information and are dropped, or perhaps the variables are combined into a single measure. In either case, the correlations among the resulting variables is less extreme.

In cases of moderate correlation structures, the distinction between the Bonferroni adjustment (2.3) and the exact adjustment (2.6) can be very slight. Miller (1977) noted this phenomenon, commenting

> ... I knew the Bonferroni inequality $\Pr\{\bigcap_1^r A_i\} \geqslant 1 - \sum_1^r \Pr\{A_i^c\}$ was very useful, but over the course of the past ten years, I have become even more impressed with the tightness of the bound if the values of $\Pr\{A_i^c\}$ are small.*

In the current setting, the events "A_i" and "A_i^c" refer to the actions "accept H_i" and "reject H_i," respectively; a small value of "$\Pr\{A_i^c\}$" corresponds to a small per-comparison error rate. The "tightness of the bound" refers to the fact that incorporating dependence structure becomes less important for smaller significance levels. Thus, the effect of incorporating dependence structures often has *little effect*, particularly for small significance levels, when there is only moderate correlation among test statistics. In these cases a Bonferroni-style adjustment provides a reasonable solution.

If a Bonferroni-style correction is reasonable, then why bother with resampling? The answer is that *other* distributional characteristics, such as discreteness and skewness, can have dramatic effects, even for small p-values. While the effects of such nonnormalities are serious in univariate situations, as indicated by Table 2.1, the effects tend to be amplified in multiple testing applications, where minimal (or maximal) statistics are considered. Given the potentially extreme effects of nonnormality, we suggest that *nonnormality is of equal or greater concern than correlation structure* in multiple testing applications. Examples given in the following sections support this claim.

2.5.1 Distribution of Extremal Statistics under Nonnormality

In this section we replicate Noreen's analysis of tests for a single lognormal mean, but use ten independent samples, such as in the fictitious "oat bran" example of Section 1.1. The data values are Y_{ij}, for sample $i = 1, \ldots, 10$, and subject $j = 1, \ldots, n$ within sample i. For convenience we have assumed that all sample sizes are identical: $n_i \equiv n$. All observations are independent and identically distributed as e^Z, where Z denotes a standard normal random variable.

The hypotheses tested are H_i: $E(Y_{ij}) = \sqrt{e}$, with upper- or lower-tailed alternatives. We know that all hypotheses are in fact true, since the data have been simulated in this fashion.

The amplified effect of nonnormality upon minimal and maximal statistics is displayed in the sampling distributions of Figures 2.2–2.4. Figure 2.2 displays the

*Reprinted from *Journal of the American Statistical Association* **72** 779–788, (1977), with permission of the American Statistical Association.

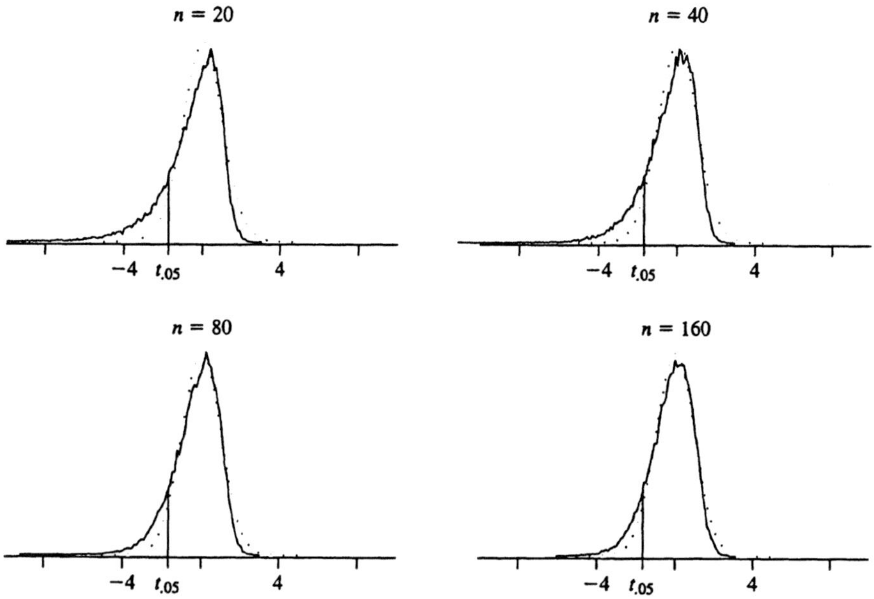

FIGURE 2.2. Distributions of t-statistics (solid line) when sampling from lognormal populations. Nominal t-distributions are superimposed (dotted line). The lower 5% critical value from the nominal t-distribution is given.

estimated sampling distributions of univariate t-statistics, defined as

$$t = \frac{\bar{y} - \sqrt{e}}{s/\sqrt{n}}$$

The solid lines show the distribution of t when sampling from lognormal populations, and the dotted lines show the distribution of t when sampling from normal populations with mean $\mu_y = \sqrt{e}$. The dotted line is the Student's t-distribution with $n - 1$ degrees of freedom.

For each graph there were 40,000 t-statistics, all simulated using lognormal y_{ij}. The sampling distributions were estimated using polygonal histograms. The range of the solid histogram line indicates extremes: for example, when $n = 20$ the smallest t-statistic was approximately -16, while the largest t-statistic was approximately 3.0. The nominal t-distributions (dotted lines) were graphed using the actual functional form of the Student's t-distribution.

Figure 2.2 corroborates the simulation study reported under the "t-test" entries of Table 2.1: since the lower tail area of the actual distribution (solid line) of the t-statistic is larger than the corresponding tail of the approximating Student's t-distribution (dotted line), the lower-tailed test rejects H more often than it should. Conversely, the upper tail area of the actual distribution is smaller than that of the approximating t-distribution, yielding fewer rejections than expected. As can be expected with larger sample sizes, the approximations become better, and the actual

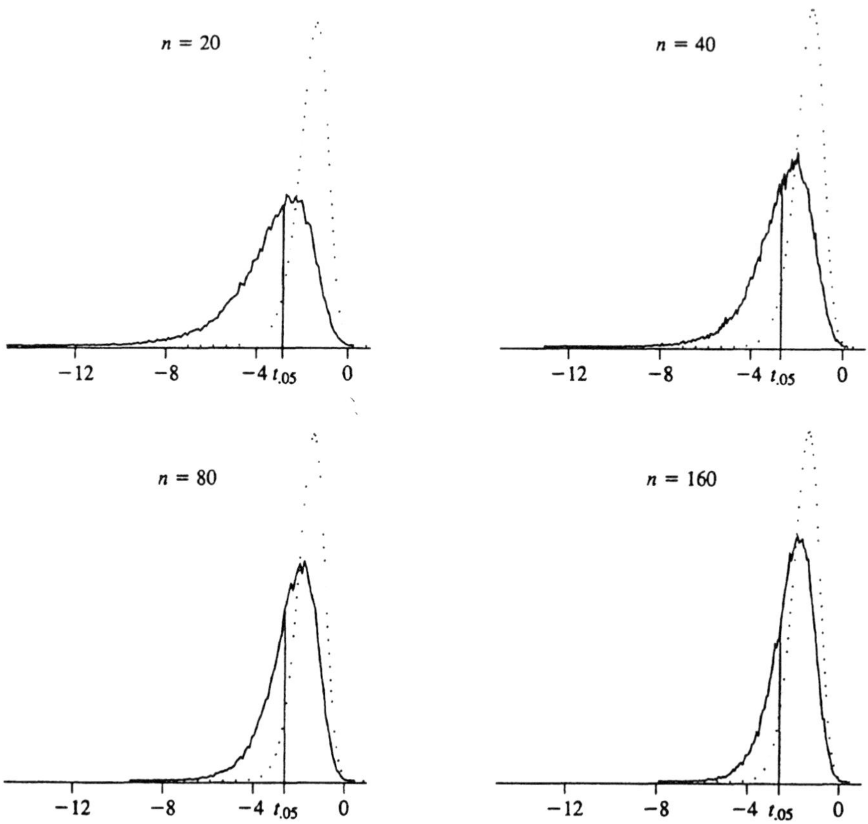

FIGURE 2.3. Distributions of *minimum* of ten independent t-statistics when sampling from lognormal (solid line) versus normal populations (dotted line). The lower 5% critical value from the nominal minimum t-distribution is given.

proportion of rejections more closely approximates the nominal proportion, as can be seen from both Figure 2.2 and the "t-test" entries of Table 2.1.

Interestingly, the actual distribution of the t-statistic when sampling from a lognormal population has *negative* skewness—exactly *opposite* the skewness of the original lognormal distribution. This result is well established (Geary, 1936), but it is still striking to see the result so graphically portrayed.

The theoretical (assuming normality) and observed distributions in Figure 2.2 do not appear to be extremely disparate, particularly for $n = 160$. This means that the actual significance levels of the t-test will not differ substantially from the nominal level α, particularly when the sample size is moderately large.

However, when one considers *maximal* and *minimal* t-statistics, the effect of the skewness is greatly amplified. As indicated in Figures 2.3 and 2.4, the theoretical and observed distributions are well separated, *even when* $n = 160$! In order to have a "significant" result at the FWE $= .05$ level using $k = 10$ lower-tailed tests and Šidák adjustments, the minimum t-statistic must be less than -2.85 when $n = 20$, less than

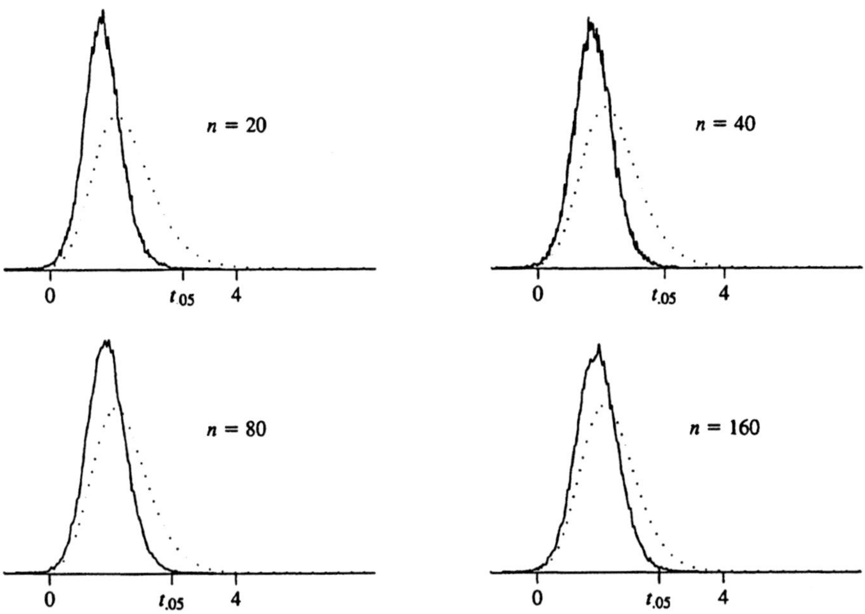

FIGURE 2.4. Distributions of *maximum* of ten independent t-statistics when sampling from lognormal (solid line) versus normal populations (dotted line). The upper 5% critical value from the nominal maximum t-distribution is given.

-2.70 when $n = 40$, -2.63 when $n = 80$, and less than -2.60 when $n = 160$. These quantities are denoted by $t_{.05}$ in Figure 2.3. Because values in the extreme lower tails of the actual distributions (solid curve) of Figure 2.3 are much more likely than under the corresponding t-distributions (dotted curve), the possibility of observing a "significant" result can be much larger than expected under the assumption of normal data. In practice, this can cause *false significances to occur*, even when a method as conservative as the Šidák method is used.

Conversely, it is quite difficult to achieve a significant upper-tailed test, since the true distributions are so sharply curtailed in the upper tails. For upper-tailed significance using the Šidák method, the critical values, $t_{.05}$, are 2.85, 2.70, 2.63, and 2.60 for the respective sample sizes. It is apparent that rejections are unlikely to occur. In practice, this means the Šidák procedure has very *low power*, and will likely fail to detect alternative hypotheses.

We can expect that these results will become worse as the number of tests (k) increases.

In fairness, we should note that the normal-based tests are much robust when testing contrasts involving two or more groups. Using analytic formulas of Geary (1947) and Gayen (1950), Miller (1986, pp. 41–42) shows how the distribution of the two-sample t-statistic

$$T = \frac{\bar{Y}_1 - \bar{Y}_2}{s\sqrt{1/n_1 + 1/n_2}}$$

Table 2.4 Rejection Proportions of Two-Sample Tests with Lognormal Data

		Nominal Level	
Tail	Method	$\alpha = .05$	$\alpha = .10$
Upper	Bootstrap	.0500	.1024
	Normal	.0562	.1124
Lower	Bootstrap	.0420	.0946
	Normal	.0299	.0852

can be robust to the skewnesses of the Y_1 and Y_2 variables, particularly when these skewnesses are roughly comparable and when the sample sizes n_1 and n_2 are approximately equal. Under these circumstances, there is an approximate cancellation of skewness terms for the distribution of T, leaving the distribution roughly symmetric. Thus, in the two-sample case, we expect the normal-based procedures to perform better than in the one-sample case.

Table 2.4 presents the results of a small simulation study in a two-sample situation involving lognormal populations. There were 10,000 data sets generated independently, from the same lognormal distribution used earlier. Within each data set, the two-sided p-value for the two-sample test was computed, as well as a bootstrapped p-value which used 1,000 bootstrap samples. The bootstrapped p-value was computed using the "homoscedastic location-shift" method presented in Chapter 3. According to the rejection proportions, both procedures perform fairly well, even though the sample sizes are small and unequal. Still, the bootstrap performs better than the normal approximation.

In Chapter 5, we discuss the effects of discreteness, particularly binary data, upon the operating characteristics of the multiple testing procedures, and note the effects are perhaps even more extreme with binary data than with continuous data. In particular, use of a Bonferroni-style adjustment for multiple tests with sparse binary data can be grossly conservative.

2.5.2 The Performance of Bootstrap Adjustments

Consider the problem of testing lognormal means from ten independent samples from Section 2.5.1. Using the definition (Equation 2.6) of an adjusted p-value, the adjusted p-values for the lower-tailed tests may be written as

$$\tilde{p}_i = \Pr\left(\min_{1 \leq j \leq 10} T_j \leq t_i \mid H_0^C\right)$$

since the p-values p_i are monotonically related to the test statistics t_i. These probabilities are obtained from the sampling distributions graphed in Figure 2.3. In practice, these functional forms of the distributions are unknown, so we will estimate them by resampling the observed data values y_{ij}.

The pivotal statistics used to test the ten hypotheses are

$$t_i = \frac{\bar{y}_i - \sqrt{e}}{s_i/\sqrt{n}}, \quad 1 \leqslant i \leqslant 10$$

Resampling-based adjustments then follow Algorithm 2.5. The independence of the Y_{ij} for different groups i is incorporated in the resampling scheme by generating bootstrap data sets Y^*_{ij} having the same independence characteristics.

Algorithm 2.7 Bootstrap Adjustments for Ten Independent Samples

0. Initialize counting variables: $COUNT_i = 0, i = 1, \ldots, 10$.
1. Let y^*_{ij} denote data which are sampled *with replacement*, and *independently*, from each group i. In other words, $y^*_{11}, \ldots, y^*_{1n}$ is a *with replacement* sample from $y_{11}, \ldots, y_{1n}$; $y^*_{21}, \ldots, y^*_{2n}$ is a *with replacement* sample from $y_{21}, \ldots, y_{2n}$, *independent* of $y^*_{11}, \ldots, y^*_{1n}$, and so on. For each bootstrap sample, compute the bootstrap test statistics

$$t^*_i = \frac{\bar{y}^*_i - \bar{y}_i}{s^*_i/\sqrt{n}}$$

2. If $\min_{1 \leqslant j \leqslant 10} t^*_j \leqslant t_i$, where the t_i are the original t-statistics, then increment the count variable: $COUNT_i \leftarrow COUNT_i + 1$. (For upper-tail tests, increment the counter if $\max_{1 \leqslant j \leqslant 10} t^*_j \geqslant t_i$.)
3. Repeat 1–2 N times. The estimated value of $\tilde{p}_i$ is $\tilde{p}^{(N)}_i = COUNT_i/N$.

While equation (2.6) provides *exact* adjusted p-values, it assumes that the underlying distributions are known. When estimated distributions are incorporated via resampling, the procedure no longer is exact, even if infinitely many resampled data sets are computed. To evaluate the approximate adjusted p-values, a simulation was performed along the lines of Algorithm 2.4 in Section 2.2.2. Specifically, the adjustment algorithm defined above in Algorithm 2.7 was placed within an "outer loop" in which the data y_{ij} were repeatedly generated, independent and identically distributed from the standard lognormal distribution. We estimated the true FWE level by generating $NSIM = 4000$ data sets, all under the complete null hypothesis. For each data set, we computed the bootstrap adjusted p-value for the *smallest* p-value in the original data using $NBOOT = 1000$ bootstrap samples. We then noted whether or not these adjusted p-values fell below the prescribed FWE level α. The proportion of the $NSIM$ samples having an adjusted p-value below α estimates the true FWE level of the method. In practice, one computes adjustments for *all* p-values; however, here we are interested in simulating the probability that *at least one* hypothesis is rejected, which happens if and only if the *minimum* p-value is less than α.

Results of this study are given in Figure 2.5. The performance of the parametric Šidák adjustment is *very* bad, corroborating Figures 2.3 and 2.4. When the sample

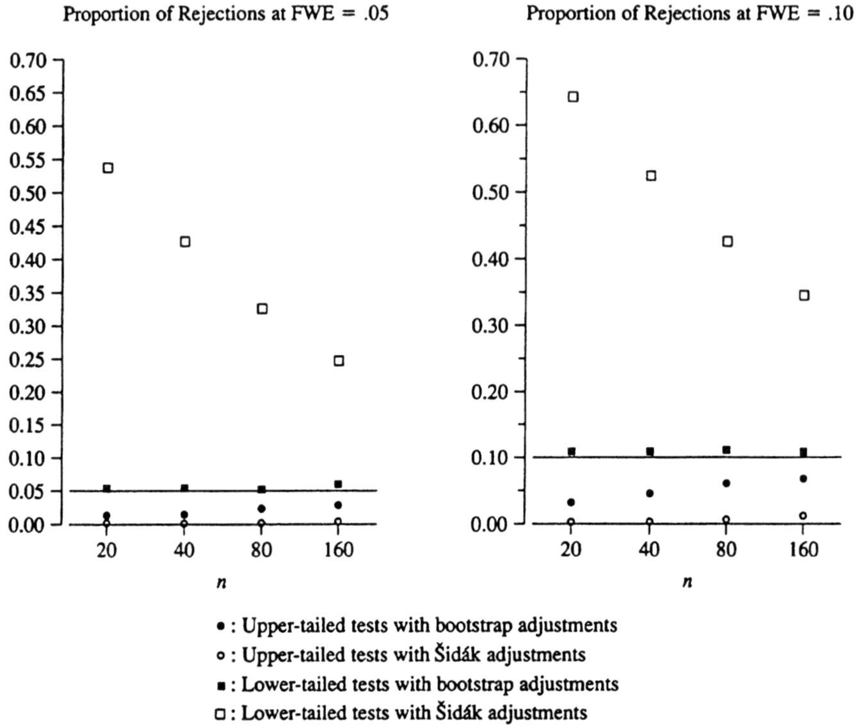

FIGURE 2.5. Rejection proportions with ten independent lognormal populations.

size is $n = 20$, there is approximately a 50% probability that at least one lower-tailed test will be declared significant at the FWE = .05 level!

The bootstrap adjustments are much better approximations. They essentially attain the nominal levels for *all* sample sizes when lower-tailed tests are used. The Šidák adjustments have Type I errors *far in excess* of the nominal .05 level, the bootstrap adjustments do not. When using upper-tailed tests, the bootstrap approximations are not as good as they are for the lower-tailed tests, but are still substantially better than the Šidák method. In the upper-tailed case, both the bootstrap and Šidák methods reject hypotheses *too rarely*, but the bootstrap has higher power.

When testing nonnormal data, the practical implications of these comparisons are (1) the bootstrap adjustments may have fewer *excess* Type I errors than the parametric Šidák adjustments (e.g., when tests are lower-tailed for lognormal data), and (2) the bootstrap adjustments may be more powerful than the parametric Šidák adjustments (e.g., when tests are upper-tailed for lognormal data).

2.6 STEP-DOWN METHODS FOR FREE COMBINATIONS

In the definition of the single-step adjustments (2.6), it is somewhat disconcerting that *all* p-values are adjusted according to the minimum p-value distribution. Shouldn't only the observed minimum p-value be adjusted according to this distribution?

As demonstrated in Section 2.3.3, the strong control of the familywise error rate *is* obtained using the single-step adjustment method, assuming subset pivotality. However, it is possible to improve the power of the procedure by making the adjusted p-values *uniformly smaller*, while retaining the same error rate protection. The method for doing this is intuitively appealing: rather than adjust *all* p-values according to the $\min P_j$ distribution, only adjust the *minimum* p-value using this distribution. Then, adjust the remaining p-values according to smaller and smaller sets of p-values. The effect of using smaller sets of p-values is to make the adjusted p-values smaller, thereby improving the power of the method. A potentially negative feature of this method is that Type III errors may be uncontrolled, but this remains an open research question.

Different methods for constructing the sets of p-values may be used, depending upon whether the hypotheses satisfy the *free combinations* or *restricted combinations* conditions. If, for every subcollection of j hypotheses $\{H_{i_1}, \ldots, H_{i_j}\}$, the simultaneous *truth* of $\{H_{i_1}, \ldots, H_{i_j}\}$ and *falsehood* of the remaining hypotheses is a plausible event, then the hypotheses satisfy the *free combinations* condition. In other words, each of the 2^k outcomes of the k-hypothesis problem is possible.

Example 2.4 Free Combinations
An animal carcinogenicity experiment is devised to detect increase in tumors with increasing doses of a compound. It is desired to test for an increase in k potential target organs. Here, H_i may refer to the hypothesis "no increase in tumors at organ i," and $H_i^!$ may refer to the hypothesis "there is an increase in tumors at organ i," $i = 1, \ldots, k$.

In this case, the hypotheses satisfy the free combinations condition: it is plausible that there is no increase in tumors for any subcollection of organs, yet there is an increase in tumors in all remaining organs. □

If the hypotheses do not satisfy the *free combinations* condition, they are *restricted* (Holm, 1979), as indicated by the following example.

Example 2.5 Restricted Combinations
An experiment is designed to compare expected yields for each of t different product formulations. It is desired to compare all $\binom{g}{2} = g(g-1)/2$ pairs of product formulations. Indexing each such comparison by i, $i = 1, \ldots, \binom{g}{2}$, H_i may refer to the hypothesis "the means are the same for comparison i," and $H_i^!$ may refer to the hypothesis "the means are different for comparison i."

The hypotheses are restricted: it is not plausible that the hypotheses $\mu_1 = \mu_2$ and $\mu_1 = \mu_3$ are true, yet $\mu_1 = \mu_3$ is false. □

Holm (1979) derived a simultaneous test procedure based on the Bonferroni inequality which strongly controls the FWE under free combinations. He also noted that the same procedure strongly controls the FWE for restricted combinations. However, with restricted combinations Holm's method may be modified to give more powerful tests (Shaffer, 1986). Holm's method remains valid, but it is potentially conservative when hypotheses are restricted.

Holm's *sequentially rejective* algorithm is based on the ordered p-values, $p_{(1)} \leqslant \cdots \leqslant p_{(k)}$, corresponding to hypotheses $H_{(1)}, \ldots, H_{(k)}$. The intuitive reasoning is as follows: once $H_{(1)}$ has been rejected using the Bonferroni critical value α/k, we should believe that $H_{(1)}$ is false. Thus, there are only $k - 1$ hypotheses which might still be true, implying the critical value $\alpha/(k-1)$ for $H_{(2)}$. If $H_{(2)}$ is rejected, use $\alpha/(k-2)$ for $H_{(3)}$.

Holm's Sequentially Rejective Algorithm

1. If $p_{(1)} > \alpha/k$, then *accept* all hypotheses $H_{(1)}, \ldots, H_{(k)}$ and stop; otherwise, *reject* $H_{(1)}$ and continue.
2. If $p_{(2)} > \alpha/(k-1)$, then *accept* all hypotheses $H_{(2)}, \ldots, H_{(k)}$ and stop; otherwise, *reject* $H_{(2)}$ and continue.

$\vdots$

j. If $p_{(j)} > \alpha/(k-j+1)$, then *accept* all hypotheses $H_{(j)}, \ldots, H_{(k)}$ and stop; otherwise, *reject* $H_{(j)}$ and continue.

$\vdots$

k. If $p_{(k)} > \alpha$, then *accept* hypothesis $H_{(k)}$; otherwise, *reject* $H_{(k)}$.

It is more convenient to display the results of this algorithm using *adjusted p-values* $\tilde{p}_{(j)}$. These may be derived using the given algorithm, along with definition (1.3):

$$\begin{aligned}
\tilde{p}_{(1)} &= kp_{(1)} \\
\tilde{p}_{(2)} &= \max(\tilde{p}_{(1)}, (k-1)p_{(2)}) \\
&\vdots \\
\tilde{p}_{(j)} &= \max(\tilde{p}_{(j-1)}, (k-j+1)p_{(j)}) \\
&\vdots \\
\tilde{p}_{(k)} &= \max(\tilde{p}_{(k-1)}, p_{(k)})
\end{aligned} \tag{2.9}$$

As is the case with the single-step Bonferroni adjusted p-values (2.3), any adjusted p-value greater than 1.0 should be set to 1.0. For any configuration of p-values $p_1, \ldots, p_k$, it may be verified that the collection of hypotheses $H_{i_1}, \ldots, H_{i_j}$ rejected using Holm's method at level α the same as the collection of hypotheses rejected using the criterion $\tilde{p}_i \leqslant \alpha$. The use of "max" at each step insures this correspondence. An additional consequence of the "max" adjustment is that the adjusted p-values have the same monotonicity as the original p-values, a highly desirable property.

To illustrate this method, consider a multiple testing situation with $k = 5$, where the ordered p-values $p_{(i)}$ are .009, .011, .012, .134, and .512. Let $H_{(1)}$ be the hypothesis corresponding to the p-value .009, $H_{(2)}$ be the hypothesis corresponding to .011, and so on. Using Holm's sequentially rejective method with $\alpha = .05$, we have the following actions:

1. Reject $H_{(1)}$ since $.009 \leqslant .05/5$, and continue.
2. Reject $H_{(2)}$ since $.011 \leqslant .05/4$, and continue.
3. Reject $H_{(3)}$ since $.012 \leqslant .05/3$, and continue.
4. Since $.134 > .05/2$, stop and accept $H_{(4)}$ and $H_{(5)}$.

The corresponding adjusted p-values are computed as follows:

1. $\tilde{p}_1 = 5(.009) = .045$.
2. $\tilde{p}_2 = \max\{.045, 4(.011)\} = .045$
3. $\tilde{p}_3 = \max\{.045, 3(.012)\} = .045$
4. $\tilde{p}_4 = \max\{.045, 2(.134)\} = .268$
5. $\tilde{p}_5 = \max\{.268, 1(.512)\} = .512$

We can interpret these adjusted p-values directly, simply by comparing them with the desired familywise significance level. For example, hypotheses $H_{(1)}$–$H_{(3)}$ are rejected at FWE = .05, but not at FWE = .01. Further, tests $H_{(4)}$ and $H_{(5)}$ are insignificant at FWE = .10. The adjusted p-values are far more convenient for reporting purposes than is the "Accept/Reject" listing that would require three separate tables, one each for FWE = .01, .05, .10, to arrive at the same conclusions.

Note that in stages 2 and 3, the adjusted p-values were set equal to the first adjusted p-value, .045. This is an example of how monotonicity enforcement works. It is also clearly needed: without such monotonicity enforcement, the adjusted p-values $\tilde{p}_2$ and $\tilde{p}_3$ both would be smaller than $\tilde{p}_1$. In this case, one might accept $H_{(1)}$ yet reject $H_{(2)}$ and $H_{(3)}$! This action clearly would run contrary to statistical intuition—smaller p-values should represent *more* evidence against the null hypothesis. It also would run contrary to Holm's algorithm: if $H_{(1)}$ is accepted, then all of $H_{(2)}$–$H_{(5)}$ must also be accepted. This concrete example illustrates how Holm's algorithm implies that monotonicity enforcement is required for adjusted p-values.

The improvement of this step-down procedure over the single-step Bonferroni procedure is clear: using the single-step method, the adjusted p-values are obtained by multiplying *every* raw p-value by five. Only the $H_{(1)}$ test would be declared significant at the FWE = .05 level in this case, whereas the step-down method allows three significant conclusions at the FWE = .05 level. Since both methods strongly control the FWE, the step-down Bonferroni method is clearly superior to the single-step Bonferroni method, if strong control of the FWE is all we require from a simultaneous test procedure.

While the FWE is strongly protected using this step-down method, it is based on the Bonferroni probability inequality, and therefore is conservative. Slightly less conservative adjustments are possible by using the Šidák inequality, taking the adjustments to be $1 - (1 - p_{(j)})^{(k-j+1)}$ at step j in (2.9). These adjustments are easily computed—both Bonferroni and Šidák step-down adjusted p-values are available in PROC MULTTEST of SAS. The Šidák adjustments are valid, but remain conservative in many cases, as discussed in Section 2.3.1.

While the Šidák adjustments yield more powerful tests than the Bonferroni adjustments, they remain conservative in that the adjusted p-values are still too large.

The adjustments may be made less conservative by incorporating the precise dependence characteristics, as with the *single-step adjusted p-values* (2.6). Let the ordered p-values have indexes $r_1, r_2, \ldots$, so that $p_{(1)} = p_{r_1}, p_{(2)} = p_{r_2}, \ldots, p_{(k)} = p_{r_k}$. The *free step-down adjusted* p-values are defined sequentially as follows:

$$\begin{aligned}
\tilde{p}_{(1)} &= \Pr\left(\min_{l \in \{r_1, r_2, \ldots, r_k\}} P_l \leqslant p_{(1)} \mid H_0^C\right) \\
\tilde{p}_{(2)} &= \max\left[\tilde{p}_{(1)}, \Pr\left(\min_{l \in \{r_2, \ldots, r_k\}} P_l \leqslant p_{(2)} \mid H_0^C\right)\right] \\
&\vdots \\
\tilde{p}_{(j)} &= \max\left[\tilde{p}_{(j-1)}, \Pr\left(\min_{l \in \{r_j, \ldots, r_k\}} P_l \leqslant p_{(j)} \mid H_0^C\right)\right] \\
&\vdots \\
\tilde{p}_{(k)} &= \max\left[\tilde{p}_{(k-1)}, \Pr(P_{r_k} \leqslant p_{(k)} \mid H_0^C)\right]
\end{aligned} \tag{2.10}$$

Note that the indexes r_j are *fixed*, not random, for the probability calculations in (2.10). Once again, use of the "max" operator insures that the order of the adjusted p-values is the same as that of the original p-values. The adjustments (2.10) are uniformly smaller than the *single-step adjusted* p-values (2.6), since the minima are taken over successively smaller sets.

While the formal mathematical definition is somewhat complex, these adjusted p-values can be simulated with relative ease. Corresponding to Algorithm 2.5 for estimating the *single-step adjusted* p-values, the following algorithm estimates the *free step-down adjusted* p-values:

Algorithm 2.8 Free Step-Down Resampling Method

0. Initialize counting variables: $COUNT_i = 0, i = 1, \ldots, k$.
1. Generate a vector $(p^*_{r_1}, \ldots, p^*_{r_k})$ from the same (or at least, approximately the same) distribution as the *original* p-values $(P_{r_1}, \ldots, P_{r_k})$ under the complete null hypothesis. Note that the sequence $\{r_j\}$ is fixed throughout the simulation. Thus the $p^*_{r_j}$ will not have the same monotonicity as the original p-values $p_{(j)}$.
2. Define the successive minima:

$$\begin{aligned}
q^*_k &= p^*_{r_k} \\
q^*_{k-1} &= \min(q^*_k, p^*_{r_{k-1}}) \\
q^*_{k-2} &= \min(q^*_{k-1}, p^*_{r_{k-2}}) \\
&\vdots \\
q^*_1 &= \min(q^*_2, p^*_{r_1}).
\end{aligned}$$

3. If $q^*_i \leqslant p_{(i)}$, then $COUNT_i \leftarrow COUNT_i + 1$.
4. Repeat 1–3 N times. Compute $\tilde{p}^{(N)}_{(i)} = COUNT_i/N$.

5. Enforce monotonicity using successive maximization:

$$\tilde{p}_{(1)}^{(N)} = \tilde{p}_{(1)}^{(N)}$$
$$\tilde{p}_{(2)}^{(N)} \leftarrow \max(\tilde{p}_{(1)}^{(N)}, \tilde{p}_{(2)}^{(N)})$$
$$\vdots$$
$$\tilde{p}_{(k)}^{(N)} \leftarrow \max(\tilde{p}_{(k-1)}^{(N)}, \tilde{p}_{(k)}^{(N)})$$

Once monotonicity has been enforced, the simulation-based estimates $\tilde{p}_{(j)}^{(N)}$ are reasonable approximations of the actual values $\tilde{p}_{(j)}$, provided N is sufficiently large. Again, $N \geqslant 10{,}000$ is recommended.

Algorithm 2.8 has two noteworthy features. First, the number of comparisons needed to find all successive minima in (2.10) has order $O(k)$ for each simulation, since only a single comparison is needed for each test. By contrast, if a crude "bubble sort" algorithm were used for each test, the number of comparisons required would be $O(k^2)$. Second, note that for all j, calculations of $\Pr(\min_{l \in \{r_j,\ldots,r_k\}} P_l \leqslant p_{(j)} \mid H_0^C)$ are done within the *same* simulation. As an alternative, one might estimate these quantities using k *separate* simulations. By using the *same* simulation, errors are reduced through the "correlated sampling" effect (Rubinstein, 1981).

Returning to the $k = 5$ scenario with p-values .009, .011, .012, .134, and .512, suppose that these correspond to the original hypotheses H_2, H_4, H_1, H_3, and H_5, respectively. Then $r_1 = 2$, $r_2 = 4$, $r_3 = 1$, $r_4 = 3$, and $r_5 = 5$. Throughout the resampling process, these indices are held fixed.

Here is how Algorithm 2.8 would operate in such a situation:

Algorithm 2.9 A Specific Step-Down Illustration

0. Initialize counting variables: $COUNT_i = 0, i = 1,\ldots,5$.
1. Generate a vector $(p_2^*, p_4^*, p_1^*, p_3^*, p_5^*)$ from the same (or at least, approximately the same) distribution as the *original* p-values $(P_2, P_4, P_1, P_3, P_5)$ under the complete null hypothesis. This can be done by generating the p-value vector $\mathbf{p}^*$ as in Algorithm 2.5, obtaining $(p_1^*, p_2^*, p_3^*, p_4^*, p_5^*)$, then simply rearranging the values to obtain $(p_2^*, p_4^*, p_1^*, p_3^*, p_5^*)$

   (Being random, the $p_{r_j}^*$ will not have the same monotonicity as the original p-values $p_{(j)}$: that is, even though the original p-values were sorted to give $p_2 \leqslant p_4 \leqslant p_1 \leqslant p_3 \leqslant p_5$, the resampled p-values most likely will *not* satisfy the ordering $p_2^* \leqslant p_4^* \leqslant p_1^* \leqslant p_3^* \leqslant p_5^*$.)
2. Define the successive minima:

$$q_5^* = p_5^*$$
$$q_4^* = \min(q_5^*, p_3^*)$$
$$q_3^* = \min(q_4^*, p_1^*)$$
$$q_2^* = \min(q_3^*, p_4^*)$$
$$q_1^* = \min(q_2^*, p_2^*).$$

3. If $q_i^* \leqslant p_{(i)}$, then $COUNT_i \leftarrow COUNT_i + 1$.
4. Repeat 1–3 N times. Compute $\tilde{p}_{(i)}^{(N)} = COUNT_i/N$.
5. Enforce monotonicity using successive maximization:

$$\begin{aligned} \tilde{p}_{(1)}^{(N)} &= \tilde{p}_{(1)}^{(N)} \\ \tilde{p}_{(2)}^{(N)} &\leftarrow \max(\tilde{p}_{(1)}^{(N)}, \tilde{p}_{(2)}^{(N)}) \\ \tilde{p}_{(3)}^{(N)} &\leftarrow \max(\tilde{p}_{(2)}^{(N)}, \tilde{p}_{(3)}^{(N)}) \\ \tilde{p}_{(4)}^{(N)} &\leftarrow \max(\tilde{p}_{(3)}^{(N)}, \tilde{p}_{(4)}^{(N)}) \\ \tilde{p}_{(5)}^{(N)} &\leftarrow \max(\tilde{p}_{(4)}^{(N)}, \tilde{p}_{(5)}^{(N)}) \end{aligned}$$

2.7 STEP-DOWN METHODS FOR RESTRICTED COMBINATIONS

When the hypotheses are *restricted*, then certain combinations of true hypotheses necessarily imply truth or falsehood of other hypotheses. In these cases, the adjustments may be made smaller than the *free step-down adjusted* p-values, while maintaining strong control of the FWE. Shaffer (1986) discusses how to incorporate such logical constraints, developing a multiplicity adjustment protocol based on the Bonferroni inequality. She comments,

> In many circumstances it may be feasible to combine logical and distributional considerations to obtain multiple testing methods better than those obtainable by either type alone.*

While Holland and Copenhaver (1987) note that the logical constraint method may be made slightly less conservative by using the Šidák inequality, we propose that the dependence structures mentioned by Shaffer be incorporated more precisely using resampling methods.

As with the *free step-down* method, the *restricted step-down* method starts with the ordered p-values, $p_{(1)} \leqslant \cdots \leqslant p_{(k)}$. Recall $p_{(j)} = p_{r_j}$. Again, hypotheses are rejected sequentially based on the magnitude of the original p-values: if $H_{(j)}$ is rejected, then $H_{(1)}, \ldots, H_{(j-1)}$ *must* have been previously rejected. However, unlike the *free step-down* method, the multiplicity adjustment for the *restricted step-down* method at stage j considers only those hypotheses that *possibly* can be true, given that the previous $j - 1$ hypotheses are all false.

Given a particular sequence $r_1, \ldots, r_k$, we define sets $\overline{S}_j$ of hypotheses *which include* $H_{(j)}$ that can be true at stage j, given that all previous hypotheses are false. Letting $S = \{r_1, \ldots, r_k\} = \{1, \ldots, k\}$, define

*Reprinted from *Journal of the American Statistical Association* **81** 826–831, (1986), with permission of the American Statistical Association.

$$
\begin{aligned}
\overline{S}_1 &= \{S\} \\
\overline{S}_2 &= \{K \subset S \mid r_2 \in K, \text{ and } (\cap_{l \in K} H_l) \bigcap (H'_{(1)}) \neq \emptyset\} \\
\overline{S}_3 &= \{K \subset S \mid r_3 \in K, \text{ and } (\cap_{l \in K} H_l) \bigcap (H'_{(1)} \cap H'_{(2)}) \neq \emptyset\} \\
&\vdots \\
\overline{S}_j &= \{K \subset S \mid r_j \in K, \text{ and } (\cap_{l \in K} H_l) \bigcap (\cap_{l=1}^{j-1} H'_{(l)}) \neq \emptyset\} \\
&\vdots \\
\overline{S}_k &= \{\{r_k\}\}
\end{aligned}
\tag{2.11}
$$

If any $\overline{S}_j = \emptyset$ in (2.11), then take $\overline{S}_j = \{\{r_j\}\}$. Also, these sets may be simplified by eliminating subsets: if $K_1 \in \overline{S}_j$ and $K_2 \in \overline{S}_j$, with $K_1 \subset K_2$, then K_1 is redundant and may be eliminated from $\overline{S}_j$.

Define

$$M_j = \max_{K \in \overline{S}_j} |K|$$

where $|K|$ = the number of elements in the set K. Step-down adjustments described in Shaffer (1986) are similar to the *free* step-down adjustments of Holm (1979), except that the divisors of the Bonferroni critical values are M_j, rather than $k - j + 1$. This adjustment strategy is more conveniently reported using adjusted p-values. Using definition (1.3), we have

$$
\begin{aligned}
\tilde{p}_{(1)} &= M_1 p_{(1)} = k p_{(1)} \\
\tilde{p}_{(2)} &= \max(\tilde{p}_{(1)}, M_2 p_{(2)}) \\
&\vdots \\
\tilde{p}_{(j)} &= \max(\tilde{p}_{(j-1)}, M_j p_{(j)}) \\
&\vdots \\
\tilde{p}_{(k)} &= \max(\tilde{p}_{(k-1)}, p_{(k)})
\end{aligned}
\tag{2.12}
$$

The adjusted p-values (2.12) can be no larger than the *free* Bonferroni adjustments (2.9), since $M_j \leqslant k - j + 1$. Note that in the case of free combinations, the truth of a collection of null hypotheses indexed by $\{r_j, \ldots, r_k\}$ cannot contradict the falsehood of all nulls indexed by $\{r_1, \ldots, r_{j-1}\}$. In this case, $\overline{S}_j = \{\{r_j, \ldots, r_k\}\}$; thus $M_j = k - j + 1$, and the *restricted* method reduces to the *free* method as a special case.

Example 2.6 Multiple Restricted Tests

Consider the case where there are 4 treatments in an experiment, with mean values μ_1, μ_2, μ_3, and μ_4. When all pairwise comparisons are considered, the hypotheses are $H_1: \mu_1 = \mu_2$, $H_2: \mu_1 = \mu_3$, $H_3: \mu_1 = \mu_4$, $H_4: \mu_2 = \mu_3$, $H_5: \mu_2 = \mu_4$, and $H_6: \mu_3 = \mu_4$. For clarity, define $H_{12} = H_1, H_{13} = H_2, \ldots, H_{34} = H_6$. Suppose that the observed p-values have the following order: $p_5 \leqslant p_2 \leqslant p_1 \leqslant p_4 \leqslant p_6 < p_3$; in other words, $r_1 = 5$, $r_2 = 2$, and so on. Obviously, these tests are quite restricted: for example, it is not possible that $\{H_1, H_2, H_3\}$ (i.e., $\{H_{12}, H_{13}, H_{14}\}$) are true while H_4 (H_{23}) is false.

The sets $\overline{S}_j$ are defined as follows:

$$\overline{S}_1 = \{\{1,2,3,4,5,6\}\} \quad (M_1 = 6)$$
$$\overline{S}_2 = \{\{1,2,4\},\{2,3,6\}\} \quad (M_2 = 3)$$
$$\overline{S}_3 = \{\{1,6\}\} \quad (M_3 = 2)$$
$$\overline{S}_4 = \{\{3,4\}\} \quad (M_4 = 2)$$
$$\overline{S}_5 = \{\{6\}\} \quad (M_5 = 1)$$
$$\overline{S}_6 = \{\{3\}\} \quad (M_6 = 1)$$

At the first stage, all hypotheses are allowed. At the second stage, allow only collections of hypotheses including $H_2: \mu_1 = \mu_3$ whose simultaneous truth does not contradict $H_5^{\prime}: \mu_2 \neq \mu_4$. For example, H_1, H_2, and H_4 simultaneously imply $\mu_1 = \mu_2 = \mu_3$, which does not contradict $\mu_2 \neq \mu_4$. The multipliers M_j $(6,3,2,2,1,1)$ are less than the corresponding free step-down multipliers $(6,5,4,3,2,1)$, resulting in smaller adjusted p-values. □

Because the sets $\overline{S}_j$ are generally difficult to determine, the associated sequence M_j is also difficult to determine. The problem of calculating the M_j may be simplified considerably by using conservative values that do not incorporate the particular ordering of the p-values. Rather than determine the *actual* sequence M_j for a given problem, one may compute the *maximum* M_j that may occur at step j for problems of the given type, resulting in a conservative adjustment. Such maximal sequences are available. Table 2.5, reproduced from Holland and Copenhaver (1987), displays maximal M_j sequences in the one-way ANOVA with t treatments, when $k = \binom{t}{2}$ pairwise tests are considered. When $t = 4$, the conservative values of Table 2.5 are $(6,3,3,3,2,1)$, only slightly larger (more conservative) than our reported sequence $(6,3,2,2,1,1)$. Yet both sequences are much less conservative than the free step-down multipliers $(6,5,4,3,2,1)$.

Shaffer (1986) discusses methods for computing maximal M_j sequences in additional applications.

Less conservative Šidák-style adjustments $1 - (1 - p_{(j)})^{M_j}$ are often valid (Holland and Copenhaver, 1987); however, the degree of conservatism can be lessened further by incorporating precise distributional characteristics, and not resorting to probability inequalities. The *restricted step-down adjusted* p-values are defined sequentially as follows:

Table 2.5 Maximal M_j for All Pairwise Tests*

j	Number of Treatments, t							
	3	4	5	6	7	8	9	10
1	3	6	10	15	21	28	36	45
2	1	3	6	10	15	21	28	36
3	1	3	6	10	15	21	28	36
4		3	6	10	15	21	28	36
5		2	6	10	15	21	28	36
6		1	4	10	15	21	28	36
7			4	7	15	21	28	36
8			3	7	11	21	28	36
9			2	7	11	16	28	36
10			1	6	11	16	22	36
11				4	11	16	22	29
12				4	10	16	22	29
13				3	9	16	22	29
14				2	7	15	22	29
15				1	7	13	22	29
16					6	13	21	29
17					5	12	18	29
18					4	11	18	28
19					3	10	18	24
20					2	9	16	24
21					1	8	16	24
22						7	15	24
23						6	13	22
24						5	13	22
25						4	12	21
26						3	11	20
27						2	10	18
28						1	9	18
29							8	17
30							7	16
31							6	15
32							5	14
33							4	13
34							3	12
35							2	11
36							1	10
37								9
38								8
39								7
40								6
41								5
42								4
43								3
44								2
45								1

* From B. S. Holland and M. D. Copenhaver, "An Improved Sequentially Rejective Bonferroni Test Procedure," *Biometrics* **43** 417–424, with permission from The Biometric Society.

$$\begin{aligned}
\tilde{p}_{(1)} &= \Pr\Big(\min_{l\in\{r_1,r_2,\ldots,r_k\}} P_l \leqslant p_{(1)} \mid H_0^C\Big)\\
\tilde{p}_{(2)} &= \max\Big[\tilde{p}_{(1)}, \max_{K\in\bar{S}_2} \Pr\Big(\min_{l\in K} P_l \leqslant p_{(2)} \mid H_0^C\Big)\Big]\\
&\;\vdots\\
\tilde{p}_{(j)} &= \max\Big[\tilde{p}_{(j-1)}, \max_{K\in\bar{S}_j} \Pr\Big(\min_{l\in K} P_l \leqslant p_{(j)} \mid H_0^C\Big)\Big]\\
&\;\vdots\\
\tilde{p}_{(k)} &= \max\left[\tilde{p}_{(k-1)}, \Pr(P_{r_k} \leqslant p_{(k)} \mid H_0^C)\right]
\end{aligned} \tag{2.13}$$

As with the *free* step-down adjustments (2.10), the indices r_j are *fixed*, not random, for the probability calculations in (2.13). At each step of (2.13), the probabilities are computed over subsets of the sets in (2.10). Thus, the restricted adjustments (2.13) can be *no larger* than the free adjustments (2.10).

In general, the sets $\bar{S}_j$ are not easily described, and construction of a simple estimation method such as Algorithm 2.8 may be difficult. We leave this problem as an open research question. Of course, brute force simulation-based calculation of the restricted adjustments are possible. This would require estimates of $\Pr(\min_{l\in K} P_l \leqslant p_{(j)} \mid H_0^C)$, for every set $K \in \bar{S}_j$.

2.8 ERROR RATE CONTROL FOR STEP-DOWN METHODS

In this section, we present details regarding control of the FWE for the free and restricted step-down adjustments. Since the *free* adjustments of Section 2.6 are a special case of the *restricted* adjustments of Section 2.7, it will suffice to demonstrate strong control of the FWE for *restricted* adjustments.

We shall distinguish between the control of the error rate under the null settings H_0^K and H^K. An example of the former case is $H_0^K = \cap_{i\in K} H_{0i}$: $\theta_i = 0$, and an example of the latter is $H^K = \cap_{i\in K} H_i$: $\theta_i \leqslant 0$. Control of the error rate under H_0^K involves reasoning under the $\theta_i \equiv 0$, for $i \in K$; in the case of H^K we wish to control the error rate for *all* $\theta_i \leqslant 0$, for $i \in K$.

2.8.1 Error Rate Control under H_0^K

Let $S = \{1, \ldots, k\}$ denote the index set for the tested hypotheses H_{0i}. To control the FWE in the strong sense, it is required that the probability of rejecting at least one H_{0i} be no larger than α, no matter what subset of the $K \subseteq S$ of hypotheses happens to be true.

Suppose that $K_0 = \{i_1, \ldots, i_j\} \subseteq S$ denotes the collection of hypotheses H_{0i} which are true, and that $K_0 \neq \varnothing$. (If $K_0 = \varnothing$, then there can be no Type I errors.)

We first note the equivalence between the adjusted p-value approach and the more customary critical value approach. Let x_K^α denote the α *quantile* of $\min_{t\in K} P_t \mid H_0^C$, discussed in Section 2.3.3 and illustrated in Figure 2.1. Further, define

$$
\begin{aligned}
x_1^\alpha &= x_S^\alpha \\
x_2^\alpha &= \min_{K \in \overline{S}_2} x_K^\alpha \\
&\vdots \\
x_j^\alpha &= \min_{K \in \overline{S}_j} x_K^\alpha \\
&\vdots \\
x_k^\alpha &= x_{\{r_k\}}^\alpha
\end{aligned}
$$

The following critical-value-based algorithm for rejecting hypotheses is equivalent to rejecting H_{0i} when $\tilde{p}_i \leqslant \alpha$:

Critical Value-Based Sequentially Rejective Algorithm

1. If $p_{(1)} > x_1^\alpha$ then *accept* all hypotheses $H_{(1)}, \ldots, H_{(k)}$ and stop; otherwise, *reject* $H_{(1)}$ and continue.
2. If $p_{(2)} > x_2^\alpha$ then *accept* all hypotheses $H_{(2)}, \ldots, H_{(k)}$ and stop; otherwise, *reject* $H_{(2)}$ and continue.

$\vdots$

j. If $p_{(j)} > x_j^\alpha$ then *accept* all hypotheses $H_{(j)}, \ldots, H_{(k)}$ and stop; otherwise, *reject* $H_{(j)}$ and continue.

$\vdots$

k. If $p_{(k)} > x_k^\alpha$ then *accept* hypothesis $H_{(k)}$; otherwise, *reject* $H_{(k)}$.

We have the following relationships:

$$
\begin{aligned}
&\{\text{There is at least one Type I Error}\} \\
&\quad = \{\text{Reject at least one } H_{0t},\ t \in K_0\} \\
&\quad \subseteq \left\{\min_{t \in K_0} P_t \leqslant x_j^\alpha\right\}
\end{aligned}
$$

where $j \leqslant k - |K_0| + 1$ is defined by $\min_{t \in K_0} P_t = P_{(j)} = P_{r_j}$.

Then

$$
\begin{aligned}
&\Pr(\text{There is at least one Type I Error}) \\
&\quad = \Pr(\text{Reject at least one } H_{0t},\ t \in K_0 \mid \cap_{i \in K_0} H_{0i}) \\
&\quad \leqslant \Pr\left(\min_{t \in K_0} P_t \leqslant x_j^\alpha \mid \cap_{i \in K_0} H_{0i}\right) \qquad (2.14) \\
&\quad = \Pr\left(\min_{t \in K_0} P_t \leqslant x_j^\alpha \mid H_0^C\right)
\end{aligned}
$$

assuming subset pivotality in the last equality. Since $K_0 \in \overline{S}_j$, we have $x_j^\alpha \leqslant x_{K_0}^\alpha$, implying

$$\Pr\Bigl(\min_{t \in K_0} P_t \leqslant x_j^{\alpha} \mid H_0^C\Bigr) \leqslant \Pr\Bigl(\min_{t \in K_0} P_t \leqslant x_{K_0}^{\alpha} \mid H_0^C\Bigr) \leqslant \alpha$$

which demonstrates that the restricted step-down adjustments strongly control the FWE.

Since the restricted step-down adjusted p-values (2.13) are everywhere smaller than the free step-down adjusted p-values (2.10), the free step-down adjustments also control the FWE in the strong sense. However, the free step-down adjustments clearly lose power relative to the restricted adjustments when there are logical constraints among hypotheses.

As in Section 2.3, it is assumed once again that the adjusted p-values are evaluated without error. When this is not true, then the resampling method approximately controls the FWE in the strong sense.

2.8.2 Error Rate Control under H^K

Suppose that H^K is true, then the distribution of $\mathbf{Y}$ is $G \in \mathcal{G}^K$. Suppose also that there exist random variables P_{i0}, defined on the same probability space as the P_i, for which $P_i \geqslant P_{i0}$ (almost everywhere) for all i; and such that the joint distribution of the $\{P_{i0}\}$ under G is identical to the joint distribution of the $\{P_i\}$ under H_0^K.

Under this condition the procedure is unbiased, and the error rate is controlled: using (2.14), note that

$$\begin{aligned}
&\Pr(\text{Reject at least one } H_{0t},\, t \in K_0 \mid G)\\
&\leqslant \Pr\Bigl(\min_{t \in K_0} P_t \leqslant x_j^{\alpha} \mid G\Bigr)\\
&\leqslant \Pr\Bigl(\min_{t \in K_0} P_{t0} \leqslant x_j^{\alpha} \mid G\Bigr)\\
&= \Pr\Bigl(\min_{t \in K_0} P_t \leqslant x_j^{\alpha} \mid H_0^K\Bigr)\\
&= \Pr\Bigl(\min_{t \in K_0} P_t \leqslant x_j^{\alpha} \mid H_0^C\Bigr) \leqslant \alpha
\end{aligned}$$

as shown in the previous section.

Such P_{i0} frequently exist in parametric analyses; for example, the two-sample t-statistic for testing $H_0\colon \mu_1 \leqslant \mu_2$ with n observations per group may be written

$$t = t_0 + \frac{\mu_1 - \mu_2}{s\sqrt{2/n}},$$

where

$$t_0 = \frac{(\bar{y}_1 - \bar{y}_2) - (\mu_1 - \mu_2)}{s\sqrt{2/n}}$$

The p-value for this test is $p = \Pr(T_{2(n-1)} \geqslant t)$, where T_ν denotes a t-distributed random variable with ν degrees of freedom. Letting p_0 be defined by $p_0 = \Pr(T_{2(n-1)} \geqslant t_0)$, $p_0 < p$ whenever $\mu_1 < \mu_2$.

In nonparametric and semiparametric situations there may be no P_{i0} which satisfy the appropriate distributional requirements *exactly*, but there may exist such P_{i0} satisfying the distributional requirements in an asymptotic sense.

CHAPTER 3

Continuous Data Applications: Univariate Analysis

When people think of multiple testing and multiple comparisons, they usually think of pairwise comparisons in the *one-way analysis of variance* (or ANOVA) model. Tukey's and Scheffé's methods, among others, are common in statistical software packages and are discussed in many texts (e.g., Neter, Wasserman, and Kutner, 1985, Ch. 17; Ott, 1988, Ch. 11; and Wilcox, 1987, Ch. 9; to name just a few). In this chapter we show how the resampling-based approach fits with classical methods, and we show its advantages and disadvantages. As a matter of fact, Tukey's, Scheffé's, and many other classical methods are equivalent, up to simulation error, to special cases of resampling-based multiple testing. For those familiar with classical multiple testing and multiple comparisons methods, these connections help clarify the resampling methodology. Others may use this chapter to understand the classical methods from the resampling perspective.

The one-way ANOVA model commonly is used to compare means of data values occurring in different groups. The model can be used whether the data are obtained *observationally* or *experimentally*. An example of observational data is the comparison of average salaries for different ethnic groups; an example of experimental data is the comparison of average cholesterol reduction in response to different medical treatments. In the experimental example the subjects are assigned (usually at random) to the groups; in the observational example the subjects themselves define the groups.

In observational studies, the treatment effects frequently are *confounded* with uncontrolled variables. In this case such effects may be "partialled out" by using the confounding variables as covariates in an analysis of covariance (ANCOVA) model. ANCOVA models also are useful for reducing error in experimental studies. Resampling methods for use with covariate models are given in Section 3.4.

From the resampling standpoint, it is no more difficult to perform multiplicity adjustments for any user-defined collection of contrasts with unbalanced analysis of covariance data, than it is to perform adjustments for all pairwise contrasts with balanced ANOVA data. Furthermore, the approximation errors are identical in these two cases: assuming that the data are normal and homoscedastic, the only source of error is from simulation. If we could simulate infinitely many data sets using a perfect random number generator, then the adjustments would be *exact* in such cases. The exactness of the resampling method is in sharp contrast to classical methods, where the distributions used are usually *approximations* in unbalanced designs, and in designs with covariates. In addition, resampling allows distributions to be nonnormal, whereas most multiple comparisons methods for ANOVA assume normality.

In Section 3.1 we show how the resampling approach works in the most familiar of multiple testing applications, namely, the problem of all pairwise comparisons in the one-way ANOVA. We consider many deviations from the usual model assumptions, including heteroscedasticity and nonparametric models, and give resampling schemes to incorporate such deviations. In Sections 3.2 we extend the ideas to more general comparisons, including general contrasts and one-sided comparisons. Two- and higher-way ANOVA models are considered in Section 3.3, and the use of covariates is considered in Section 3.4.

3.1 ONE-WAY ANOVA MODEL: PAIRWISE COMPARISONS

The commonly used one-way ANOVA model is as given in Example 2.3 of Section 2.3.2:

$$Y_{ij} = \mu_i + \epsilon_{ij}$$

where $i = 1,\ldots,g; j = 1,\ldots,n_i$; the μ_i are fixed unknown constants; and the ϵ_{ij} are random variables with mean zero. Specific probability models for the ϵ_{ij} include:

Homoscedastic Location Shift Model: The ϵ_{ij} are independent and identically distributed as some distribution function G. A special case is the *homoscedastic normal model*, where the ϵ_{ij} are independent and identically distributed as $\mathcal{N}(0,\sigma^2)$.

Heteroscedastic Location Shift Model: The ϵ_{ij} are independently distributed as G_i, for some distributions G_i. A special case is the *heteroscedastic normal model*, where the ϵ_{ij} are independently distributed as $\mathcal{N}(0,\sigma_i^2)$.

Randomization Model: The probability model for the observable data is determined by the random assignment of subjects to groups. Following the notation of Rao (1973), p. 503, let the experimental units in the combined sample be numbered from 1 to N, where $N = \sum n_i$. Conceptually, each experimental unit might respond differently to each treatment. Denote such a "true response" of unit u to treatment i by R_{iu}, $i = 1,\ldots g$, $u = 1,\ldots,N$. If the N units are randomly divided into g groups, with n_i within each group, then the observed data values $y_{11},\ldots,y_{1n_1}$ are a random sample from $R_{11},\ldots,R_{1N}$; $y_{21},\ldots,y_{2n_2}$ are a random sample from $R_{21},\ldots,R_{2N}$ (excluding all units u previously selected for group 1), and so on. Using this probability model, the mean value for Y_{ij} is $\mu_i = \bar{R}_i = \sum_u R_{iu}/N$, and the residual is $\epsilon_{ij} = Y_{ij} - \bar{R}_i$. The residuals have mean zero, but are neither independent nor homoscedastic; expressions for their variances and covariances are given by Rao (1973), p. 504.

Even though the three preceding types of models are quite different, they are all easily analyzed using resampling methods.

Consider the pairwise comparisons problem. In the one-way ANOVA model, these comparisons are

$$\theta_{ii'} = \mu_i - \mu_{i'}$$

for $1 \leqslant i < i' \leqslant g$. There are $\binom{g}{2} = g(g-1)/2$ contrasts $\theta_{ii'}$, so for five treatments there are 10 pairwise comparisons; for 10 treatments there are 45. The estimates of these comparisons are not mutually independent: two distinct comparisons having a particular treatment group in common will be correlated.

3.1.1 Homoscedastic Model: Single-Step Analysis

When the homoscedastic normal model is valid, the usual statistic for testing $H_{ii'}\colon \theta_{ii'} = 0$ is

$$T_{ii'} = \frac{\bar{Y}_i - \bar{Y}_{i'}}{s(1/n_i + 1/n_{i'})^{1/2}} \tag{3.1}$$

where $s^2 = \sum\sum(Y_{ij} - \bar{Y}_i)^2/\sum(n_i - 1)$. The p-values are computed as

$$p_{ii'} = \Pr(|T_\nu| \geqslant |t_{ii'}|)$$

where T_ν denotes a random variable having the central t-distribution with ν degrees of freedom. In this example $\nu = \sum(n_i - 1)$. The lower-case "t" indicates that the value is fixed for the probability calculation. Under the homoscedastic normal model, these statistics are exact, meaning that the nominal and actual Type I error probabilities coincide. Under the homoscedastic location shift and randomization models, the statistics are approximately valid for large sample sizes, meaning that the actual Type I error level converges to the nominal level as the n_i tend to infinity. If the sample sizes n_i all are equal, then these statistics asymptotically valid under the heteroscedastic models as well.

In the pairwise comparisons problem, the complete null hypothesis is $H_0^C\colon \mu_i \equiv \mu$; i.e., all means are equal. In the homoscedastic normal model with the t statistics (3.1), the *single-step adjusted* p-values are, from (2.6) in Section 2.3.2,

$$\tilde{p}_{ii'} = \Pr\left(\min_{1 \leqslant j < j' \leqslant g} P_{jj'} \leqslant p_{ii'} \mid H_0^C\right)$$

Since the p-values are strictly monotonically decreasing in the absolute values of the test statistics, the adjusted p-values may be reformulated as follows:

$$\tilde{p}_{ii'} = \Pr\left(\max_{1 \leqslant j < j' \leqslant g} |T_{jj'}| \geqslant |t_{ii'}| \mid H_0^C\right) \tag{3.2}$$

Note that the subset pivotality condition (Condition 2.1) holds for the p-values. This is clear since the test statistics $t_{ii'}$ depend only on the ϵ's when $H_{ii'}$ is true. Because the subset pivotality condition is very important in resampling-based multiple testing, it is worth demonstrating this property here, in more detail. Note that

$$\bar{Y}_i = \mu_i + \bar{\epsilon}_i$$

where $\bar{\epsilon}_i = \sum_j \epsilon_{ij}/n_i$. Thus, the residual mean square is invariant to (does not depend on) the means μ_i:

$$s^2 = \sum\sum(\epsilon_{ij} - \bar{\epsilon}_i)^2 \Big/ \sum(n_i - 1) = s^2(\epsilon)$$

where ϵ is the vector of residuals ϵ_{ij}. The test statistic for testing $H_{ii'}$: $\theta_{ii'} = 0$ is

$$T_{ii'} = \frac{(\mu_i - \mu_{i'}) + (\bar{\epsilon}_i - \bar{\epsilon}_{i'})}{s(\epsilon)(1/n_i + 1/n_{i'})^{1/2}} \tag{3.3}$$

which clearly does not depend on *any* μ_i when $H_{ii'}$ is true. Therefore, the test statistics possess subset pivotality. Notice that this property does not depend on normality of the residuals.

Since the distribution of the p-values $P_{ii'}$ is completely determined by the distribution of the test statistics $T_{ii'}$, the p-values have subset pivotality as well: the distribution for any subset K is identical under the restrictions H_0^K and H_0^C. Therefore, it will suffice to use the complete null distribution for multiplicity adjustments. Given the homoscedastic model with the ϵ's independent and identically distributed according to a common distribution G, equation (3.2) becomes

$$\tilde{p}_{G,ii'} = \Pr\left(\max_{1 \leqslant j < j' \leqslant g} |T_{jj'}| \geqslant |t_{ii'}| \;\Big|\; G \right) \tag{3.4}$$

Because G is unknown, it must be estimated in order to use (3.4). A consistent estimate $\hat{G}$ of G, as discussed by Freedman (1981), is the empirical distribution of the *residuals* from the one-way ANOVA model. The single-step resampling-based adjusted p-values are then computed using $\hat{G}$ in place of G in (3.4).

Resampling from the residuals, rather than the actual data values, is important. If the actual data are pooled and resampled, then the underlying residual distribution will be estimated using a *multi-modal* empirical distribution function, when the means differ. Such resampling violates the first principle of Hall and Wilson (1991), "... even if the data might be drawn in a way that fails to satisfy H_0, resampling should be done in a way that reflects H_0." Johnsson (1988) misses this point, and proposes using bootstrap resampling with the pooled, uncentered data.

So the adjusted p-values may be computed from the bootstrap data sets

$$Y_{ij}^* = \hat{\epsilon}_{ij}^* \tag{3.5}$$

where the $\hat{\epsilon}_{ij}^*$ are a *with replacement* sample from the residuals $\hat{\epsilon}_{ij} = Y_{ij} - \bar{Y}_{i}$. The following modification of Algorithm 2.5 describes this procedure:

Algorithm 3.1 Single-Step Homoscedastic Bootstrap Pairwise Comparisons

0. Initialize counting variables: $COUNT_{ii'} = 0,\ 1 \leqslant i < i' \leqslant g$.
1. Compute the sample means $\bar{y}_i^*$ ($i = 1, \ldots, g$) and the residual mean square $(s^*)^2$ from the bootstrap data (3.5).
2. Compute the statistics

$$t_{ii'}^* = \frac{\bar{y}_i^* - \bar{y}_{i'}^*}{s^*(1/n_i + 1/n_{i'})^{1/2}}$$

3. If $\max_{1 \leqslant j < j' \leqslant g} |t_{jj'}^*| \geqslant |t_{ii'}|$, where the $t_{ii'}$ are computed using (3.1), then increment the count variable: $COUNT_{ii'} \leftarrow COUNT_{ii'} + 1$.
4. Repeat 1–3 N times. The estimated value of $\tilde{p}_{ii'}$ is $\tilde{p}_{ii'}^{(N)} = COUNT_{ii'}/N$.

A potential problem with resampling the residuals $\hat{\epsilon}_{ij} = Y_{ij} - \bar{Y}_i$ is that residuals from groups with smaller sample sizes have less variability: $\mathrm{Var}(\hat{\epsilon}_{ij}) = \sigma^2(1 - 1/n_i)$. When the residuals are resampled from the pooled set, the resulting distribution might, incorrectly, appear to be a *mixture* distribution. To alleviate this problem in ANOVA and in more general regression applications, "weighted residuals" might be resampled, as discussed by Wu (1986).

Example 3.1 Effectiveness of Weight-Reducing Agents
Table 3.1 is reproduced from Ott (1988), with summary statistics given in Table 3.2. For ease of comparison, the data in Table 3.2 are arranged in ascending order of mean values. Data values represent weight losses using five different weight-reducing agents. There are $n = 10$ male subjects per group, so the balanced one-way ANOVA model is appropriate. Different types of adjusted and unadjusted p-values are given in Table 3.3, again with column and row headings arranged in ascending order of the observed sample means.

The *Single-Step Homoscedastic Bootstrap* (subtable 3.3c) adjustments are computed as in Algorithm 3.1, using PROC MULTTEST with $N = 10{,}000$ bootstrap

Table 3.1 Weight Loss Data* for Reducing Agents A, ..., E

Agent	Observations									
A	12.4	10.7	11.9	11.0	12.4	12.3	13.0	12.5	11.2	13.1
B	9.1	11.5	11.3	9.7	13.2	10.7	10.6	11.3	11.1	11.7
C	8.5	11.6	10.2	10.9	9.0	9.6	9.9	11.3	10.5	11.2
D	8.7	9.3	8.2	8.3	9.0	9.4	9.2	12.2	8.5	9.9
E	12.7	13.2	11.8	11.9	12.2	11.2	13.7	11.8	11.5	11.7

*Lyman Ott, *An Introduction to Statistical Methods and Data Analysis*, 3rd ed., (1988), 463, PWS-Kent Publishing Co., Boston.

Table 3.2 Summary Statistics for Weight Loss Data

	Agent				
Statistic	D	C	B	A	E
Mean	9.27	10.27	11.02	12.05	12.17
SD	1.16	1.03	1.12	.83	.79
SE	.37	.32	.35	.26	.25
N	10	10	10	10	10

Table 3.3 *p*-Values for Weight Loss Data

3.3a: Unadjusted Homoscedastic Normal

	D	C	B	A	E
D	—				
C	.030	—			
B	.000	.099	—		
A	.000	.000	.025	—	
E	.000	.000	.013	.789	—

3.3b: Unadjusted Heteroscedastic Normal

	D	C	B	A	E
D	—				
C	.056	—			
B	.003	.136	—		
A	.000	.001	.032	—	
E	.000	.000	.017	.744	—

3.3c: Single-Step Homoscedastic Bootstrap (*N* = 10,000)

	D	C	B	A	E
D	—				
C	.181	—			
B	.003	.451	—		
A	.000	.002	.160	—	
E	.000	.001	.092	.999	—

3.3d: Single-Step Homoscedastic Normal (*N* = 100,000)

	D	C	B	A	E
D	—				
C	.184	—			
B	.002	.458	—		
A	.000	.002	.161	—	
E	.000	.001	.093	.999	—

3.3e: Free Step-Down Homoscedastic Bootstrap (*N* = 10,000)

	D	C	B	A	E
D	—				
C	.088	—			
B	.002	.180	—		
A	.000	.001	.088	—	
E	.000	.000	.058	.784	—

3.3f: Free Step-Down Homoscedastic Normal (*N* = 40,000)

	D	C	B	A	E
D	—				
C	.086	—			
B	.001	.186	—		
A	.000	.001	.086	—	
E	.000	.001	.056	.790	—

3.3g: REGWQ

	D	C	B	A	E
D	—				
C	.073	—			
B	.002	.230	—		
A	.000	.001	.063	—	
E	.000	.001	.057	.980	—

3.3h: Free Step-Down Heteroscedastic Bootstrap (*N* = 10,000)

	D	C	B	A	E
D	—				
C	.166	—			
B	.027	.257	—		
A	.002	.008	.125	—	
E	.001	.006	.087	.740	—

samples. Comparing the *Unadjusted Homoscedastic Normal* p-values (subtable 3.3d) with the *Single-Step Homoscedastic Bootstrap* p-values, it is clear that the adjusted p-values are larger, making it more difficult to detect significant differences. On the other hand, those pairwise differences which remain significant after adjustment (comparisons (A, C), (A, D), (B, D), (C, E), and (D, E)) are not likely to be artifacts resulting from excessive data dredging. □

3.1.2 Simultaneous Confidence Intervals and Directional Decisions

While the *single-step* adjustments are not as powerful as their step-down counterparts, they have the attractive property that *directional* decisions may be claimed whenever an adjusted p-value is significant. In particular, the method controls (at least asymptotically) the *familywise directional error rate*, or *Type III FWE* (see Hochberg and Tamhane, 1987, p. 39). As pointed out by Tukey (1991), the ability to make directional decisions is highly desirable:

> ... asking "Are the effects different?" is foolish.
>
> What we should be answering first is "Can we tell the direction in which the effects of A differ from the effects of B?" Is it "up," "down," or "uncertain"?*

Using the *Single-Step Homoscedastic Bootstrap* (subtable 3.3c) adjusted p-values in Table 3.3, determination that the direction is "uncertain" is made whenever $\tilde{p}_{ii'} \geqslant \alpha$, where α is the familywise significance level. If α is chosen to be .10, then the directional determination is "uncertain" for comparisons of (μ_A, μ_B), (μ_A, μ_E), (μ_B, μ_C), and (μ_C, μ_D); all other pairs of means $(\mu_i, \mu_{i'})$ can confidently be declared to have the same ordering as their sample counterparts $(\bar{y}_i, \bar{y}_{i'})$.

To demonstrate control of the Type III FWE, we relate the adjusted p-values to simultaneous confidence intervals. Let

$$M_0 = \max_{1 \leqslant j < j' \leqslant g} |T_{jj'}|$$

Under the complete null hypothesis, the distribution of M_0 depends on the error distribution G, and not on the μ's. For convenience, let us assume that this error distribution is continuous. Let Q_G denote the distribution of $M_0 \mid H_0^C$, and let $Q_G^{(\alpha)}$ denote its $(1-\alpha)$ quantile. It follows from equation (3.3) that the pivotal statistic

$$\max_{1 \leqslant j < j' \leqslant g} \frac{|(\bar{y}_j - \mu_j) - (\bar{y}_{j'} - \mu_{j'})|}{s(1/n_j + 1/n_{j'})^{1/2}}$$

has distribution Q_G as well, for all configurations of the μ_i. Hence, *exact* $100(1-\alpha)\%$ simultaneous confidence intervals for $\mu_i - \mu_{i'}$ are given by

$$\bar{y}_i - \bar{y}_{i'} \pm Q_G^{(\alpha)} s(1/n_i + 1/n_{i'})^{1/2} \tag{3.6}$$

*With permission of the Institute of Mathematical Statistics.

Zero falls outside this interval if and only if $|t_{ii'}| > Q_G^{(\alpha)}$. Defining adjusted p-values in terms of the test statistics, $\tilde{p}_{G,ii'} = \Pr(M_0 \geqslant |t_{ii'}| \mid H_0^C)$. Under H_0^C, the distribution of M_0 depends only on G, so $\tilde{p}_{G,ii'} = \Pr(M_0 \geqslant |t_{ii'}| \mid G)$. The condition $\tilde{p}_{G,ii'} < \alpha$ is therefore equivalent to the condition $|t_{ii'}| > Q_G^{(\alpha)}$.

An estimate of $Q_G^{(\alpha)}$ is easily obtained by bootstrapping the statistic M_0, giving estimated quantiles $Q_{\hat{G}}^{(\alpha)}$. Because bootstrap estimates are asymptotically consistent (e.g., Beran, 1988a) we can anticipate that the simultaneous bootstrap confidence intervals have coverage probability approaching $100(1 - \alpha)\%$ in large samples. In this case the bootstrap-based criterion $\tilde{p}_{\hat{G},ii'} < \alpha$ also approximately controls the Type III FWE in large samples: we can have a Type III error only when at least one simultaneous confidence interval fails to contain the true parameter value.

The following algorithm may be used to construct bootstrap simultaneous confidence intervals in the homoscedastic ANOVA model.

Algorithm 3.2 Bootstrap Simultaneous Pairwise Intervals

1. Compute the sample means $\bar{y}_i^*, i = 1, \ldots, g$, as well as the residual mean square $(s^*)^2$ from the bootstrap data (3.5).
2. Compute the statistics

$$t_{ii'}^* = \frac{\bar{y}_i^* - \bar{y}_{i'}^*}{s^*(1/n_i + 1/n_{i'})^{1/2}}$$

3. Compute and store $\max_{1 \leqslant i < i' \leqslant g} |t_{ii'}^*|$.
4. Repeat 1–3 N times. The estimated value of $Q_{\hat{G}}^{(\alpha)}$ is $\hat{Q}_{\hat{G}}^{(\alpha)}$, the $1 - \alpha$ empirical quantile of the N values $\max_{1 \leqslant i < i' \leqslant g} |t_{ii'}^*|$.

This method may be used to construct simultaneous confidence intervals of the form

$$\bar{y}_i - \bar{y}_{i'} \pm \hat{Q}_{\hat{G}}^{(\alpha)} s(1/n_i + 1/n_{i'})^{1/2}$$

Note that these intervals are symmetric. In cases of strongly skewed distributions, it may be more reasonable to construct asymmetric confidence intervals. However, as discussed in Chapter 2, Section 2.5, skewnesses of individual distributions may approximately cancel when considering two-sample comparisons. In such cases, the sampling distribution of the pivotal statistic is approximately symmetric, suggesting the use of a symmetric interval. See Beran (1988a, 1988b) for further refinements of bootstrap simultaneous confidence sets and tests of hypotheses.

The simplest possible homoscedastic location shift model is the normal model with equal sample sizes $n_i \equiv n$. In this case the following simplification is possible:

$$\tilde{p}_{ii'} = \Pr\left(\frac{1}{\sqrt{2}} \max_{1 \leqslant j < j' \leqslant g} \frac{|\bar{Y}_j - \bar{Y}_{j'}|}{s/\sqrt{n}} \geqslant |t_{ii'}| \;\Big|\; H_0^C\right)$$

Recognizing that

$$\max_{1\leqslant j<j'\leqslant g} \frac{|\bar{Y}_j - \bar{Y}_{j'}|}{s/\sqrt{n}} \mid H_0^C$$

has the *Studentized range distribution* $Q_{g,\nu}$, it follows that $\tilde{p}_{ii'} \leqslant \alpha$ if and only if $t_{ii'} \geqslant \frac{1}{\sqrt{2}} Q_{g,\nu}^{(\alpha)}$, where $Q_{g,\nu}^{(\alpha)}$ denotes the $(1 - \alpha)$ quantile of the given Studentized range distribution. Equivalently, $\tilde{p}_{ii'} \leqslant \alpha$ if and only if $|\bar{y}_i - \bar{y}_{i'}| \geqslant Q_{g,\nu}^{(\alpha)} s/\sqrt{n}$.

Thus, in the homoscedastic normal model with equal sample sizes, the single-step resampling-based adjustment is equivalent to Tukey's (1953) *T procedure* (e.g., Hochberg and Tamhane, 1987, p. 31.) In this case, simulation analysis is not required to estimate the adjusted p-values, since the Studentized range distribution is well known. However, one still might use simulation methods if it is inconvenient to integrate the Studentized range distribution numerically. As is discussed later in this section, the use of simulation methods is even more appealing in designs with unequal sample sizes, since tabled values of the appropriate range distributions are virtually nonexistent, and since the analytic evaluation of the distribution function can be extremely difficult.

The Tukey *T procedure* may be used either for multiple tests or for simultaneous intervals. The simultaneous intervals are *exact*; therefore, the criterion $\tilde{p}_{ii'} \leqslant \alpha$ controls the Type III FWE *exactly* in the case of balanced homoscedastic normal data. The *Single-Step Homoscedastic Normal* entries of Table 3.3 are the single-step adjusted p-values under the assumption of a normal residual distribution. Comparing these with the *Single-Step Homoscedastic Bootstrap* p-values shows little difference; this is to be expected for approximately normal data. The central limit theorem guarantees that the results are asymptotically equivalent under normal sampling and bootstrap sampling; however, as indicated in Section 2.5, the results may differ dramatically in small to moderate sample sizes with skewed distributions.

When the normality assumption holds, the bootstrap samples can be generated from the $\mathcal{N}(0, s^2)$ distribution, then proceeding as in Algorithm 3.1. The calculations simplify immensely in this special case. The *Single-Step Homoscedastic Normal* adjusted p-values in Table 3.3 (subtable 3.3d) were computed as follows: let $Z_A, \ldots, Z_E$ denote independent standard normal variables and V be distributed independently as χ^2_{45}. The adjusted p-values in (3.2) may be simulated as

$$\Pr\left(\frac{\max(Z_j) - \min(Z_j)}{(2V/45)^{1/2}} \geqslant |t_{ii'}|\right)$$

Standard normal and chi-squared (gamma) random number generators are available in many software packages, making this simulation easy. The entries in Table 3.3 were computed using 100,000 such simulations with handwritten code using the RANNOR and RANGAM random number functions of the SAS (1990) software. With 100,000 simulations, the binomial simulation standard error is .0016 when the true adjusted p-value of .5; the simulation standard error is .0007 when true adjusted p-value is .05.

In this case the computations can be performed analytically using numerical integration of the Studentized range distribution (e.g., Hochberg and Tamhane, 1987, p. 376). The Studentized range distribution is available directly in the SYSTAT® software package as the "SCF" function (Wilkinson, 1988); comparison of such values with the simulated values in Table 3.3 shows no differences larger than .003.

What about unbalanced designs? Even in the homoscedastic one-way ANOVA model, this simple difference has caused great problems for classical multiple comparisons procedures. The main problem is that the distribution of M_0 is not simply related to the Studentized range distribution for general n_i. There is a large literature concerning this distribution: Spurrier and Isham (1985), and Hayter (1985) develop analytic representations for this distribution, but numerical evaluation of its quantiles involves substantial computational difficulty.

The resampling approach provides a simple solution to the problem of imbalance, since the distribution of $M_0 = \max_{1 \leq j < j' \leq g} |T_{jj'}|$ is estimated with arbitrary precision, given sufficient computing resources and accurate random number generators. The simultaneous directional error rate is likewise controlled for single-step analysis of two-sided tests.

3.1.3 Homoscedastic Model: Step-Down Analysis

We will now demonstrate that the computationally convenient *free step-down method* compares favorably with commonly used step-down methods. When all pairwise comparisons are considered, the hypotheses are *restricted*, as discussed in Section 2.7. Nevertheless, as given in Section 2.8, the general resampling method for *free* combinations may be used to control the FWE. The computations are much simpler for the *free* step-down adjustments than they are for the *restricted* adjustments, as seen by comparing (2.10) with (2.13).

For the homoscedastic location shift model, the bootstrap data sets are obtained by sampling the residuals with replacement, as given in (3.4). Algorithm 3.1 then can be modified to perform the free step-down adjustments. Using the weight loss data of Table 3.1 as an example, the pooled residual mean square is $s = .997$, the t-statistics are computed as

$$t_{ii'} = \frac{\bar{y}_i - \bar{y}_{i'}}{.997 \times .447}$$

and the ordering of the observed absolute t-statistics is as follows: $|t_{AE}| = .269$, $|t_{BC}| = 1.683$, $|t_{CD}| = 2.243$, $|t_{AB}| = 2.311$, $|t_{BE}| = 2.580$, $|t_{BD}| = 3.926$, $|t_{AC}| = 3.993$, $|t_{CE}| = 4.263$, $|t_{AD}| = 6.237$, and $|t_{DE}| = 6.506$. Combining elements of Algorithms 2.8 and 3.1, the following algorithm computes the free step-down adjusted p-values for the weight loss data. (Note that this algorithm is specific to this particular application only. The more general form is given in Algorithm 2.8.)

Algorithm 3.3 Free Step-Down Homoscedastic Bootstrap Pairwise Comparisons

0. Initialize counting variables: $COUNT_{ii'} = 0, 1 \leq i < i' \leq g$.

1. Compute the sample means $\bar{y}_i^*, i = 1, \ldots, g$, as well as the residual mean square $(s^*)^2$ from the bootstrap data (3.5).
2. Compute the statistics

$$t_{ii'}^* = \frac{\bar{y}_i^* - \bar{y}_{i'}^*}{s^*(1/n_i + 1/n_{i'})^{1/2}}$$

   Define the successive *maxima*:

$$\begin{aligned} u_{AE}^* &= |t_{AE}^*| \\ u_{BC}^* &= \max(u_{AE}^*, |t_{BC}^*|) \\ &\vdots \\ u_{DE}^* &= \max(u_{AD}^*, |t_{DE}^*|) \end{aligned}$$

3. If $u_{ii'}^* \geqslant |t_{ii'}|$, where the $t_{ii'}$ are computed using (3.1), then increment the count variable: $COUNT_{ii'} \leftarrow COUNT_{ii'} + 1$.
4. Repeat 1–3 N times. Compute $\tilde{p}_{ii'}^{(N)} = COUNT_{ii'}/N$.
5. Enforce monotonicity using successive maximization:

$$\begin{aligned} \tilde{p}_{DE}^{(N)} &= \tilde{p}_{DE}^{(N)} \\ \tilde{p}_{AD}^{(N)} &\leftarrow \max(\tilde{p}_{DE}^{(N)}, \tilde{p}_{AD}^{(N)}) \\ &\vdots \\ \tilde{p}_{AE}^{(N)} &\leftarrow \max(\tilde{p}_{BC}^{(N)}, \tilde{p}_{AE}^{(N)}) \end{aligned}$$

Table 3.3 gives the resulting *Free Step-Down Homoscedastic Bootstrap* adjusted p-values (subtable 3.3e), computed using PROC MULTTEST with $N = 10{,}000$ simulations. Unlike the single-step adjustments given in Table 3.3, subtable 3.3c, directional associations are not declared with the step-down adjustments. However, it is clear that the differences are "more significant" using the step-down method since the adjusted p-values are much smaller.

As a point of comparison, this method may also be used with normally distributed data. For this particular application, the normality and balance conditions simplify the computations, and steps 1. and 2. of Algorithm 3.3 may be replaced by

1. Generate $z_A^*, \ldots, z_E^*$ independently from the standard normal distribution, and generate v^* independently from the χ_{45}^2 distribution. Compute the pseudo-test statistics

$$t_{ii'}^* = \frac{(z_i^* - z_{i'}^*)}{(2v^*/45)^{1/2}}$$

2. Define the successive *maxima*:

$$
\begin{aligned}
u_{AE}^* &= |t_{AE}^*| \\
u_{BC}^* &= \max(u_{AE}^*, |t_{BC}^*|) \\
&\vdots \\
u_{DE}^* &= \max(u_{AD}^*, |t_{DE}^*|)
\end{aligned}
$$

3. If $u_{ii'}^* \geqslant |t_{ii'}|$, then $COUNT_{ii'} \leftarrow COUNT_{ii'} + 1$.

The remaining steps are as before. The *Free Step-Down Homoscedastic Normal* entries of Table 3.3, subtable 3.3f, show the resulting adjusted p-values from 40,000 simulations using the RANNOR and RANGAM random number generators of SAS. As in the case of the single-step adjustments, there is little discrepancy between normal sampling and bootstrap sampling in this application.

Use of adjusted p-values facilitates comparison with alternative step-down multiple testing methods. A method developed by Ryan (1960), Einot and Gabriel (1975), and Welsch (1977), called the "REGWQ Method" in the GLM procedure of SAS/STAT (SAS/STAT® User's Guide, Version 6, Fourth Edition, Volume 2, 1989, p. 947) is comparable to the minimum p-value technique. Letting the ordered means be $\bar{y}_1 \leqslant \cdots \leqslant \bar{y}_g$, the test $H_{ii'}$: $\mu_i = \mu_{i'}$ is rejected at FWE level α when

$$t_{ii'} \geqslant Q_{r,\nu}^{(\alpha_r)}/\sqrt{2}$$

where $r = |i - i'| + 1$ and

$$\alpha_r = \begin{cases} 1 - (1 - \alpha)^{r/g} & \text{for } r < g - 1 \\ \alpha & \text{for } r \geqslant g - 1 \end{cases}$$

Thus, the adjusted p-value for test $H_{ii'}$ is

$$\tilde{p}_{ii'} = \begin{cases} 1 - (\Pr(Q_{r,\nu} \leqslant \sqrt{2}t_{ii'}))^{g/r} & \text{for } r < g - 1 \\ 1 - \Pr(Q_{r,\nu} \leqslant \sqrt{2}t_{ii'}) & \text{for } r \geqslant g - 1 \end{cases}$$

where $Q_{r,\nu}$ denotes a random variable having the Studentized range distribution with range r and degrees of freedom ν.

The resulting adjusted p-values are displayed under the *REGWQ* heading of Table 3.3, subtable 3.3g; the Studentized range probabilities were calculated using the "SCF" function of SYSTAT® (Wilkinson, 1988). By and large, the results are similar to the free step-down adjustments of Table 3.3. Neither procedure dominates the other in the sense of having uniformly smaller p-values. On the other hand, *both* methods dominate the single-step adjustments.

At the expense of increased computational difficulty, it is possible to improve upon these adjustments uniformly (that is, by making the adjusted p-values uniformly smaller) when logical constraints are considered. This resampling-based technique is discussed in Section 2.7. Analytic range based methods are discussed by Marcus, Peritz, and Gabriel (1976), Begun and Gabriel (1981), and Welsch (1977). Bonferroni- and independence-based methods for incorporating logical constraints (as discussed by Shaffer, 1986; Holland and Copenhaver, 1987; and in Section 2.8), also may give smaller adjustments than either the free step-down approach or the REGWQ method.

3.1.4 Heteroscedastic Model

The heteroscedastic model allows different distributions G_i for residuals ϵ_{ij} occurring in different groups. Under heteroscedasticity, the following statistics are commonly used for pairwise testing:

$$t_{ii'} = \frac{\bar{y}_i - \bar{y}_{i'}}{(s_i^2/n_i + s_{i'}^2/n_{i'})^{1/2}} \tag{3.7}$$

where $s_i^2 = \sum_j (y_{ij} - \bar{y}_i)^2/(n_i - 1)$. Approximately valid p-values are computed using the Satterthwaite (1946) method: the degrees of freedom are given by

$$\nu_{ii'} = \frac{(s_i^2/n_i + s_{i'}^2/n_{i'})^2}{(s_i^2/n_i)^2/(n_i - 1) + (s_{i'}^2/n_{i'})^2/(n_{i'} - 1)}$$

and the unadjusted p-values are given by

$$p_{ii'} = \Pr(|T_{\nu_{ii'}}| > |t_{ii'}|) \tag{3.8}$$

To judge the extremity of these p-values, it is no longer sufficient to consider distributions of the maxima of test statistics $|t_{ii'}|$: since the degrees of freedom are variable, the strict inverse monotonicity between $p_{ii'}$ and $|t_{ii'}|$ no longer holds. This is a situation where the adjustments should be defined directly in terms of minimum p-value distributions, as the general method of Section 2.3.2 suggests.

Using the test statistics (3.7), we note again that the subset pivotality condition holds under the heteroscedastic location shift model, since the distribution of any test statistic depends only on the ϵ's when the corresponding hypothesis is true. Thus, it will suffice to compute adjustments under the complete null model $\mu_i \equiv \mu$. Because the value μ cancels from the test statistics, we can take $\mu = 0$ in our resampled data sets.

In this case, the adjusted p-values (either the single-step, Equation 2.6, or the step-down, Equation 2.10) depend on the unknown distributions $G_1, \ldots, G_g$. For the step-down adjustments (2.10),

$$\Pr\left(\min_{l \in \{r_j, \ldots, r_k\}} P_l \leqslant p_{(j)} \mid H_0^C\right) = \Pr\left(\min_{l \in \{r_j, \ldots, r_k\}} P_l \leqslant p_{(j)} \mid G_1, \ldots, G_g\right)$$

The bootstrap method for calculating such adjustments involves replacing the unknown G_i with their empirical distribution function estimates $\hat{G}_i$:

$$\Pr\left(\min_{l\in\{r_j,\ldots,r_k\}} P_l \leqslant p_{(j)} \mid H_0^C\right) \approx \Pr\left(\min_{l\in\{r_j,\ldots,r_k\}} P_l \leqslant p_{(j)} \mid \hat{G}_1,\ldots,\hat{G}_g\right)$$

The adjusted p-values may be computed using the bootstrap data sets

$$Y_{ij}^* = \hat{\epsilon}_{ij}^*$$

where the $\hat{\epsilon}_{ij}^*$ are obtained using g *distinct* samples drawn *with replacement*: $\hat{\epsilon}_{11}^*,\ldots,\hat{\epsilon}_{1n_1}^*$ are a *with replacement* sample from the residuals $\hat{\epsilon}_{11},\ldots,\hat{\epsilon}_{1n_1}$; $\hat{\epsilon}_{21}^*,\ldots,\hat{\epsilon}_{2n_2}^*$ are a *with replacement* sample from the residuals $\hat{\epsilon}_{21},\ldots,\hat{\epsilon}_{2n_2}$, and so on. As before, the residuals are defined by $\hat{\epsilon}_{ij} = Y_{ij} - \bar{Y}_i$.

The unadjusted p-values for the weight loss data using Satterthwaite's approximation, as well as the free step-down adjustments using heteroscedastic bootstrap resampling, are given in Table 3.3, subtable 3.3b. The adjusted p-values were programmed and calculated using GAUSS386 (© Aptech Systems, Inc., 1992). Comparing the *homoscedastic bootstrap* adjustments (subtable 3.3e) with the *heteroscedastic bootstrap* adjustments (subtable 3.3h), it is clear that the heteroscedastic resampling of p-values (3.8) results in a more conservative multiple test procedure. This is not surprising, given that the unadjusted Satterthwaite p-values are somewhat larger than the unadjusted homoscedastic normal p-values. In addition, there is greater instability of the test statistics (3.7). This greater instability may result in a greater probability of observing an "extreme" value of the form (3.7) than of the "pooled variance" variety (3.1), when many such statistics are considered jointly.

Resampling also may be performed using the normal model, with variances estimated from the data. In this case, the residuals may be simulated as $\hat{\epsilon}_{ij}^* \sim \mathcal{N}(0, s_i^2)$. Further simplifications in the resampling algorithm are possible with normal resampling.

It should be noted, when performing heteroscedastic resampling, that there are many more parameters (distributions or variances) that must be estimated for use in the resampling model. Thus, the resulting adjusted p-values are likely to have greater instability than those resulting from homoscedastic resampling, particularly when sample sizes are small. In small sample situations, it may be prudent to calculate multiplicity adjustments that are probabilistically guaranteed to be conservative. The resampling method may then be a useful adjunct to estimate the *degree* of conservatism of the given method. An example of such a conservative methods for normal heteroscedastic models is given by Hochberg (1976).

Example 3.2 Fat Absorption Data

The data in Table 3.4 are taken from Snedecor and Cochran (1976). Unadjusted Satterthwaite p-values (subtable 3.5a) and free step-down adjusted p-values based upon heteroscedastic normal resampling are (subtable 3.5c) displayed in Table 3.5. The calculations were programmed and executed using the GAUSS386 software.

Table 3.4 Fat Absorbed by Doughnuts*

Cooking Method			
1	2	3	4
164	178	175	155
172	191	193	166
168	197	178	149
177	182	171	164
156	185	163	170
195	177	176	168

*From *Statistical Methods*, 8th ed. (1989) Snedecor and Cochran, with permission of Iowa State University Press.

Table 3.5 *p*-Values for Fat Absorption Data

3.5a: Unadjusted Heteroscedastic Normal (Satterthwaite)

	4	1	3	2
4	—			
1	.155	—		
3	.024	.569	—	
2	.001	.073	.112	—

3.5b: Unadjusted Homoscedastic Normal

	4	1	3	2
4	—			
1	.100	—		
3	.026	.498	—	
2	.001	.037	.136	—

3.5c: Free Step-Down Heteroscedastic Normal (N = 10,000)

	4	1	3	2
4	—			
1	.254	—		
3	.080	.563	—	
2	.003	.198	.254	—

3.5d: Free Step-Down Homoscedastic Bootstrap (N = 10,000)

	4	1	3	2
4	—			
1	.243	—		
3	.094	.450	—	
2	.004	.111	.243	—

3.5e: Unadjusted Rerandomization (N = 10,000)

	4	1	3	2
4	—			
1	.158	—		
3	.024	.563	—	
2	.002	.070	.129	—

3.5f: Single-Step Rerandomization (N = 10,000)

	4	1	3	2
4	—			
1	.338	—		
3	.107	.905	—	
2	.004	.147	.424	—

3.5g: Free Step-Down Rerandomization (N = 10,000)

	4	1	3	2
4	—			
1	.238	—		
3	.096	.501	—	
2	.004	.114	.238	—

3.5h: Restricted Step-Down Rerandomization (N = 10,000)

	4	1	3	2
4	—			
1	.158	—		
3	.083	.563	—	
2	.004	.117	*	—

*Reference distribution unavailable using PROC MULTTEST.

Similar to the analysis of Hochberg and Tamhane (1987, p. 191), who analyzed the data using a different method, Table 3.5 shows only one significance, the (μ_2, μ_4) comparison. □

As an aside, heteroscedastic resampling using the normal distribution model seems preferable to heteroscedastic bootstrap sampling in this application because the sample sizes are uncomfortably small. The empirical distribution functions do not estimate the true functions very well with only 6 observations per group. One measure of this problem is the probability that the bootstrap will produce samples having a sample variance of zero. We wish bootstrap samples to mimic the data-generating process, but with continuous data we expect that sample variances will be positive. If the bootstrap model is likely to generate zero variances, then the bootstrap model clearly is much different from the real model.

To illustrate this phenomenon, note that within group 1 there is a $(\frac{1}{6})^6$ probability that all bootstrap residuals will be set to $\hat{e}_{11}$. There is similarly a $(\frac{1}{6})^6$ chance that all bootstrap residuals are set to $\hat{e}_{12}$, and so on, so the probability that all residuals are set to a common value is at least $6(\frac{1}{6})^6 = (\frac{1}{6})^5$. (This probability is higher if the residuals $\hat{e}_{11}, \ldots, \hat{e}_{16}$ are not all distinct.)

Now, when all four groups are considered, the probability that *at least one* resampled data set has zero variance is at least $1 - (1 - (\frac{1}{6})^5)^4 = .0005143$, a small probability. However, when *NBOOT* = 10,000 bootstrap data sets are considered, the probability that at least one will contain a group with zero variance is at least $1 - (1 - .0005143)^{10,000} = .994$, which is a fairly certain chance.

This situation is less of a problem in homoscedastic bootstrap sampling. Here, there is a combined pool of 24 residuals to resample from, and the probability of having all bootstrap residuals within a group equal to a common value is $24(\frac{1}{24})^6$, assuming all $\hat{e}_{ij}$ are distinct. The probability of *at least one* group having zero variance is then $1 - (1 - (\frac{1}{24})^5)^4$, and the probability of this occurring in at least one of 10,000 bootstrap data sets is only .005.

3.1.5 Randomization Model

Under the randomization ANOVA model, the "complete null hypothesis" means that all treatments have identical effects (possibly no effect) on all experimental units. Using the notation R_{iu} of Section 3.1 to denote the response of experimental unit u to treatment i, the randomization null hypothesis is $H: R_{iu} \equiv R_u$. If this hypothesis is true, the N observed data values y_{ij} are simply a random arrangement of $R_1, \ldots, R_N$ into g groups of size n_i each. Critical values and p-values therefore are obtained easily via resampling: simply "shuffle" the N data values y_{ij}, placing the first n_1 values y^*_{1j} into group 1, the second n_2 values y^*_{2j} into group 2, and so on. Each similarly constructed data set is equally likely to have occurred under the complete randomization null hypothesis.

The randomization model poses a challenge in that the subset pivotality condition of Section 2.2.3 is not valid. That is, the joint distribution of a subcollection of test statistics under its partial randomization null hypothesis is not identical to its joint distribution under the complete randomization null hypothesis. For example, suppose that treatments $i = 1$ and $i = 2$ have identical effects, but that the remaining treatments all affect the units in different ways. Then $R_{1u} \equiv R_{2u}$, but all remaining R_{iu} may differ. In this case we may state that, conditional upon the combined collection of experimental units selected for groups 1 and 2, *all random shuffles of* $y_{11}, \ldots, y_{1n_1}, y_{21}, \ldots, y_{2n_2}$ *into groups 1 and 2 only are equally likely to have occurred.* We may not say that all random permutations of the *entire collection* of observed values $\{y_{ij}; i = 1, \ldots, g, j = 1, \ldots, n_i\}$ are equally likely to have occurred. Petrondas and Gabriel (1983) discuss this phenomenon in more detail, noting that use of the complete null randomization distribution can result in "excess Type I errors": for particular partial null hypotheses, the probability of wrongly declaring a particular comparison significant may exceed α.

Rerandomization (or permutation) analysis is inherently a "resampling method," since, by definition, the observed data values define the reference distributions. If one uses the complete null distribution for all calculations (admittedly risking excess Type I errors, as discussed by Petrondas and Gabriel, 1983), the single-step adjusted p-values may be defined as

$$\begin{aligned}\tilde{p}_{ii'} &= \Pr\left(\min_{1 \leq j < j' \leq g} P_{jj'} \leq p_{ii'} \mid H_0^C\right) \\ &= \Pr\left(\min_{1 \leq j < j' \leq g} P_{jj'} \leq p_{ii'} \mid \text{all permutations of the data are equally likely}\right) \qquad (3.9)\end{aligned}$$

Often, the statistics $\bar{y}_i - \bar{y}_{i'}$ are chosen in place of p-value statistics in randomization analysis. Using the usual t-statistics and their resulting p-values is an alternative method. One motivation for using t-statistics and their resulting p-values in randomization analysis is that they provide an approximation of the corresponding randomization-based probabilities (Pitman, 1938). In addition, the use of the pivotal statistics may result in a more "balanced" multiple testing analysis (Beran, 1988a, 1988b).

To compute the adjustments in (3.9) via simulation, generate *random permutations* of the observed data values y_{ij}. This differs from the bootstrap method in two ways: first, the data values are *not centered*; second, sampling will be performed *without replacement* instead of *with replacement.*

Algorithm 3.4 Single-Step Rerandomized Pairwise Comparisons

0. Initialize counting variables: $COUNT_{ii'} = 0$, $1 \leq i < i' \leq g$.
1. Let y_{ij}^* denote a *without replacement sample* (or permutation) of the y_{ij}. That is, the values $y_{11}, \ldots, y_{1n_1}, \ldots, y_{gn_g}$ simply are *shuffled*, and the first n_1 elements of the resulting shuffle are $y_{11}^*, \ldots, y_{1n_1}^*$, the second n_2 elements are $y_{21}^*, \ldots, y_{2n_2}^*$, and so on.

   Compute the sample means $\bar{y}_i^*, i = 1, \ldots, g$, as well as the residual mean square $(s^*)^2$ from the permuted data y_{ij}^*.

2. Compute the statistics

$$t^*_{ii'} = \frac{\bar{y}^*_i - \bar{y}^*_{i'}}{s^*(1/n_i + 1/n_{i'})^{1/2}}$$

3. If $\max_{1\leqslant j<j'\leqslant g} |t^*_{jj'}| \geqslant |t_{ii'}|$, where the $t_{ii'}$ are the two-sample t-statistics (3.1), then increment the count variable: $COUNT_{ii'} \leftarrow COUNT_{ii'} + 1$.
4. Repeat 1–3 N times. The estimated value of $\tilde{p}_{ii'}$ is $\tilde{p}^{(N)}_{ii'} = COUNT_{ii'}/N$.

This algorithm is available from PROC MULTTEST. Analysis of the "fat data" of Example 3.2 gives the *Single-Step Rerandomization* p-values of Table 3.5, subtable 3.5f. Since the subset pivotality condition is violated, all p-value adjustments other than that of the minimum, (the (2,4) comparison), strictly speaking, are not valid. Furthermore, the adjusted p-value for the (2,4) comparison, $\tilde{p}_{24} = .004$, should properly be considered a test of the complete null randomization hypothesis.

On the other hand, if one simply considered simple rerandomization tests, without adjustment, the resulting p-values are much smaller than indicated by the single-step adjustments. Such p-values appear in Table 3.5, under the heading *Unadjusted Rerandomization*, subtable 3.5e. These were computed by applying PROC MULTTEST to the specific groups to be compared, and specifying only one test. In this case, the "adjusted p-value" of the PROC MULTTEST output is in fact the exact rerandomization p-value. These were computed using permutation resampling, with the two-sample t-statistic as the base test. In this example, the risk of "excess Type I errors" is certainly less if one performs a single-step adjustment than if one performs no adjustment: in Table 3.5, the p-values under the "Unadjusted Rerandomization" heading are all smaller than they are under the "Single-Step Rerandomization" heading.

As suggested by Petrondas and Gabriel (1983), proper rerandomization-based multiplicity adjustments use the reference distributions of the subset hypotheses, rather than the complete null hypothesis. These adjustments are also defined in Section 2.7, equation (2.13), except the probability calculations are performed under the particular null hypothesis corresponding to the subset considered at that particular step. In the fat absorption data, the ordering of the hypotheses, from most significantly different to least significantly different, is as follows: $H_1: \mu_2 = \mu_4$, $H_2: \mu_3 = \mu_4$, $H_3: \mu_1 = \mu_2$, $H_4: \mu_2 = \mu_3$, $H_5: \mu_1 = \mu_4$, and $H_6: \mu_1 = \mu_3$. The maximal subcollections of null hypotheses allowed at each step ("K," from Equation 2.13) are as follows:

$K_1 = \{1,2,3,4,5,6\}$ all treatments identical

$K_2 = \{2,3,6\}$ treatments 1, 3, and 4 identical

$K_3 = \{1,2,4\}$ treatments 1, 2, and 3 identical

$K_4 = \{3,4\}$ treatments 2 and 3 identical; treatments 1 and 4 identical

$K_5 = \{3\}$ treatments 1 and 4 identical

$K_6 = \{2\}$ treatments 1 and 3 identical

Under the heading *Restricted Step-Down Rerandomization*, Table 3.5 (subtable 3.5h) displays the results of using PROC MULTTEST repeatedly, eliminating particular groups not included in the given set K, and adjusting the minimum p-value over the given sets K. For example, for K_2, data from treatment group 2 was deleted, and the multiplicity adjustment was obtained by randomizing the observations among the remaining three treatment groups. The separate permutation distributions needed for K_4 pose a unique problem (see the discussion of "closure" by Einot and Gabriel, 1975); this particular reference distribution is not available using PROC MULTTEST since the distribution cannot be obtained by simply eliminating particular groups of data. One would need to retain all groups in the data sets, and shuffle data from treatment groups 2 and 3 within those groups only, and shuffle data from treatment groups 1 and 4 within those groups only. Comparing the *Restricted Step-Down Rerandomization* adjustments (3.5h) with *Single-Step Rerandomization* adjustments (3.5f) and *Free Step-Down Rerandomization* adjustments (3.5g, both single-step and step-down adjustments available directly from PROC MULTTEST), we believe that "excess Type I errors" are not of concern in this example. However, in comparing the *Free Step-Down Rerandomization* adjusted p-value .501 with the larger *Restricted Step-Down Rerandomization* adjustment .563, for the (3, 1) comparison, the potential for such errors to occur becomes apparent.

3.2 ONE-WAY ANOVA MODEL: OTHER COMPARISONS

It should be clear that resampling-based multiplicity adjustment can apply to any collection of tests, not just pairwise comparisons. Other possibilities are comparisons of all treated groups against a common control group, and orthogonal contrasts to measure linear, quadratic, and higher order trends in the dosage levels.

3.2.1 Comparing All Treatments against a Control

The comparison of all treatments against a control often is done using Dunnett's one-sided tests (Dunnett, 1955). In balanced designs, homoscedastic normal resampling is equivalent to Dunnett's test. Further, it models the underlying dependence structure exactly in unbalanced designs and/or cases involving covariates.

Example 3.3 Uterine Weights of Mice

An example where all pairwise comparisons against a control is desired is given by Steel and Torrie (1980, p. 144), with data in Table 3.6. We are interested in lower-tailed comparisons. Unadjusted p-values are given by

$$p_{ic} = \Pr(T_\nu < t_{ic})$$

where t_{ic} is the pooled-variance t statistic for comparing group i with the control, and where the degrees of freedom are $\nu = 21$. Table 3.7 reports the unadjusted p-values, the *free* step-down adjusted p-values using 40,000 bootstrap simulations (homosce-

Table 3.6 Uterine Weights of Mice*

	Solutions						
	1	2	3	4	5	6	Control
	84.4	64.4	75.2	88.4	56.4	65.6	89.8
	116.0	79.8	62.4	90.2	83.2	79.4	93.8
	84.0	88.0	62.4	73.2	90.4	65.6	88.4
	68.6	69.4	73.8	87.8	85.6	70.2	112.6
$\bar{y}$	88.25	75.40	68.45	84.90	78.90	70.20	96.15

*From Steel and Torrie, *Principles and Procedures of Statistics*, (1980), with permission of McGraw-Hill, Inc.

dastic model), and the *free* step-down Bonferroni and Šidák adjusted p-values. All calculations were performed using PROC MULTTEST. At the $\alpha = .10$ familywise significance level, treatments 2, 3, 5, and 6 have significantly lower uterine weights; treatments 2, 3, and 6 remain significant at the $\alpha = .05$ level. These conclusions are unchanged even when the more conservative Bonferroni adjustments are used.

□

Side-by-side comparison of the resampling-based adjustments with the Bonferroni adjustments shows the effects of dependence structure on the multiplicity adjustments. As suggested in Section 2.5 (see the Miller, 1977, quote), these differences are not large, particularly for small adjusted p-values.

If sampling is performed from a normal homoscedastic model, then this approach is equivalent to the step-down analysis described by Hochberg and Tamhane (1987), p. 149, except for the *display* of the results, which uses adjusted p-values, and except for the Monte Carlo error of the resampling-based approach. Since the tests are *free* combinations in this example (called "nonhierarchical" by Hochberg and Tamhane), the adjustments may not be improved by incorporating logical constraints.

The comparison of the single-step bootstrap adjustments with the single-step homoscedastic normal adjustments given by Hochberg and Tamhane shows an interesting feature: the bootstrap adjustment for the comparison (5 vs. control) is signifi-

Table 3.7 p-Values for Uterine Weight Data

		Step-Down			Single-Step		
Test	Raw_p	Adj_p	Sid_p	Bon_p	Adj_p	Sid_p	Bon_p
1 vs. control	.1827	.1794	.1917	.2019	.4605	.7020	1.0000
4 vs. control	.1009	.1605	.1917	.2019	.2908	.4719	0.6056
5 vs. control	.0282*	.0631	.0821	.0845	.0957	.1575	0.1689
2 vs. control	.0121*	.0344*	.0474*	.0483*	.0443*	.0703	0.0725
6 vs. control	.0031*	.0108*	.0155*	.0156*	.0122*	.0186*	0.0187*
3 vs. control	.0019*	.0080*	.0116*	.0117*	.0080*	.0116*	0.0117*

*p-value $< .05$.

cant at the $\alpha = .10$ level, while the corresponding adjustment based on the normality assumption is insignificant. (Hochberg and Tamhane show that the appropriate test statistic does not exceed the $\alpha = .10$ quantile of the relevant distribution; equivalently, the theoretical homoscedastic normal resampling-based adjusted p-value is greater than .10.) This discrepancy is due solely to the fact that resampling was done from a nonnormal distribution: given the large simulation sample size of 40,000, it is unlikely that the difference is attributable to simulation error. Since the normality of the residual distribution is in question, it is prudent to examine this distribution via normal quantile–quantile plot, shown in Figure 3.1. The residual distribution does not deviate markedly from normal, although there are some interesting "kinks" that may suggest nonnormal characteristics.

In the face of this discrepancy between normal and bootstrap sampling, three comments are in order. First, the use of adjusted p-values rather than formal "reject/accept" decisions alleviates the problem. If we believe that a p-value of .095 is not dramatically different from .105, then the bootstrap analysis and the normality-assumption analysis are quite similar. The second comment concerns the validity of the two adjustments: since the normality assumption is always suspect, and can never be considered to hold *exactly*, the normality-assumption adjusted p-value is, in reality, always an approximation. Thus, we have two approximations: one based on the normality assumption, and the other based on a bootstrap estimate of the underlying error distribution. Determining which adjustment is more accurate is nearly impossible, since the true underlying distribution is unknown. The third comment relates to the second: in extremely small sample sizes, the normality-based adjustments may be preferable because of the instability of the estimated empirical distribution function, even if the normality assumption is obviously in error. In moderate to large sample sizes, there is less instability of the bootstrap estimates, and

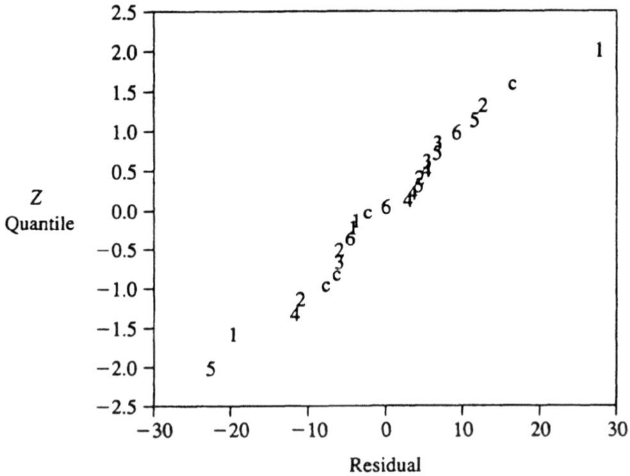

FIGURE 3.1. Normal Q–Q plot of residuals from uterine weight data. Symbol plotted is treatment group.

the bootstrap adjustments will tend to be more reliable in markedly non-normal situations, as indicated in Section 2.5. Of course, if the assumption of homoscedasticity is questionable, then a heteroscedastic bootstrap may be preferable, provided that the sample sizes are large enough to estimate the individual distribution functions reliably.

The uterine weight data of Example 3.3 has extremely small sample sizes, with only four observations per group. In this case, the bootstrap estimates of the within-group residual distributions are suspect. However, the bootstrap has proved itself reliable for moderately small sample sizes. Freedman and Peters (1984) use the bootstrap in a complicated regression involving estimation of many parameters with only $n = 18$. By contrast, the uterine weight example, when viewed as a regression problem with homoscedastic resampling, has $n = 28$. Even though the bootstrap can be useful in such small sample situations, extra caution is urged. To assess the stability of the method, one might try different resampling methods, for example, normal and location shift, and take the analysis yielding the largest adjusted p-values. This method will err on the conservative side, in other words, on the side of not claiming too many significant results. Simulation analyses, such as those performed in Section 2.5, would be helpful to identify situations where particular resampling plans are preferred.

3.2.2 Directional Decisions for One-Sided Comparisons with a Control

When comparing all treatments against a common control with lower-tailed tests, the rule $\tilde{p}_i \leqslant \alpha$ may be used to declare $H_i^l: \mu_i < \mu_c$ with FWE $= \alpha$. As discussed in Section 2.8.2, the rule $\tilde{p}_i \leqslant \alpha$ controls the probability of declaring *at least one* $H_i^l: \mu_i < \mu_c$ when in fact $H_i: \mu_i \geqslant \mu_c$, at or below the level α. Thus, it is safe to make directional conclusions—we may claim specifically that $\mu_i < \mu_c$, rather than simply $\mu_i \neq \mu_c$, given a small $\tilde{p}_i$.

Since directional decisions are allowed with this *free* step-down approach, and since these adjusted p-values are uniformly smaller than the *single-step* adjustments, the step-down adjustments are preferred. Table 3.7 also contains single-step adjustments, which are clearly more conservative.

This directional property of one-sided tests may be exploited to obtain improved one-sided confidence bounds. Given a particular collection of hypothesized difference δ_{0i}, the hypotheses $H_i: \mu_i - \mu_c \geqslant \delta_{0i}$ may be tested using the free step-down approach. The simultaneous $100(1 - \alpha)\%$ confidence set for the true values δ_i is then the set of all δ_{0i} for which all hypotheses $H_i: \mu_i - \mu_c \geqslant \delta_{0i}$ are accepted at FWE $= \alpha$. See Hochberg and Tamhane (1987, pp. 59–62) or Kim, Stefánsson and Hsu (1987) for further details.

3.2.3 General Contrasts

As demonstrated by Example (2.3) in Section 2.3.2, contrasts of interest in an ANOVA setting need not be restricted to pairwise comparisons: for example, we may wish to compare the average of two selected treated groups against the control, or we may wish to test contrasts measuring linear and/or quadratic dose/response relationships. A general contrast of the means in an ANOVA model is $\theta = c_1\mu_1 + \cdots + c_g\mu_g$, where $\sum_i c_i = 0$. Often many such contrasts will be of interest, for example, $\theta_l = c_{l1}\mu_1 + \cdots + c_{lg}\mu_g$, for $l = 1, \ldots, k$.

In a balanced homoscedastic normal ANOVA, the Tukey method for multiple pairwise comparisons can be extended to tests for multiple contrasts. A contrast θ_l is significantly different from zero if

$$\frac{2\sqrt{n}\,|\sum_i c_{li}\bar{y}_i|}{s\sum_i |c_{li}|} > Q_{g,\nu}^{(\alpha)}$$

where n is the common sample size, $Q_{g,\nu}^{(\alpha)}$ is the $1 - \alpha$ quantile of the Studentized range distribution with g means and degrees of freedom $\nu = g(n - 1)$.

Note that the ordinary t-statistic for testing θ_l is

$$t_l = \frac{\sqrt{n}\sum_i c_{li}\bar{y}_i}{s\sqrt{\sum_i c_{li}^2}}$$

Thus, when using Tukey's method for testing contrasts in a balanced homoscedastic normal ANOVA, the adjusted p-value for any given contrast is

$$\tilde{p}_l = \Pr\left(Q_{g,\nu} > 2|t_l|\sqrt{\textstyle\sum_i c_{li}^2}\Big/\textstyle\sum_i |c_{li}|\right) \tag{3.10}$$

where $Q_{g,\nu}$ denotes a random variable with the Studentized range distribution. Useful features of these adjusted p-values are (1) there is a direct correspondence with simultaneous confidence intervals, so directional decisions are allowed, and (2) the "family" of tests may include infinitely many contrasts while maintaining (at least conservative) control of the FWE. Thus, tests need not be pre-planned. On the negative side, the method requires balanced designs.

An alternative for testing infinitely many contrasts is Scheffé's method, which does not require balanced designs. In this case the ordinary t-statistic for testing θ_l is

$$t_l = \frac{\sum_i c_{li}\bar{y}_i}{s\sqrt{\sum_i c_{li}^2/n_i}}$$

A contrast θ_l is significantly different from zero if

$$|t_l| > \sqrt{(g-1)F_{g-1,\nu}^{(\alpha)}}$$

where $F_{g-1,\nu}^{(\alpha)}$ is the $1 - \alpha$ quantile of the F distribution with $g - 1$ numerator degrees of freedom and $\nu = \sum(n_i - 1)$ denominator degrees of freedom. Thus, the corresponding adjusted p-values are

$$\tilde{p}_l = \Pr(F_{g-1,\nu} > t_l^2/(g-1)) \tag{3.11}$$

where $F_{g-1,\nu}$ denotes a random variable with the F distribution. Like the Tukey adjusted p-values, the Scheffé adjustments allow directional decisions, and the family may include infinitely many contrasts. However, the Scheffé adjustments are valid for unbalanced designs. Even for balanced designs, the Scheffé method may produce smaller adjusted p-values than the Tukey adjustments for particular types of contrasts.

Example 3.4 Litter Weight Dose Response Tests

A data set analyzed by Westfall and Rom (1990), displayed in Table 3.8, involves litter weights of mice born from mothers (dams) assigned to three different dosage groups and a control. For the low dose group the dose metameter is 5 (units of a study compound), for the medium dose group it is 50, and for the high dose group it is 500. The criterion variable Y is the average non-birth litter weight for the first three weeks of the study, obtained by averaging weight data for weeks 1–3 over both males and females. This criterion variable Y is an example of a selected $\mathbf{Y}_2$ variable in the data analysis paradigm of Section 1.6. We could have used *all* weekly weights in the analysis, but selecting a *particular combined variable* of interest greatly reduces the multiplicity problem.

We are interested in the dose–response relationship. A contrast that is sensitive to a linear dose–response relationship is $\theta_1 = -3\mu_1 - \mu_2 + \mu_3 + 3\mu_4$. However, it is not known whether the dose–response function is linear, so other contrasts should be considered. By mean-centering the dose metameter 0, 5, 50, 500, the following contrast is obtained: $\theta_2 = -138.75\mu_1 - 133.75\mu_2 - 88.75\mu_3 + 361.25\mu_4$. Contrast θ_2 is sensitive to a dose–response relationship that mimics the metameter, in this case an exponential function. At the other extreme, one might consider a dose–response relationship that is logarithmic, indicating decreasing effects at larger doses. Defining the metameter $\log(i)$, $i = 1, 2, 3, 4$, and mean-centering, we obtain the following contrast: $\theta_3 = -.795\mu_1 - .105\mu_2 + .305\mu_3 + .595\mu_4$. All three contrasts are sensitive to monotonic dose–response relationships, but differ in the specific functional forms that they attempt to mimic. Use of these particular contrasts for detecting a dose-response relationship is advocated by Mantel et al. (1982).

We are also interested in the quadratic dose–response function. The contrast $\theta_4 = \mu_1 - \mu_2 - \mu_3 + \mu_4$ is sensitive to quadratic "up-then-down" or "down-then-up" behavior, often called an "umbrella" shape.

Summary statistics are as follows:

	Control	5	50	500
$\bar{y}_i$	32.309	29.310	29.866	29.646
n_i	20	19	18	17

The residual root mean square is $s = 4.33066$.

The t-statistics for testing $H_{0i} : \theta_i = 0$ are $t_1 = -1.6458$, $t_2 = -.7936$, $t_3 = -1.9289$, and $t_4 = 1.3779$. Given the pattern of the sample averages $\bar{y}_i$, it is not surprising that the contrast mimicking the logarithmic dose-response relationship is most significant.

Table 3.9 reports the results of the tests, where the adjustments use the Tukey method (3.10), the Scheffé method (3.11), and the homoscedastic normal resampling-

Table 3.8 Litter Weight Data*

		Gestation	Litter	Ave. Female Weight in Week				Ave. Male Weight in Week			
Dose	ID	Time	Size	0	1	2	3	0	1	2	3
0	2911	22.5	15	6.4	12.4	27.7	44.6	6.8	12.5	27.2	43.9
0	2912	22.5	14	6.2	16.5	33.6	50.8	6.5	16.6	32.1	50.4
0	2913	22.0	14	6.0	18.0	35.2	54.0	6.6	18.7	36.9	55.4
0	2914	22.0	13	5.1	14.8	31.9	48.3	6.0	19.0	39.2	59.9
0	2915	21.5	15	5.8	16.3	37.2	57.4	6.0	16.2	36.0	57.5
0	2916	23.0	5	5.4	13.6	30.0	44.4	5.9	13.4	29.4	46.8
0	2917	21.5	16	5.6	11.3	25.9	41.4	5.8	12.5	28.8	46.4
0	2918	22.5	15	6.0	14.9	30.9	53.8	6.5	16.1	31.7	54.6
0	2919	22.5	14	6.2	15.6	31.2	46.2	6.6	17.2	33.5	51.6
0	2920	21.5	15	5.3	15.3	32.0	49.5	5.8	16.0	32.7	51.2
0	2921	22.0	12	5.7	14.1	28.8	47.2	6.0	15.7	30.1	50.4
0	2922	22.5	15	6.0	15.8	31.2	52.2	6.4	16.3	31.6	53.3
0	2923	22.0	16	5.6	14.3	30.4	44.2	6.0	15.2	31.3	45.8
0	2924	21.5	7	6.3	14.0	26.6	43.4	6.6	14.4	27.4	46.0
0	2925	22.0	15	6.1	16.0	32.6	52.8	6.1	15.6	31.8	51.5
0	2926	22.0	13	6.0	15.6	32.1	51.8	6.0	15.8	32.7	52.6
0	2927	21.5	14	5.6	13.2	29.5	46.8	5.8	13.6	29.1	45.6
0	2928	22.0	15	5.3	14.6	32.5	48.3	5.7	16.4	35.0	51.7
0	2929	22.5	16	5.6	14.2	29.5	48.4	6.1	15.4	30.9	50.8
0	2930	22.0	9	5.9	16.7	32.8	52.0	6.6	18.8	36.2	56.4
5	2931	22.5	15	6.6	16.0	33.4	53.8	7.1	16.8	34.2	54.8
5	2932	22.5	7	6.2	11.7	24.0	40.9	7.4	12.5	26.6	42.3
5	2933	22.5	15	6.4	11.0	23.6	38.2	6.7	11.6	23.5	37.8
5	2934	23.0	9	7.8	18.4	33.7	58.1	8.0	20.1	37.1	64.4
5	2935	22.0	13	5.5	12.5	24.4	39.0	6.2	14.9	29.9	46.8
5	2936	22.5	13	5.9	15.3	30.9	51.7	6.6	17.3	32.5	55.4
5	2937	22.5	17	5.9	11.9	24.8	36.3	6.2	12.6	26.2	37.9
5	2938	22.5	15	5.9	14.1	30.0	48.5	6.6	16.6	34.2	55.8
5	2939	22.5	12	6.1	17.1	33.7	56.1	6.8	17.0	34.5	57.8
5	2940	22.0	13	5.2	10.1	24.3	39.6	5.8	12.1	28.1	46.6
5	2941	22.0	14	5.3	14.8	29.4	48.9	5.0	16.3	30.0	50.6
5	2942	21.5	12	5.5	11.9	27.7	44.3	5.9	14.6	32.9	50.6
5	2943	22.5	14	6.2	13.2	25.6	39.5	6.7	13.7	27.2	41.8
5	2944	22.0	13	5.7	14.1	30.7	48.4	6.3	16.1	33.0	50.8
5	2945	22.5	16	6.1	16.6	33.2	52.8	6.2	16.2	32.1	51.7
5	2946	21.5	9	5.4	9.0	18.8	32.3	5.9	10.1	21.4	36.2
5	2947	21.5	14	5.3	11.0	23.0	37.2	5.8	12.8	27.6	43.1
5	2949	21.5	12	5.5	9.3	19.6	31.8	5.8	9.1	18.9	30.7
5	2950	22.5	16	6.0	12.4	25.8	42.4	6.3	14.0	28.1	47.0

*From Dr. D. Rom.

Table 3.8 ***(Continued)*** **Litter Weight Data**

Dose	ID	Gestation Time	Litter Size	Ave. Female Weight in Week 0	1	2	3	Ave. Male Weight in Week 0	1	2	3
50	2951	22.0	16	5.2	14.7	30.8	47.4	5.5	14.5	30.5	49.8
50	2952	21.5	16	5.5	16.1	37.1	51.5	5.7	17.0	38.2	54.9
50	2953	21.5	14	5.8	13.7	27.8	42.8	5.9	14.1	27.8	41.6
50	2954	22.5	15	6.1	15.3	30.8	50.2	6.5	16.9	32.4	53.2
50	2955	22.5	15	6.0	13.4	29.4	47.2	6.4	14.4	30.6	48.6
50	2956	21.5	15	5.5	13.2	28.6	45.2	5.8	15.0	30.9	48.1
50	2957	21.5	14	5.6	12.3	27.3	41.3	5.7	12.4	27.6	41.5
50	2958	22.0	17	5.5	13.3	30.1	46.8	5.8	15.8	33.4	52.7
50	2959	22.5	13	6.6	17.1	34.6	55.8	6.8	18.3	36.0	58.5
50	2960	21.5	14	5.7	12.6	28.4	45.0	6.0	13.1	27.9	44.0
50	2961	21.5	16	5.3	8.5	18.9	32.2	5.5	10.2	22.6	37.1
50	2962	22.5	17	6.1	14.2	29.1	45.9	6.5	15.2	30.2	50.3
50	2963	22.0	14	6.0	15.7	29.0	48.6	6.2	14.9	28.0	47.1
50	2964	22.0	14	5.6	12.7	27.0	41.8	6.0	13.5	28.0	42.8
50	2965	22.0	12	5.9	10.6	20.7	34.4	6.6	11.9	22.5	37.7
50	2966	21.5	12	5.5	11.4	27.4	44.2	5.8	13.8	31.4	49.1
50	2968	22.0	14	5.6	14.2	29.4	48.4	6.2	15.7	31.6	52.3
50	2970	21.5	16	5.2	13.1	28.4	43.2	5.4	14.1	31.2	45.8
500	2972	22.0	7	4.3	9.0	20.0	35.2	4.8	12.0	25.6	45.5
500	2973	22.5	13	5.8	14.4	32.2	51.2	6.7	15.8	34.2	54.9
500	2974	22.0	10	6.3	14.0	29.4	50.2	6.8	15.4	32.8	56.1
500	2975	21.5	11	5.8	10.9	23.8	38.6	6.1	12.0	26.0	41.0
500	2976	22.0	15	5.3	13.2	28.8	46.6	5.6	14.6	29.9	48.8
500	2977	22.5	11	6.6	8.7	19.4	32.6	7.1	8.8	17.1	28.7
500	2978	21.5	14	5.3	11.0	24.8	36.8	5.4	13.4	28.6	43.6
500	2979	22.5	9	6.5	13.4	27.6	46.2	6.1	12.9	27.8	43.7
500	2980	22.0	11	5.8	7.7	17.9	29.9	6.1	8.5	20.5	33.7
500	2981	22.5	15	5.9	14.8	31.0	49.7	6.3	16.4	32.2	53.2
500	2982	22.5	13	5.3	11.4	24.8	39.1	5.9	12.4	26.2	42.8
500	2983	22.0	12	5.5	14.4	31.9	52.6	5.9	14.4	31.8	54.1
500	2984	22.5	14	6.1	13.9	30.8	48.3	6.6	16.4	34.2	54.2
500	2985	23.0	16	5.8	15.3	34.6	58.7	6.6	17.6	39.7	66.6
500	2987	22.5	12	6.3	15.2	31.6	48.6	6.6	16.9	34.3	52.3
500	2990	21.5	13	5.2	14.4	28.2	47.8	5.1	14.7	29.5	49.6
500	2991	22.0	17	5.7	15.9	33.2	54.3	6.2	17.0	35.2	56.3

Table 3.9 Adjusted Two-Tailed *p*-Values for Contrasts

Contrast	*t*-Value	Raw *p*-Value	Adjusted *p*-Values: Resampling	Scheffé	Tukey
Linear	−1.6458	.1043	.2323	.4443	.5657
Exponential	−0.7936	.4301	.7209	.8892	.9150
Logarithmic	−1.9289	.0578	.1372	.3017	.3017
Quadratic	1.3779	.1726	.3579	.5963	.5963

based single-step method using 20,000 simulations. Note that the Tukey method does not strictly apply because the sample sizes are unbalanced. However, as the sample sizes are approximately equal, the use of the Studentized range distribution with $4 \times (17 - 1) = 64$ rather than $\nu = \sum(n_i - 1) = 70$ denominator degrees of freedom provides a reasonable approximation. For the Scheffé adjustments, $\nu = 70$ is used. The resampling method provides the smallest *p*-values, but no test shows significance even at the FWE = .10 level for this data. □

Even though no results achieve statistical significance, the resampling-based method is superior to the Scheffé and Tukey methods. This comparison is slightly unfair, since the Scheffé and Tukey methods allow for infinitely many contrasts. Even contrasts *suggested by the data* (called *post-hoc* comparisons) are allowed with the Scheffé and extended Tukey methods, while the resampling method requires that *all* contrasts be specified in advance. Nevertheless, these comparisons reinforce the idea that analysts should pre-specify a collection of interesting contrasts if they wish to claim statistical significance.

Resampling-based adjusted *p*-values also may be computed for infinitely many contrasts. Similar to the case where resampling is done for finitely many contrasts, the advantage of using the resampling approach for infinitely many contrasts is that specific distributional characteristics are incorporated. Guidelines for the construction of such resampling-based adjustments are given in Chapter 6, Section 6.4.

3.3 TWO- AND HIGHER-WAY ANOVA MODELS

In two-way ANOVA models, one frequently tests hypotheses concerning main effects and interactions. Similar to Tukey (1991), we claim that these tests ought to be formulated using meaningful contrasts, and that multiplicity adjustment should be considered for the resulting collection of contrasts. In the one-way ANOVA, the global test $H: \mu_1 = \cdots = \mu_g$ is equivalent to the simultaneous truth of all $H_{ii'}: \mu_i = \mu_{i'}$; evaluation of these pairwise hypotheses or other specified contrasts is often much more useful than the evaluation of the global hypothesis, especially if directional decisions are allowed.

The two-way ANOVA model is

$$Y_{ijk} = \mu_{ij} + \epsilon_{ijk}$$

where $i = 1, \ldots, g$; $j = 1, \ldots, b$; $k = 1, \ldots, n_{ij}$; the μ_{ij} are fixed unknown constants; and the ϵ_{ijk} are random variables with mean zero. Frequently, the following decomposition of the μ_{ij} is considered:

$$\mu_{ij} = \mu + \alpha_i + \beta_j + \gamma_{ij}$$

Side conditions $\sum_i \alpha_i = \sum_j \beta_j = \sum_i \gamma_{ij} = \sum_j \gamma_{ij} = 0$ are often imposed to make the parameters uniquely defined. Without these side conditions, μ, α_i, β_j, and γ_{ij} are not uniquely determined; however, the interaction hypothesis H_γ: $\gamma_{ij} \equiv 0$ is meaningful: it states that the cell means profile $\mu_{ij}, j = 1, \ldots, b$ is *parallel* to the profile $\mu_{i'j}, j = 1, \ldots, b$, for all $1 \leqslant i < i' \leqslant g$ pairs of factor levels. This condition implies in turn that parallelism exists in each (2×2) *subtable*, in other words, that

$$\tau_{(i,i'),(j,j')} = (\mu_{ij} - \mu_{ij'}) - (\mu_{i'j} - \mu_{i'j'}) = 0$$

for every (2×2) subtable defined using rows (i, i') and columns (j, j'). The differences $\tau_{(i,i'),(j,j')}$ are called *tetrad differences*; an important problem in simultaneous inference is the analysis of all such differences in a two-way ANOVA (Tukey, 1991; Bradu and Gabriel, 1974).

The marginal test statistic for each tetrad difference is

$$t_{(i,i'),(j,j')} = \frac{(\bar{y}_{ij} - \bar{y}_{ij'}) - (\bar{y}_{i'j} - \bar{y}_{i'j'})}{s(1/n_{ij} + 1/n_{ij'} + 1/n_{i'j} + 1/n_{i'j'})^{1/2}}$$

where $s^2 = \sum_i \sum_j \sum_k (y_{ijk} - \bar{y}_{ij})^2/\nu$, and $\nu = \sum_i \sum_j (n_{ij} - 1)$ is the error degrees of freedom. Unadjusted p-values for individual tetrad differences are given by

$$p_{(i,i'),(j,j')} = \Pr\left(|T_\nu| > |t_{(i,i'),(j,j')}|\right)$$

The exact distribution of the minimum p-value distribution is available in simple cases: for example, in balanced models with $g = 2$ or $b = 2$ it is simply related to the Studentized range distribution (Bradu and Gabriel, 1974). However, in more general circumstances, exact determination of the minimum p-value distribution over all tetrad tests is *extremely* complicated. Nevertheless, resampling-based analysis provides a simple solution. The resampling-based analysis yields an estimate of the *exact* underlying distribution in homoscedastic normal analysis, and yields an estimate of the approximate distribution in general location-shift error models.

As suggested by Tukey (1991),

> ... it is unlikely that we will wish to spend 5% error rate on each main effect ..., 5% on each two-factor interaction ..., and so on. Some more sensible budgeting of error rate is quite likely to be needed.*

One way to address Professor Tukey's concern is to place main effects contrasts *into the same family*, alongside the interaction tetrad contrasts. It is reasonable to define main effect contrasts in terms of pairwise comparisons $\bar{\mu}_{i.} - \bar{\mu}_{i'.}$ and $\bar{\mu}_{.j} - \bar{\mu}_{.j'}$. Indeed, under the usual side conditions on the parameters α_i, β_j, and γ_{ij}, the typical ANOVA "main effects" test of $H: \alpha_i \equiv 0$ is equivalent to $H_{ii'}: \bar{\mu}_{i.} - \bar{\mu}_{i'.} = 0$, for all $1 \leq i < i' \leq g$; and the "main effects" test of $H: \beta_j \equiv 0$ is equivalent to $H_{jj'}: \bar{\mu}_{.j} - \bar{\mu}_{.j'} = 0$, for all $1 \leq j < j' \leq b$.

To test for main effects and interaction contrasts, one may treat the two-way model as if it were a one-way model, then formulate a family of linear contrasts of interest. Resampling-based multiplicity adjustments are available using the methods of Sections 3.1 and 3.2.

In higher-way ANOVA models, the same basic methods apply. To analyze three-factor interactions, one might consider comparisons of tetrad contrasts for differing levels of the third factor. For example, in the model

$$Y_{ijkl} = \mu_{ijk} + \epsilon_{ijkl}$$

specific linear contrasts measuring three-factor interactions are given by

$$\big((\mu_{ijk} - \mu_{ij'k}) - (\mu_{i'jk} - \mu_{i'j'k})\big) - \big((\mu_{ijk'} - \mu_{ij'k'}) - (\mu_{i'jk'} - \mu_{i'j'k'})\big)$$

Since the resampling approach easily applies to any collection of linear contrasts, there are essentially no new difficulties.

Example 3.5 Productivity Improvement

Table 3.10 and Figure 3.2 display data on productivity improvement taken from Neter, Wasserman, and Kutner (1985, pp. 559 and 629). Letting μ_{ij} be the mean value for R&D expense level i, i = L, M, or H and country j, j = E or U, all possible main effects and interaction contrasts, using cell means, are given in Table 3.11. Multiplicity adjustment is performed over the resulting collection ($k = 7$) of tests. The unadjusted two-sided p-values, and the single-step adjusted p-values (as computed by PROC MULTTEST), are also given in Table 3.11. Since directional decisions are allowed, we conclude that the "Medium" and "High" research expenditure firms have significantly higher productivity improvements than "Low" research

Table 3.10 Annual Percent Increase in Productivity*

	Geographic Location	
R&D Expense	Europe	U.S.
Low	8.2, 6.8, 5.8, 6.9	7.6, 6.6, 6.3, 7.7, 6.0
Moderate	6.7, 8.1, 9.4, 8.6, 8.7, 7.1, 8.4	7.8, 7.7, 8.9, 7.9, 8.3
High	8.5, 10.1, 9.5	9.7, 7.8, 9.6

*From Neter, Wasserman, and Kutner, *Applied Linear Statistical Models*, 2nd ed. (1985), pp. 559–560, with permission of Richard Irwin, Inc.

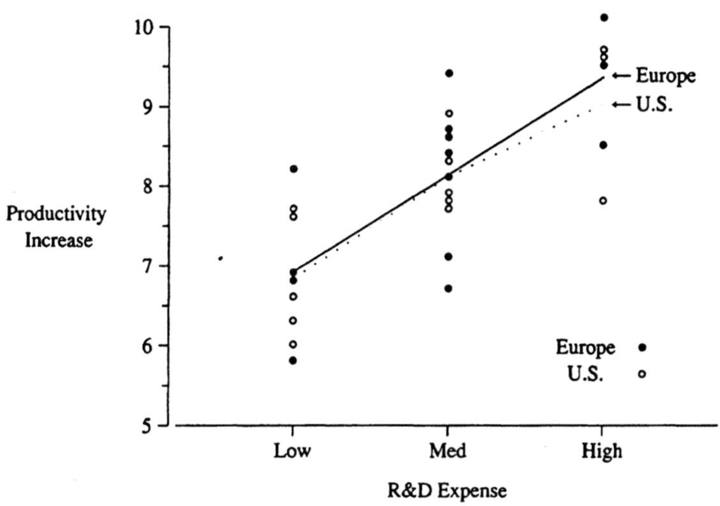

FIGURE 3.2. Percent increase in productivity data. Cell means profiles are solid for Europe, dotted for U.S.

expenditure firms, when the cell means are averaged over U.S. and European firms. All other comparisons are insignificant. The *unadjusted* p-values suggest that the "High" research expenditure is an improvement over "Medium" research expenditure ($p = .0207$), but the corresponding *adjusted* p-value ($\tilde{p} = .1136$) would ordinarily be considered "insignificant." If step-down analysis is performed, this particular comparison becomes "significant" at the $\alpha = .10$ level ($\tilde{p} = .0858$ based on 10,000 simulations), but directional decisions are not allowed. □

Often, detecting significant interaction contrasts is considered less important than detecting significant main-effects contrasts. In this case, one may wish not to "spend"

Table 3.11 Analysis of Productivity Improvement Data

	Cell Means						p-Values	
Effects	μ_{LE}	μ_{LU}	μ_{ME}	μ_{MU}	μ_{HE}	μ_{HU}	Raw_p	Adj_p
Main								
L vs. M	+1	+1	−1	−1	0	0	.0034	.0204
L vs. H	+1	+1	0	0	−1	−1	.0001	.0002
M vs. H	0	0	+1	+1	−1	−1	.0207	.1136
Europe vs. U.S.	+1	−1	+1	−1	+1	−1	.6719	.9971
Interaction								
(LM) × (EU)	+1	−1	−1	+1	0	0	.9354	1.0000
(LH) × (EU)	+1	−1	0	0	−1	+1	.7849	.9997
(MH) × (EU)	0	0	+1	−1	−1	+1	.7199	.9989

the same amount of the FWE on an interaction contrast that one "spends" on a main effect contrast. It may be more sensible to "budget" (in the words of Tukey, 1991) more for the detection of main effect contrasts than for the detection of interaction contrasts. Such methods also may be accomplished via resampling. In Chapter 6, Section 6.1, the "weighted adjusted p-value" is presented for this particular application.

3.4 ALLOWING FOR COVARIATES

The general resampling methods of Sections 3.1–3.3 naturally extend to applications with covariates. The key idea is that analysis of covariance models can be viewed as regression models, and resampling analysis can be performed using the residuals. The analyses given here are *conditional* on the observed covariates. If the covariates are random variables, then the usual condition that the covariates are independent of the residuals is assumed.

3.4.1 Continuous Covariates

The following example illustrates multiplicity adjustment with continuous covariates.

Example 3.6 Litter Weights with Covariates
The data set of Westfall and Rom (1990) described in Section 3.2.3 involving litter weights of mice resulting from dams assigned to different dosage groups contained covariates X_1 = litter size, and X_2 = length of gestation. We wish to assess whether any treatment reduces litter weight, controlling for litter size and length of gestation.

A model for the data is

$$Y_{ij} = \mu_i + \beta_1 X_{1ij} + \beta_2 X_{2ij} + \epsilon_{ij}$$

where $i = 1, \ldots, 4$ (group 1 represents control and group 4 represents high dose); $j = 1, \ldots, n_i$; $\mu_1, \ldots, \mu_4, \beta_1, \beta_2$ are fixed unknown constants; and the ϵ_{ij} are random variables with mean zero, assumed to be independent of one another and of the covariates. For now, we also assume that the residuals are homoscedastic, having common variance σ^2. In matrix form, the model is

$$\mathbf{Y} = \mathbf{D}\boldsymbol{\gamma} + \boldsymbol{\epsilon}$$

where $\mathbf{Y}$ is a lexicographic arrangement $\mathbf{Y}' = (Y_{11}, Y_{12}, \ldots, Y_{4n_4})$, $\mathbf{D}$ is the corresponding design matrix

$$\mathbf{D} = \begin{pmatrix} 1 & 0 & 0 & 0 & X_{111} & X_{211} \\ 1 & 0 & 0 & 0 & X_{112} & X_{212} \\ \vdots & \vdots & \vdots & \vdots & \vdots & \vdots \\ 0 & 0 & 0 & 1 & X_{14n_4} & X_{24n_4} \end{pmatrix}$$

$\boldsymbol{\gamma}' = (\mu_1, \mu_2, \mu_3, \mu_4, \beta_1, \beta_2)$ is the parameter vector, and $\boldsymbol{\epsilon}$ is the lexicographic arrangement of the ϵ_{ij} corresponding to $\mathbf{Y}$.

The estimated parameter vector is

$$\hat{\boldsymbol{\gamma}} = (\mathbf{D}'\mathbf{D})^{-1}\mathbf{D}'\mathbf{Y}$$

with covariance matrix $\sigma^2(\mathbf{D}'\mathbf{D})^{-1}$. Linear contrasts measuring covariate-adjusted differences from the control are

$$\boldsymbol{\theta} = \begin{pmatrix} \theta_1 \\ \theta_2 \\ \theta_3 \end{pmatrix} = \begin{pmatrix} -1 & 1 & 0 & 0 & 0 & 0 \\ -1 & 0 & 1 & 0 & 0 & 0 \\ -1 & 0 & 0 & 1 & 0 & 0 \end{pmatrix} \boldsymbol{\gamma} = \begin{pmatrix} \mathbf{c}_1' \\ \mathbf{c}_2' \\ \mathbf{c}_3' \end{pmatrix} \boldsymbol{\gamma} = \mathbf{C}'\boldsymbol{\gamma}$$

The estimated contrasts are $\hat{\theta}_i = \mathbf{c}_i'\hat{\boldsymbol{\gamma}}$, having variance

$$\mathrm{Var}(\mathbf{c}_i'\hat{\boldsymbol{\gamma}}) = \sigma^2\mathbf{c}_i'(\mathbf{D}'\mathbf{D})^{-1}\mathbf{c}_i$$

The usual test statistic for comparing the adjusted treatment means is

$$t_i = \frac{\mathbf{c}_i'\hat{\boldsymbol{\gamma}}}{s[\mathbf{c}_i'(\mathbf{D}'\mathbf{D})^{-1}\mathbf{c}_i]^{1/2}}$$

where s^2 is the usual "mean square for error" $s^2 = (\mathbf{Y} - \mathbf{D}\hat{\boldsymbol{\gamma}})'(\mathbf{Y} - \mathbf{D}\hat{\boldsymbol{\gamma}})/\nu$, and ν is the degrees of freedom $\nu = \#$ rows of $\mathbf{D} - \#$ columns of $\mathbf{D} = 68$. ($\mathbf{D}$ has full column rank in this example; in general the degrees of freedom are $\nu = n - \mathrm{rank}(\mathbf{D})$.)

Since we wish to detect only *decreases* in weights, all tests are *lower-tailed*, with unadjusted p-values

$$p_i = \Pr(T_\nu < t_i)$$

The p-values are monotonically related to the t-statistics in this application, and adjusted p-values may be estimated using the distribution of the minimum test statistic. As in Example 3.3, there are no logical restrictions among these hypotheses, suggesting the use of the *free* step-down method.

Under homoscedasticity, the error distribution is assumed to be G, which may be estimated consistently using the empirical distribution function of the residuals $\hat{\epsilon}_{ij} = Y_{ij} - \mathbf{d}_{ij}'\hat{\boldsymbol{\gamma}}$, where $\mathbf{d}_{ij}'$ is the row of $\mathbf{D}$ corresponding to the (i, j) element of $\mathbf{Y}$. Under any partial null hypothesis the *subset pivotality* condition holds: the joint distribution of any subset K of test statistics depends only upon G whenever the corresponding intersection null hypothesis H_0^K is true. Thus, all free step-down adjusted p-values depend only upon G.

To generate resampled data in this application, use

$$Y_{ij}^* = \hat{\epsilon}_{ij}^* \tag{3.12}$$

where the $\hat{\epsilon}_{ij}^*$ are a *with replacement* sample from the residuals $\hat{\epsilon}_{ij}$.

In this particular instance, the unadjusted p-values (lower-tailed) are $p_1 = .0058$, $p_2 = .0457$, and $p_3 = .0244$. Thus, the "Low vs. Control" comparison is most significant, the "High vs. Control" comparison is second most significant, and the "Medium vs. Control" comparison is least significant of the three tests.

Algorithm 2.8 shows how to calculate the free step-down adjusted p-values, for general situations. For this example, the algorithm is as follows:

Algorithm 3.5 Step-Down Bootstrap Adjustments with ANCOVA Data

0. Initialize counting variables: $COUNT_i = 0, i = 1, 2, 3$.
1. Compute the parameter estimates $\hat{\boldsymbol{\gamma}}^* = (\mathbf{D}'\mathbf{D})^{-1}\mathbf{D}'\mathbf{Y}^*$ and the residual mean square error $s^* = (\mathbf{Y}^* - \mathbf{D}\hat{\boldsymbol{\gamma}}^*)'(\mathbf{Y}^* - \mathbf{D}\hat{\boldsymbol{\gamma}}^*)/\nu$ from the bootstrap data (3.12).
2. Compute the statistics

$$t_i^* = \frac{\mathbf{c}_i'\hat{\boldsymbol{\gamma}}^*}{s^*[\mathbf{c}_i'(\mathbf{D}'\mathbf{D})^{-1}\mathbf{c}_i]^{1/2}}$$

3. Corresponding to the ordering of the lower-tailed p-values $p_1 < p_3 < p_2$ for this particular application, $p_{(1)} = p_1$, $p_{(2)} = p_3$, and $p_{(3)} = p_2$. The t-values are correspondingly ordered as $t_1 < t_3 < t_2$, $t_{(1)} = t_1$, $t_{(2)} = t_3$, and $t_{(3)} = t_2$. Using step 2 of Algorithm 2.8 with t-statistics rather than p-values, define the successive *minima*:

$$u_3^* = t_2^*$$
$$u_2^* = \min(u_3^*, t_3^*)$$
$$u_1^* = \min(u_2^*, t_1^*)$$

4. If $u_i^* \leqslant t_{(i)}$, then $COUNT_i \leftarrow COUNT_i + 1$.
5. Repeat 1–3 N times. Compute $\tilde{p}_i^{(N)} = COUNT_i/N$. Note that $\tilde{p}_{(1)}^{(N)} = \tilde{p}_1^{(N)}$, $\tilde{p}_{(2)}^{(N)} = \tilde{p}_3^{(N)}$, and $\tilde{p}_{(3)}^{(N)} = \tilde{p}_2^{(N)}$.
6. Enforce monotonicity using successive maximization:

$$\tilde{p}_1^{(N)} = \tilde{p}_1^{(N)}$$
$$\tilde{p}_3^{(N)} \leftarrow \max(\tilde{p}_1^{(N)}, \tilde{p}_3^{(N)})$$
$$\tilde{p}_2^{(N)} \leftarrow \max(\tilde{p}_3^{(N)}, \tilde{p}_2^{(N)})$$

The resulting adjusted p-values, using 10,000 bootstrap simulations, are $\tilde{p}_1 = .015$, $\tilde{p}_2 = .0457$, and $\tilde{p}_3 = .0444$, all significant at the $\alpha = .05$ familywise level. Since all tests are lower-tailed, directional decisions are allowed: average body weight is *reduced* by all treatments, controlling for covariates. □

The calculations were performed using program written in the SAS/IML matrix language. While PROC MULTTEST does perform step-down adjustments, it currently does not allow ANCOVA models with continuous covariates. At this time, no general software exists to perform such adjustments.

The bootstrap analysis is quite similar to the normal-based resampling analysis in this problem. When the residuals are assumed to be homoscedastic normal, the adjustments are somewhat more easily calculated. We require the ability to generate multivariate normal vectors with covariance matrix equal to the correlation matrix of $\hat{\boldsymbol{\theta}}$, as well as the ability to generate an independently distributed χ^2 variate with ν degrees of freedom. In this application, the correlation matrix of $\hat{\boldsymbol{\theta}}$ is given by

$$\begin{aligned}\boldsymbol{\rho} &= \mathrm{Corr}(\hat{\boldsymbol{\theta}}) \\ &= \{\mathrm{diag}(\mathbf{C}'(\mathbf{D}'\mathbf{D})^{-1}\mathbf{C})\}^{-1/2}\mathbf{C}'(\mathbf{D}'\mathbf{D})^{-1}\mathbf{C}\{\mathrm{diag}(\mathbf{C}'(\mathbf{D}'\mathbf{D})^{-1}\mathbf{C})\}^{-1/2} \\ &= \begin{pmatrix} 1 & .4356 & .4797 \\ .4356 & 1 & .4105 \\ .4797 & .4105 & 1 \end{pmatrix}\end{aligned} \tag{3.13}$$

Note that the "diag" operator creates a diagonal matrix:

$$\mathrm{diag}\begin{pmatrix} a_{11} & a_{12} & \dots & a_{1n} \\ a_{21} & a_{22} & \dots & a_{2n} \\ \vdots & \vdots & \ddots & \vdots \\ a_{n1} & a_{n2} & \dots & a_{nn} \end{pmatrix} = \begin{pmatrix} a_{11} & 0 & \dots & 0 \\ 0 & a_{22} & \dots & 0 \\ \vdots & \vdots & \ddots & \vdots \\ 0 & 0 & \dots & a_{nn} \end{pmatrix}$$

For normal resampling, steps 1–2 of Algorithm 3.5 are replaced by

1. Generate the (3×1) vector $\mathbf{U} \sim \mathcal{N}(0, \boldsymbol{\rho})$, where $\boldsymbol{\rho}$ is given by (3.13); and generate V independently as χ^2_ν.
2. Compute the statistics

$$t_i^* = \frac{U_i}{(V/\nu)^{1/2}}$$

The resulting adjusted p-values, computed using 20,000 homoscedastic normal simulations, are $\tilde{p}_1 = .0161$, $\tilde{p}_2 = .0465$, and $\tilde{p}_3 = .0465$. The unadjusted p-value for "Medium vs. Control" is .0457 when computed analytically from the t-distribution. The value .0465 results from the simulation-based adjustment for the "High vs. Control" test; the p-value for the "Medium vs. Control" test was assigned the same value to force the same order for the adjusted p-values as exists for the unadjusted p-values (step 6 in Algorithm 3.5).

3.4.2 Discrete Covariates: "Block Variables" and Meta-Analysis

The hypothetical "oat bran study" (Example 1.1 of Chapter 1) is a typical situation where the use of "blocking" is appropriate. A "block" denotes a collection of observations that exhibit similar characteristics, and the goal of using a blocking factor usually is to reduce the experimental variation. In multicenter clinical studies, each center is its own "block," the sense that patient populations within particular centers have similar characteristics. Thus, the variation within centers tends to be smaller than variation in the entire population.

Frequently, an experiment with multiple treatments is carried out in several centers (or blocks), resulting in data

$$y_{ijk} = \mu_{ij} + \epsilon_{ijk}$$

where $i = 1, \ldots, g$ denote treatment groups; $j = 1, \ldots, b$ denote blocks; $k = 1, \ldots, n_{ij}$ denote replications of treatment i within block j; the μ_{ij} are fixed unknown constants; and the ϵ_{ijk} are random variables with mean zero.

In such analyses, it often is the case that similar hypotheses are tested within each block. In the "oat bran" study, whether the oat bran effect is real should not be overly dependent upon particular treatment centers. Thus, it is often considered more appropriate to *combine* the data from all blocks to test a single hypothesis, rather than test all hypotheses individually. This is an example of *meta-analysis* (Iyengar, 1991; Hedges and Olkin, 1985), referred to in Section 1.5.3.

When there is a small but consistent effect at all locations, the meta-analytic approach has substantially higher power to detect a significant difference than does the minimum p-value statistic. On the other hand, when there is little or no effect at most locations, but a pronounced effect at one or a few locations, then the minimum p-value method will have higher power. Further discussions of these concepts are given by Louv and Littell (1986), Hedges and Olkin (1985, Chapter 3), and in Chapter 6 of this book, where we compare the power of meta-analytic tests with multiplicity-adjusted tests.

Combining data over all centers, or blocks, provides a partial solution to the multiplicity dilemma. However, meta-analysis does not provide a complete solution to the multiple testing problem for two reasons: first, it is doubtful that researchers are willing to ignore significant results found at individual centers. Second, typical experiments are more complex with several meta-analytic tests computed within various strata, resulting in another multiplicity problem. For example, had the analyses of Tables 1.2 or 1.3 been analyzed by combining over centers, it remains likely that subdivisions by age, sex, and other categories would have been tested. It often is argued that the treatment affects different subtypes differently, so multiple meta-analytic tests are recommended. Thus, while meta-analysis can reduce the dimension of the multiplicity problem, experimenters are unlikely to so completely restrict the examination of their data.

While the block model appears identical to the two-way ANOVA model, design and randomization considerations may suggest alternate methods of analysis. In particular, we may wish to perform resampling *within blocks* to preserve possible heteroscedastic and/or randomized assignment characteristics.

As base test statistics, one may use treatment contrasts that are *combined* over the block variables, either via simple summate or weighted average. Such combined contrasts are similar in spirit to regression-style covariate adjustments. Another option is to combine the tests using a function of the p-values for each site. A simple test, known as the *Fisher test* (Fisher, 1958, Koziol and Perlman, 1978), takes $\chi^2 = -2\sum_i \ln(p_i)$, where the p_i are the individual p-values for each site. If the p-values are independent and uniformly distributed under their respective null hypothesis, then the chi-squared distribution may be used to determine significance, where the degrees of freedom are twice the number of sites.

For example, using the data from Table 1.2, $\chi^2 = 19.78$, with a p-value of $p = .472$ computed from the χ^2 distribution with 20 degrees of freedom. The data from Table 1.3 similarly shows lack of significance: $\chi^2 = 16.57$, with $p = .680$. Since both tests indicate insignificant effects, the meta-analysis has given the correct conclusion in both cases.

When there are multiple meta-analytic tests, multiplicity adjustments that preserve heteroscedastic block characteristics, nonnormalities, and dependence structures may be performed by resampling data independently within blocks. The combined (meta-analytic) p-values are computed, and the minimum of these p-values may be used for multiplicity adjustment. There are many ways to combine data from different blocks, including the Fisher test and other tests which combine means from various blocks. PROC MULTTEST currently tests combinations of means (and/or proportions, when data are binary), combined over a single block variable (or STRATA variable, in the PROC MULTTEST syntax). This software also performs multiplicity adjustments over all such tests when many variables and/or between-group contrasts are specified. Appendix III lists options that are available in this procedure at the time this book is being written.

CHAPTER 4

Continuous Data Applications: Multivariate Analysis

As demonstrated in Chapter 3, the resampling approach to multiple testing works well in univariate applications. The method is often equivalent to standard methods used in homoscedastic normal applications. Further, resampling offers a means to assess robustness in more general models.

While there is much theory and literature devoted to univariate linear model applications, there has been much less development of multiple testing in multivariate applications. This is unfortunate, since multivariate data predominate real applications: in most studies, more than just one characteristic of the experimental unit is measured. The general form of such data is presented in Chapter 1, Section 1.6. Multivariate data make the multiplicity problem worse, because analysts typically study all measured variables, and wish to claim significance whenever a particular $p < .05$ result is found. As the potential multiplicity is the number of variables times the number of comparisons within a variable, the multiplicity problem is magnified with multivariate data. Thus, it is especially important that sound, easy to use multiplicity adjustments are available for multivariate applications.

Unknown correlation structures present an additional problem for multiple tests with multivariate data. The proper multiplicity adjustment for any collection of tests depends on underlying correlation structures, which are usually known in univariate applications, but are usually unknown in multivariate applications. To handle unknown correlations, the methods that have been suggested (such as Bonferroni-type methods) are conservative for *all* correlation structures. Usually, less adjustment is needed when the correlations are stronger; therefore, the Bonferroni-type analyses are *more* conservative in these cases.

Resampling methods are particularly useful in these multivariate applications, since they provide multiplicity adjustments which automatically correct for *all* relevant correlation structures, both among the variables and among the tests over groups. In multivariate location-shift models (including multivariate normal), properly constructed resampling methods will perform well for large experimental sample sizes. However, in small-sample situations it is prudent to perform sensitivity analysis, perhaps using alternative adjustment methods (Section 3.2).

On the other hand, when performing *one test* per variable, the resampling method is exact (up to simulation error) for *all* sample sizes in multivariate randomization models. We begin by discussing the application to randomization models in Section 4.1. In Section 4.2, the resampling method is presented and developed in the general multivariate regression model. Applications of the general method are developed in Section 4.3, which discusses multiple regression analysis, and in Section 4.4, which discusses repeated-measures models. Alternative methods, including choice of resampling plan and choice of base test statistics, are compared in Section 4.5.

4.1 MULTIVARIATE RANDOMIZATION ANALYSIS

As in the univariate randomization model discussed in Section 3.1, suppose there are N experimental units, which are randomly assigned to g groups having n_i observations per group. Further, let us suppose there are v measurements taken on each unit. In the data structure of Section 1.6, v is the number of columns of the $\mathbf{Y}_2$ data matrix. The "true response" of experimental unit u to treatment i is now a vector:

$$\mathbf{R}_{iu} = (R_{iu1} \quad R_{iu2} \quad \dots \quad R_{iuv})$$

for $i = 1, \dots, g$, and $u = 1, \dots, N$. If the N units are randomly divided into g groups, with n_i within each group, then the observed data vectors $\mathbf{y}_{11}, \dots, \mathbf{y}_{1n_1}$ are a random sample from $\mathbf{R}_{11}, \dots, \mathbf{R}_{1N}$; $\mathbf{y}_{21}, \dots, \mathbf{y}_{2n_2}$ are a random sample from $\mathbf{R}_{21}, \dots, \mathbf{R}_{2N}$ (excluding all units previously selected for group 1) and so on. The observed data vectors $\mathbf{y}_{ij}$ are

$$\mathbf{y}_{ij} = (y_{ij1} \quad y_{ij2} \quad \dots \quad y_{ijv})$$

If a single test is considered for each variable, the null hypotheses of interest are ordinarily

H_k: the treatments identically affect (or have no effect on)
the responses for variable k

There are v such hypotheses. If H_k is true, then the "true" responses for variable k are equal for all treatment groups:

$$(R_{i1k} \quad \dots \quad R_{iNk}) \equiv (R_{1k} \quad \dots \quad R_{Nk}) \tag{4.1}$$

Using the randomized assignment of units to treatment groups, equation (4.1) leads to the restatement of the kth hypothesis as

H_k: All random permutations of the observed data $y_{11k}, y_{12k}, \dots, y_{gn_gk}$
into the g groups, with sample sizes n_i, are equally likely

To measure evidence against H_k, a test statistic t_k is used. For now, the actual form of t_k is not important—it may be an F-test statistic, or some other statistic measuring differences between sample means. Let us suppose that larger values of t_k offer greater evidence against H_k. Unadjusted p-values for testing H_k are then

$$p_k = \Pr(T_k \geqslant t_k \mid H_k) \tag{4.2}$$

Calculation of such unadjusted p-values presents a large computational burden. When the analytic distribution of t_k under H_k is intractable, these p-values are often computed using resampling methods.

If the p-values *themselves* are computed using resampling methods, the additional use of resampling to compute the *adjusted* p-values can present an enormous computational burden. For example, suppose all experimental p-values p_k are computed using simulation sample sizes of 10,000. Then suppose that we wish to compute *adjusted* p-values, using 4,000 simulations. Within each of the 4,000 simulated data sets, we would have to re-compute all p-values p_k using 10,000 simulations. The total number of simulations is very large here: $10{,}000 \times 4{,}000 = 40{,}000{,}000$! Beran (1988a, 1988b) notes this problem in the contexts of simultaneous confidence sets and univariate hypothesis tests.

An interesting research question is: Can we obtain both the unadjusted and the adjusted p-values within a single simulation analysis, using a number of simulations on the order of *NRESAMPLE* rather than $(NRESAMPLE)^2$? This problem may be solved when performing multiple randomization tests with multivariate binomial data (Chapter 5), since the original permutation-based p-values can be computed analytically, without resorting to resampling. However, the problem remains to be solved in general continuous applications.

Rather than use the exact permutation distributions to obtain all unadjusted p-values, we suggest that these p-values be approximated using standard normal-theory approximations, which are easily calculated. Then resampling be used to evaluate the permutation distribution of the resulting approximated p-values. With such an analysis, the multiplicity-adjusted p-values are *exact* from a randomization standpoint, even though the unadjusted p-values are not.

If the unadjusted p-values are not exact, the marginal distributions of the p_k may differ. This can happen if, for example, one variable has an approximately normal distribution, and another variable has a highly skewed distribution. Use of normal-theory test statistics for both variables will lead to a p-value distribution that is approximately uniform in the former case, but perhaps very nonuniform in the latter case. To see whether there is a potential problem, one might compare the rerandomization-based p-values with the normal-theory p-values. If there are any marked discrepancies, one might consider some data transformation procedures to bring the two closer together before performing multiplicity adjustments.

The resampling method is not invalidated by disagreement between rerandomized p-values and normal-theory p-values; however, if the marginal randomization distributions of the normal-theory p-values differ greatly from test to test, the net result may be that particular tests are given *more weight*, or emphasis, than other tests. As discussed in Chapter 2, Section 2.3.2, a researcher may or may not wish that distributional characteristics determine such weights. In Chapter 5 we note that such implicit distributional weighting is sensible with sparse binomial data; in Chapter 6, we discuss a resampling-based adjustment procedure that allows the researcher to input his/her own weights explicitly, depending upon subject matter considerations.

Multivariate randomization models are an important application where the subset pivotality condition holds *exactly*, provided that only one test per variable is performed. A particular subset hypothesis H^K corresponding to a set of variables $K = \{k_1, \ldots, k_h\}$ is defined as

H^K: the treatments identically affect (or have no effect on) the responses for variables $k_1, \ldots, k_h$

Defining the response vectors

$$\mathbf{y}_{ijK} = (y_{ijk_1} \quad y_{ijk_2} \quad \cdots \quad y_{ijk_h})$$

the partial null hypothesis can be restated as

H^K: All random permutations of the observed vectors $\mathbf{y}_{11K}, \mathbf{y}_{12K}, \ldots, \mathbf{y}_{gn_gK}$ into the g groups, with sample sizes n_i, are equally likely

Assuming the test statistics t_k depend only upon the kth variable (as is commonly true), the marginal distribution of *any* subcollection $\{T_l;\ l \in K\}$ of test statistics is the same under the partial null hypothesis H^K, and the complete null hypothesis H^C. In either case the marginal distribution simply refers to the distribution induced by the random assignment of the observed complete data vectors $\mathbf{y}_{11}, \ldots, \mathbf{y}_{gn_g}$ into the g groups.

Example 4.1 Percentage Increase over Baseline Measurements

Table 4.1 presents artificial data (with realistic characteristics) from measurements on human subjects in Control, Low, and High dosage groups of a particular compound. To control the variation between subjects, each observation has been calculated as a percentage increase over a baseline measurement. The data set in Table 4.1 is therefore the "$\mathbf{Y}_2$" matrix of the data structure presented in Section 1.6. The raw data matrix "$\mathbf{Y}_1$" contains initial as well as final measurements, but the percentage increases were selected as the primary variables for analysis.

We are interested in determining whether the treatments cause an *increase* in the percentage measurements, for *any* of the variables. A possible test statistic is

$$t_k = \frac{(\bar{y}_{2k} + \bar{y}_{3k})/2 - \bar{y}_{1k}}{s_k\left(\frac{1}{n_1} + \frac{1}{4n_2} + \frac{1}{4n_3}\right)^{1/2}}$$

where $\bar{y}_{ik}$ is the mean of variable k in group i, and s_k^2 is the pooled within-group mean square error for variable k. The statistic t_k assesses whether the average of the treatment means is larger than the control mean for variable k.

Table 4.1 Percentage Increase Data

	Control ($n_1 = 11$)			Low Dose ($n_2 = 9$)			High Dose ($n_3 = 11$)		
	y_1	y_2	y_3	y_1	y_2	y_3	y_1	y_2	y_3
	−5	6	−2	−5	−34	−52	−10	−12	−27
	13	11	1	15	6	−27	4	−1	20
	15	22	9	5	1	14	29	26	−36
	−5	−3	10	3	−1	22	3	1	58
	−1	2	−27	35	23	30	7	5	47
	9	12	−6	12	14	15	31	27	66
	5	7	−2	4	9	66	−5	11	61
	4	9	14	−18	4	44	21	25	−41
	−30	−16	−3	−1	3	45	35	25	8
	−2	−23	−14				30	16	52
	−8	−3	5				−36	−25	69
Mean	−0.45	2.18	−1.36	5.78	2.78	17.44	9.91	8.91	25.18
SD	12.38	12.92	11.66	14.55	15.63	36.77	21.87	17.19	42.79
y_1	1.00			1.00			1.00		
Corr y_2	.75	1.00		.58	1.00		.93	1.00	
y_3	.14	.36	1.00	−.04	.65	1.00	−.29	−.28	1.00

For moderate sample sizes, the true randomized p-value (4.2) should be well approximated by the normal-theory value $p_k^{\mathcal{N}} = \Pr(T_\nu > t_k)$, where T_ν denotes a t-distributed random variable with $\nu = n_1 + n_2 + n_3 - 3$ degrees of freedom. Here, the test statistics and upper-tailed normal-theory p-values are $t_1 = 1.3048, p_1 = .1013$; $t_2 = .6353, p_2 = .2652$; $t_3 = 1.82815, p_3 = .0391$. Based on 100,000 simulations using PROC MULTTEST, the corresponding *unadjusted* rerandomized p-values are $p_{1r} = .1023, p_{2r} = .2624$, and $p_{3r} = .0387$. Because the true rerandomized p-values and their normal-theory approximations are close, it is reasonable to resample the normal-theory p-values to obtain the multiplicity adjustments: we see no evidence that different tests might be "weighted" differently. As previously mentioned, resampling the rerandomized p-values is an ungainly simulation problem; thus, practical considerations favor resampling the normal-theory p-values. In any case, the resampled multiplicity-adjusted p-values are exact (up to simulation error) even if the unadjusted p-values are not.

The resampling-based adjustments for this application may be performed by permuting, or "shuffling," the data as indicated in Section 3.1.4; however, in this case, *entire vectors* are shuffled to preserve the dependence characteristics. In this example, the following algorithm was used:

Algorithm 4.1 Multivariate Permutation Resampling of t-Statistics

0. Initialize counting variables: $COUNT_i = 0, i = 1, 2, 3$.
1. Let $\mathbf{y}_{ij}^*$ denote a *without replacement sample* (or permutation) of the observed data vectors $\mathbf{y}_{ij}$. That is, the vectors $\mathbf{y}_{11}, \ldots, \mathbf{y}_{1n_1}, \ldots, \mathbf{y}_{3n_3}$ simply are *shuffled*,

the first n_1 elements of the resulting shuffle are the vectors $\mathbf{y}_{11}^*, \ldots, \mathbf{y}_{1n_1}^*$; the second n_2 elements are $\mathbf{y}_{21}^*, \ldots, \mathbf{y}_{2n_2}^*$, and so on.

Compute the marginal sample means $\bar{y}_{ik}^*$, for $i = 1, 2, 3$ and $k = 1, 2, 3$, as well as the residual mean squares $(s_k^*)^2$ from the rerandomized vectors $\mathbf{y}_{ij}^*$.

2. Compute the statistics

$$t_k^* = \frac{(\bar{y}_{2k}^* + \bar{y}_{3k}^*)/2 - \bar{y}_{1k}^*}{s_k^*\left(\frac{1}{n_1} + \frac{1}{4n_2} + \frac{1}{4n_3}\right)^{1/2}}$$

3. Using Algorithm 2.8, the unadjusted p-values are ordered from *largest* to *smallest* to compute the successive extrema in a computationally convenient fashion. When defining the adjustment using upper-tailed t-statistics, the corresponding ordering is *smallest* t-statistic to *largest* t-statistic.

With the ordering $t_2 = .6353 < t_1 = 1.3048 < t_3 = 1.82815$, we have $t_{(1)} = t_3$, $t_{(2)} = t_1$, and $t_{(3)} = t_2$. Defining the adjustments using t-statistics for this particular application, define the successive *maxima*:

$$\begin{aligned} u_3^* &= t_2^* \\ u_2^* &= \max(u_3^*, t_1^*) \\ u_1^* &= \max(u_2^*, t_3^*) \end{aligned}$$

4. If $u_i^* \geqslant t_{(i)}$, then $COUNT_i \leftarrow COUNT_i + 1$.
5. Repeat 1–3 N times. Compute $\tilde{p}_i^{(N)} = COUNT_i/N$. Note that $\tilde{p}_{(1)}^{(N)} = \tilde{p}_3^{(N)}$, $\tilde{p}_{(2)}^{(N)} = \tilde{p}_1^{(N)}$, and $\tilde{p}_{(3)}^{(N)} = \tilde{p}_2^{(N)}$.
6. Enforce monotonicity using successive maximization:

$$\begin{aligned} \tilde{p}_{(1)}^{(N)} &= \tilde{p}_{(1)}^{(N)} \\ \tilde{p}_{(2)}^{(N)} &\leftarrow \max(\tilde{p}_{(1)}^{(N)}, \tilde{p}_{(2)}^{(N)}) \\ \tilde{p}_{(3)}^{(N)} &\leftarrow \max(\tilde{p}_{(2)}^{(N)}, \tilde{p}_{(3)}^{(N)}) \end{aligned}$$

Simultaneous permutation-based inference for multivariate applications involving linear contrasts can be done using PROC MULTTEST. The p-values for this application are given in Table 4.2. There is only marginal evidence (adjusted $p = .0997$) of a significant increase over control, occurring for the y_3 measurement. □

Several interesting features are brought out by this example. First, the large simulation size (100,000) was used in an attempt to reconcile the smallest adjusted p-value (for the y_3 test) with significance or non-significance at the $\alpha = .10$ FWE level. Even with this large simulation size, the attempt failed, because the standard error of the adjusted p-value estimate is $\{.0997(1 - .0997)/100{,}000\}^{1/2} = .00095$, whereas the adjusted p-value differs from .10 by .0003, a lesser amount. Depending upon how important significance or non-significance at the $\alpha = .10$ level is perceived to be, the

Table 4.2 Step-Down Rerandomization Tests for Percentage Increase Data

	p-Values			
Variable	Raw_p	Adj_p	Sid_p	Bon_p
y_1	.1013	.1499	.1924	.2026
y_2	.2652	.2624	.2652	.2652
y_3	.0391*	.0997	.1129	.1174

*p-value < .05.

experimenter may wish to try even larger simulation sample sizes to reconcile the question. We are content to say that this adjusted p-value is "about .10."

A second point concerning this analysis is that there is little difference between the largest unadjusted p-value (.2652, for y_2) and the corresponding adjusted p-value .2624. Since this is the largest p-value and since there was no need to enforce monotonicity in this example, there is no multiplicity adjustment for this largest p-value using the step-down method. This resampled value is in fact the true rerandomized p-value, modulo simulation error.

A third point concerns the disparity between the adjusted p-value for y_1 (.1499) and the independence-assuming Šidák adjustment (.1924): at the second stage of the step-down analysis, the distribution of the minimum of (P_1, P_2) is considered. There is a large correlation between Y_1 and Y_2 (Table 4.1). Because vectors are resampled, the adjusted p-value (.1499) incorporates this dependence structure, making the adjustment less conservative.

The idea for Example 4.1 was taken from Conover and Salsburg (1988), who discussed methods for detecting treatment effects with similar data using rank transformation methods. The data for this example suggest nonnormality: Figure 4.1 shows a predominantly left-skewed appearance in the distributions, with low outliers. These features suggest that rank-based tests might have more power than their parametric counterparts.

An alternative test procedure described by Conover and Salsburg involves rank-transforming each variable and computing some statistics upon the rank data. When the distributions satisfy the "Lehmann Alternatives" condition, the locally most powerful rank test is obtained using a certain *exponent* of the ranks in the test statistic. The optimal value of this exponent is unknown. It is reasonable to try the analysis using several different exponents; however, this method results in increased multiplicity, with an attendant increase in the probability of Type I errors. A simple solution is to use several different exponents, and to include all resulting tests into the "family" for multiplicity adjustment. The test statistics corresponding to different exponents are highly dependent, but this dependence structure is naturally accommodated when one computes all test statistics in each resampled data set.

In a re-analysis of the percentage increase data, we used the rank transformation on each variable, considered four different exponents (1, 2, 4, and 8), and considered one-tailed two-sample t-tests (with pooled variance from all three groups) for two

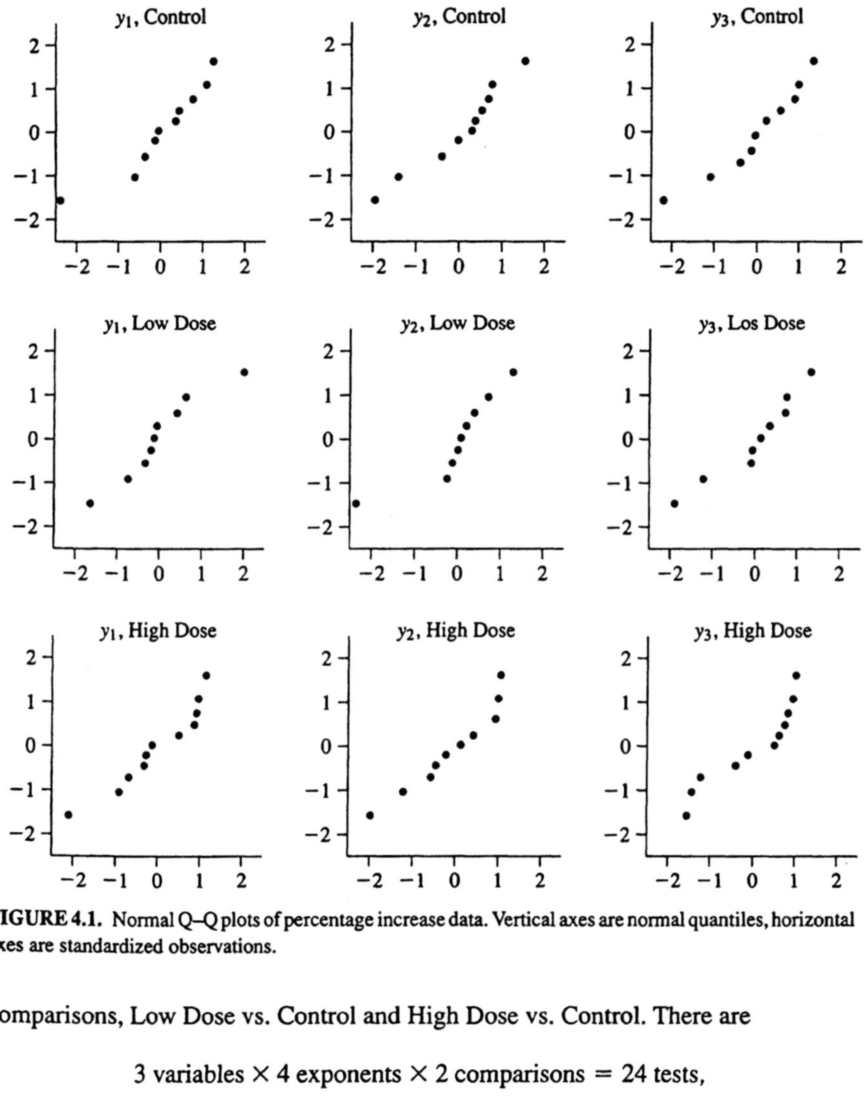

FIGURE 4.1. Normal Q–Q plots of percentage increase data. Vertical axes are normal quantiles, horizontal axes are standardized observations.

comparisons, Low Dose vs. Control and High Dose vs. Control. There are

$$3 \text{ variables} \times 4 \text{ exponents} \times 2 \text{ comparisons} = 24 \text{ tests},$$

many more than the previous analysis, which used only 3 tests.

Permutation resampling is better than bootstrap resampling in this example. Bootstrapping the rank-transformed data will lead to many "tied" ranks; thus the bootstrap distribution will reflect more discreteness than actually exists in reality. On the other hand, the permutation distribution is appropriate for evaluation of rank-based tests (e.g., Hájek and Šidák, 1967, Chapter IV).

This analysis was performed by Westfall, Lin and Young (1990) using PROC MTEST (the predecessor to PROC MULTTEST). Table 4.3 presents the results of a single-step analysis of all raw and adjusted p-values for this re-analysis. In this particular problem and in any problem where multiplicity of tests is under

Table 4.3 Single-Step Rerandomization Rank Tests Using Percentage Increase Data

		Low vs. Control				High vs. Control			
Variable	Exponent	Raw_p	Adj_p	Sid_p	Bon_p	Raw_p	Adj_p	Sid_p	Bon_p
y_1	1	0.2265	0.7849	0.9979	1.0000	0.0984	0.5154	0.9168	1.0000
	2	0.2207	0.7748	0.9975	1.0000	0.0511	0.3374	0.7163	1.0000
	4	0.2139	0.7645	0.9969	1.0000	0.0294*	0.2168	0.5112	0.7052
	8	0.1888	0.7252	0.9934	1.0000	0.0289*	0.2149	0.5058	0.6946
y_2	1	0.4368	0.9496	1.0000	1.0000	0.1320	0.6086	0.9665	1.0000
	2	0.4553	0.9560	1.0000	1.0000	0.0695	0.4123	0.8227	1.0000
	4	0.4602	0.9578	1.0000	1.0000	0.0256*	0.1968	0.4630	0.6137
	8	0.4412	0.9512	1.0000	1.0000	0.0114*	0.0969	0.2399	0.2727
y_3	1	0.0457*	0.3105	0.6750	1.0000	0.0201*	0.1615	0.3860	0.4828
	2	0.0200*	0.1607	0.3845	0.4804	0.0030*	0.0267*	0.0691	0.0715
	4	0.0359*	0.2574	0.5838	0.8608	0.0015*	0.0134*	0.0358*	0.0365*
	8	0.1397	0.6253	0.9730	1.0000	0.0058*	0.0495*	0.1309	0.1399

*p-value < .05.

consideration, it is very useful to compare the raw p-values to the adjusted p-values. The columnar arrangement of Table 4.3 facilitates these comparisons. The smallest observed raw p-value that is adjusted to above .05 is .0114, the high vs. control comparison with a rank exponent of eight. These results are interpreted as follows: there is only a 1% chance of observing a result as extreme as the observed result for the high vs. control comparison with rank exponent equal to eight, but there is *greater than a 5% chance* that a result this extreme will be observed *somewhere* among this collection of 24 tests. Raw p-values give a numerical measure of evidence concerning a hypothesis, when only one hypothesis is considered; adjusted p-values offer a measure of evidence concerning a particular hypothesis selected from many hypotheses.

It is quite intriguing that the global null hypothesis can be rejected when one is allowed to use many exponents. Even though the multiplicity problem is worse in this analysis, significance is still attained, amounting to "having one's cake and eating it too" (assuming that statistical significance is sought). Although not presented here, the adjusted p-values are *insignificant* at the $\alpha = .05$ FWE level when the *raw* ranks are used, without exponents. Even though the number of tests k over which multiplicity is performed is smaller with only the *raw* rank tests (6 vs. 24), we have significance when multiplicity adjustment is performed with all 24 tests, yet no significances when multiplicity adjustment is performed over the 6 tests! Generally, one does not expect that a *more* significant result can be found using a *larger* family of tests—from a crude Bonferroni perspective, the multiplier k is larger when the family size is larger, making it more difficult to obtain a significant result. Although this example is somewhat anomalous, it illustrates the potential power and utility of resampling methods for multiplicity adjustment.

Using the 24 *unadjusted* p-values, the p-value plot of Schweder and Spjøtvoll (1982) is presented in Figure 4.2. A very useful plot for assessing multiplicity, this

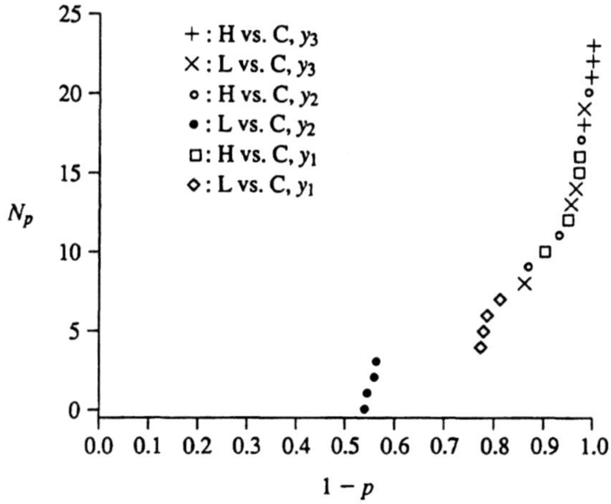

FIGURE 4.2. p-value plot for percentage increase analysis using rank tests with different exponents.

plot graphs N_p, the number of p-values greater than p, against the values $1-p$. Deviations from a straight line, particularly those curving "upward" in the upper right-hand area of the plot, are interpretable as deviations from the complete null hypothesis.

There is substantial deviation from linearity in Figure 4.2, but some of this departure can be attributed to the extremely large correlations between the test statistics for the various powers. There are some "clusters" in the plot corresponding to p-values which test the same hypothesis, but use different rank exponents. These large correlations are incorporated into the resampling-based adjusted p-values. Comparing the resampling-based "Adj_p" entries with the independence-assuming "Sid_p" entries in Table 4.3, it is clear that incorporating these correlations reduces the p-values.

Multivariate randomization analysis is straightforward when the data structure includes a block (or strata) variable. Once the various blocks of experimental units have been defined, the observation vectors may be shuffled independently within blocks to simulate the complete null hypothesis of no treatment effect for any block. The test statistics are then recomputed by combining over blocks. There will be one such combined test statistic for each variable/contrast combination. For specified contrasts among treatments, resampling-based multiplicity adjustment for multiple tests that have been combined over blocks can be done using PROC MULTTEST.

4.2 MULTIVARIATE REGRESSION MODELS

While the use of a block variable allows for the possibility of covariate-adjusted analysis, it is often desirable to use continuous variables, particularly if there are several covariates. Permutation-based resampling methods become somewhat complicated in such applications; however, the bootstrap provides a simple, elegant solution to this general class of problems.

4.2.1 The Model and Hypotheses

Consider the multivariate regression model

$$\mathbf{Y} = \mathbf{X}\boldsymbol{\beta} + \boldsymbol{\epsilon} \tag{4.3}$$

where $\mathbf{Y}$ is a $(n \times v)$ matrix of response variables; $\mathbf{X}$ is a full rank $(n \times b)$ design matrix; $\boldsymbol{\beta}$ is a $(b \times v)$ matrix of regression parameters, and $\boldsymbol{\epsilon}$ is a $(n \times v)$ matrix of random error terms, all with mean zero. This model subsumes univariate ANOVA as well as multivariate ANOVA (MANOVA) models, univariate and multivariate analysis of covariance models (ANCOVA and MANCOVA), and univariate and multivariate regression models. These models also apply to nonexperimental situations such as the analysis of sample surveys and the analysis of financial markets. In this section we assume that the rows $\boldsymbol{\epsilon}_i$ of $\boldsymbol{\epsilon}$ are independent and identically distributed according to some multivariate distribution G; however, as discussed in Section 3.1.3, it is possible to expand this model to include heteroscedastic distributions. We also assume that any variables in the $\mathbf{X}$ matrix that are considered random (e.g., covariates) are independent of the random matrix $\boldsymbol{\epsilon}$.

Specific tests of interest in such models are usually one-dimensional (or single-degree-of-freedom) tests. As a matter of statistical convenience, omnibus hypotheses are often tested, but if they are rejected, it is most important to identify which particular subcomponent is significant. Such subcomponent tests may be formulated as

$$H_0: \mathbf{c}'\boldsymbol{\beta}\mathbf{d} = f$$

Here, $\mathbf{c}'$ and $\mathbf{d}$ are $(1 \times b)$ and $(v \times 1)$ vectors of user-specified constants, and f is a user-specified scalar. When k such tests are specified, these become

$$H_{0i}: \mathbf{c}_i'\boldsymbol{\beta}\mathbf{d}_i = f_i, \; i = 1, \ldots, k$$

Test statistics and unadjusted p-values for each H_i are easily obtained: let

$$\hat{\boldsymbol{\beta}} = (\mathbf{X}'\mathbf{X})^{-1}\mathbf{X}'\mathbf{Y}$$

be the least squares matrix of parameter estimates, and let

$$\mathbf{S} = (\mathbf{Y} - \mathbf{X}\hat{\boldsymbol{\beta}})'(\mathbf{Y} - \mathbf{X}\hat{\boldsymbol{\beta}})/(n - b)$$

be the unbiased residual variance-covariance matrix estimate. The statistic

$$T_i = \frac{\mathbf{c}_i'\hat{\boldsymbol{\beta}}\mathbf{d}_i - f_i}{(\mathbf{d}_i'\mathbf{S}\mathbf{d}_i)^{1/2}[\mathbf{c}_i'(\mathbf{X}'\mathbf{X})^{-1}\mathbf{c}_i]^{1/2}} \tag{4.4}$$

may be used to test H_i. When the residual distribution G is multivariate normal and H_{0i} is true, then T_i has the t distribution with $\nu = n - b$ degrees of freedom.

The raw (unadjusted) p-values are either upper-, lower-, or two-tailed, depending upon the alternative of interest. These p-values are asymptotically valid even when the multivariate distribution function G is not multivariate normal, provided that all variances are finite.

4.2.2 Bootstrap Resampling

Note that

$$\mathbf{c}_i'\hat{\boldsymbol{\beta}}\mathbf{d}_i - f_i = \mathbf{c}_i'(\mathbf{X}'\mathbf{X})^{-1}\mathbf{X}'\mathbf{Y}\mathbf{d}_i - f_i = (\mathbf{c}_i'\boldsymbol{\beta}\mathbf{d}_i - f_i) + \mathbf{c}_i'(\mathbf{X}'\mathbf{X})^{-1}\mathbf{X}'\boldsymbol{\epsilon}\mathbf{d}_i$$

Thus, whenever H_{0i} is true, the distribution of the test statistic T_i of (4.4) depends only on the distribution of the random error matrix $\boldsymbol{\epsilon}$. We may use this fact to estimate the joint sampling distribution of the test statistics—using bootstrap sampling, replace $\boldsymbol{\epsilon}$ by a resampled version $\boldsymbol{\epsilon}^*$. If the matrix $\mathbf{X}$ has random elements, then we consider all sampling distributions to be conditional upon the realized values in $\mathbf{X}$.

The test statistics also are invariant to the unknown $\boldsymbol{\beta}$, so the subset pivotality condition holds, and we may resample under the complete null hypothesis. In this case, the simplest complete null model sets $\boldsymbol{\beta} = \mathbf{0}$, and resampling is performed using an estimate of the residual distribution. As in Section 3.1.1, equation (3.5), the required bootstrap data sets are

$$\mathbf{Y}_i^* = \boldsymbol{\epsilon}_i^*,\ i = 1, \ldots, n$$

In contrast to the univariate case, the $\boldsymbol{\epsilon}_i^*$ are now $(1 \times v)$ *vectors*, rather than scalars. Again, they are defined as a *with replacement* sample of the sample residual vectors:

$$\hat{\boldsymbol{\epsilon}}_i = \mathbf{Y}_i - \mathbf{X}_i\hat{\boldsymbol{\beta}}$$

Here, $\mathbf{Y}_i$ and $\mathbf{X}_i$ are the ith rows of $\mathbf{Y}$ and $\mathbf{X}$, respectively. By resampling *vectors*, the correlations among variables are incorporated into the multiplicity adjustments.

Residual vectors may be "weighted" before resampling (Wu, 1986) to correct for heteroscedasticity induced by the design matrix $\mathbf{X}$ (see the discussion following Algorithm 3.1 in Section 3.1.1).

Resampling-based multiple testing may be performed using the resampled data matrices $\mathbf{Y}^*$ to compute the test statistics t_i^* of (4.4), with $\hat{\boldsymbol{\beta}}$ and $\mathbf{S}$ replaced by $\hat{\boldsymbol{\beta}}^*$ and $\mathbf{S}^*$, and with the f_i set to zero. Adjusted p-values may be computed using Algorithm 2.5 in Section 2.3.2 (for single-step analysis), or Algorithm 2.8 in Section 2.6 (for step-down analysis). In most cases, t-values may be used instead of p-values in Algorithms 2.5 or 2.8, with appropriate modifications to account for one- or two-sided testing.

4.2.3 Multivariate Normal Resampling

Resampling analysis in the multivariate normal case is identical to bootstrap resampling, except that the $(1 \times v)$ residual row vectors $\boldsymbol{\epsilon}_i^*$ are generated from a multivariate normal distribution: $\epsilon_i^* \sim \mathcal{N}_v(\mathbf{0}, \mathbf{S})$. Note that the covariance matrix $\mathbf{S}$ is the covariance matrix estimated from the original data. Using special properties of the multivariate normal distribution allows more convenient generation of the resampled test statistics t_i^*. In particular, the resampled residual covariance estimate $\mathbf{S}^*$ is distributed as $1/(n-b)$ times a Wishart random matrix $\mathcal{W}(n-b, \mathbf{S})$ (see Anderson, 1984, Section 7.2). Furthermore, the resampling distribution of the vector of estimates $\{\mathbf{c}_i'\hat{\boldsymbol{\beta}}^*\mathbf{d}_i\,;\, i = 1, \ldots, k\}$ is multivariate normal with mean zero, and with the (i,j) covariance given by

$$\mathrm{Cov}(\mathbf{c}_i'\hat{\boldsymbol{\beta}}^*\mathbf{d}_i,\ \mathbf{c}_j'\hat{\boldsymbol{\beta}}^*\mathbf{d}_j) = \mathbf{c}_i'(\mathbf{X}'\mathbf{X})^{-1}\mathbf{c}_j\mathbf{d}_i'\mathbf{S}\mathbf{d}_j \tag{4.5}$$

for $1 \leqslant i,j \leqslant k$ (Anderson, 1984, pp. 287–291.) Letting $\mathbf{C} = [\mathbf{c}_1 : \cdots : \mathbf{c}_k]$ and $\mathbf{D} = [\mathbf{d}_1 : \cdots : \mathbf{d}_k]$, a computationally convenient expression for the covariance matrix of $\{\mathbf{c}_i'\hat{\boldsymbol{\beta}}^*\mathbf{d}_i\,;\, i = 1, \ldots, k\}$ is

$$\mathrm{Cov}\begin{pmatrix} \mathbf{c}_1'\hat{\boldsymbol{\beta}}^*\mathbf{d}_1 \\ \vdots \\ \mathbf{c}_k'\hat{\boldsymbol{\beta}}^*\mathbf{d}_k \end{pmatrix} = \mathbf{C}'(\mathbf{X}'\mathbf{X})^{-1}\mathbf{C} \odot \mathbf{D}'\mathbf{S}\mathbf{D} \tag{4.6}$$

where "$\odot$" refers to the elementwise matrix multiplication operator. The $\odot$ operator is also called the *Hadamard product*: given $(m \times n)$ matrices $\mathbf{A} = \{a_{ij}\}$ and $\mathbf{B} = \{b_{ij}\}$, $\mathbf{A} \odot \mathbf{B} = \{a_{ij}b_{ij}\}$ (Rao, 1973, p. 30).

Therefore, rather than compute the test statistics t_i^* from resampled data, one may generate the test statistics directly from the Wishart and multivariate normal distributions. If $\mathbf{Z}$ is a $(k \times 1)$ vector of independent standard normal variables, and if $\mathbf{H}$ is a $(k \times r)$ matrix of constants, then

$$\mathbf{H}'\mathbf{Z} \sim \mathcal{N}_r(\mathbf{0}, \mathbf{H}'\mathbf{H})$$

Many mathematical and statistical software packages have a "square root" function for a given matrix. Using the covariance matrix of (4.6) one may then generate vectors having the same distribution as the vector of estimates $\{\mathbf{c}_i'\hat{\boldsymbol{\beta}}^*\mathbf{d}_i\,;\, i = 1, \ldots, k\}$ as follows: (1) generate a $(k \times 1)$ vector $\mathbf{Z}$ of independent standard normal variates, (2) compute the "square root" matrix $\mathbf{H}$ of the covariance matrix in (4.6), then (3) compute the vector $\mathbf{H}'\mathbf{Z}$, which has the required distribution.

If the "square root" function is not available, use the eigenvalue/eigenvector decomposition of the covariance matrix (4.6). Let $\mathbf{A}$ denote a positive semidefinite $(k \times k)$ matrix with eigenvalues λ_i, $i = 1, \ldots, k$, and corresponding orthonormal eigenvectors $\mathbf{E}_i$, $\mathbf{A}\mathbf{E}_i = \lambda_i\mathbf{E}_i$. Defining the matrix

$$\mathbf{L}^{1/2} = \begin{pmatrix} \lambda_1^{1/2} & 0 & \ldots & 0 \\ 0 & \lambda_2^{1/2} & \ldots & 0 \\ \vdots & \vdots & \ddots & \vdots \\ 0 & 0 & \ldots & \lambda_k^{1/2} \end{pmatrix}$$

and the matrix

$$\mathbf{E} = [\mathbf{E}_1 : \mathbf{E}_2 : \cdots : \mathbf{E}_k]$$

a "square root" matrix $\mathbf{H}$ for $\mathbf{A}$ such that $\mathbf{A} = \mathbf{H}'\mathbf{H}$ is then

$$\mathbf{H} = \mathbf{L}^{1/2}\mathbf{E}'$$

To generate the resampled matrix $\mathbf{S}^*$, $(1 \times v)$ vectors of pseudo-residuals $\boldsymbol{\epsilon}_i^* \sim \mathcal{N}_v(\mathbf{0}, \mathbf{S})$ may be generated using the square root matrix for $\mathbf{S}$. Then $\mathbf{S}^* = \sum(\boldsymbol{\epsilon}_i^*)'\boldsymbol{\epsilon}_i^*/(n - b)$ has the required distribution.

Efficient algorithms generating such random vectors and matrices are discussed by Johnson (1987, pp. 49–55, 203–204). The method given here for generating $\mathbf{S}^*$ is inefficient but easy to program.

As discussed in Chapter 3, the resampling method often provides exact analyses (modulo Monte Carlo error) when the assumed model is homoscedastic normal, and when sampling is done from the normal distribution. In multivariate models (that is, when $v > 1$), resampling-based analysis provides only an *approximate* solution when the distribution G is assumed to be multivariate normal and resampling is done from the multivariate normal distribution, even when the adjusted p-values are evaluated without Monte Carlo error. This is a consequence of the fact that the correlations between columns of the $\boldsymbol{\epsilon}$ matrix are unknown; the resampling method uses estimates of these correlations, determined from the sample covariance matrix $\mathbf{S}$, in the multiplicity adjustments.

4.2.4 Control of Type III Errors

When single-step analysis is performed for multiple two-sided tests in this general framework, the Type III directional error rate is asymptotically controlled, and directional decisions are allowed. As in the univariate ANOVA application of Section 3.1.2, let

$$M_0 = \max_{1 \leqslant i \leqslant k} |T_i|$$

Under the complete null hypothesis, the distribution of M_0 depends on the error distribution G, assumed continuous for convenience, and not on $\boldsymbol{\beta}$. Let Q_G denote the distribution of $M_0 \mid H_0^C$, and let $Q_G^{(\alpha)}$ denote its $(1 - \alpha)$ quantile. Also having distribution Q_G is the pivotal statistic

$$\max_{1 \leqslant i \leqslant k} \left| \frac{(\mathbf{c}_i'\hat{\boldsymbol{\beta}}\mathbf{d}_i - f_i) - (\mathbf{c}_i'\boldsymbol{\beta}\mathbf{d}_i - f_i)}{(\mathbf{d}_i'\mathbf{S}\mathbf{d}_i)^{1/2}[\mathbf{c}_i'(\mathbf{X}'\mathbf{X})^{-1}\mathbf{c}_i]^{1/2}} \right|$$

which yields *exact* $100(1 - \alpha)\%$ simultaneous confidence intervals

$$(\mathbf{c}_i'\hat{\boldsymbol{\beta}}\mathbf{d}_i - f_i) \pm Q_G^{(\alpha)} (\mathbf{d}_i'\mathbf{S}\mathbf{d}_i)^{1/2}[\mathbf{c}_i'(\mathbf{X}'\mathbf{X})^{-1}\mathbf{c}_i]^{1/2}$$

An interval does not contain zero if and only if $|t_i| > Q_G^{(\alpha)}$ or, equivalently, when $\tilde{p}_{G,i} < \alpha$.

The quantiles $Q_G^{(\alpha)}$ may be estimated via bootstrap resampling, or by multivariate normal resampling if a multivariate normal model is assumed. Given the asymptotic consistency of the bootstrap estimate $Q_{\hat{G}}^{(\alpha)}$, we can anticipate that the intervals based on $Q_{\hat{G}}^{(\alpha)}$ have coverage probability approaching $100(1-\alpha)\%$ in large samples, and therefore that the criterion $\tilde{p}_{\hat{G},i} < \alpha$ approximately controls the Type III FWE in large samples.

4.2.5 The Effect of Using Estimated Correlations

Simulation-based simultaneous confidence sets have been studied by Edwards and Berry (1987), who assume all correlations are known (e.g., when $v = 1$ in an ANOVA or regression model). These authors are somewhat leery of its use in cases of unknown correlation structures, since the operating characteristics of the method are unknown.

Westfall and LuValle (1985a, 1985b) investigate such methods when correlation structures are unknown. For specific examples, they find that using estimated correlations can substantially shorten the intervals, while maintaining the simultaneous coverage levels.

As discussed by Westfall and LuValle, there is a potential problem with using estimated correlations in simultaneous inference. Suppose that $b = 1$, and that the **X** matrix consists of a single column of 1's. In this case the regression parameter matrix $\boldsymbol{\beta}$ is simply the row vector of variable means:

$$\boldsymbol{\beta} = (\mu_1 \quad \mu_2 \quad \ldots \quad \mu_t)$$

Suppose further that the data are multivariate normal, but with *zero correlations*, so the variables $Y_1, \ldots, Y_t$ are independent. The tests to be considered are H_{0i}: $\mu_i = 0$. A researcher ordinarily is not aware that the variables are independent, and may therefore attempt to incorporate the correlation structure using a resampling analysis. Incorporating correlations makes the adjusted p-values *smaller* than the Šidák adjustments, which are *exact* for the complete null hypothesis owing to independence among variables. Therefore, the resampling-based method must be *anticonservative* in this case, yielding a greater proportion of false positives than the nominally stated FWE level.

To assess the potential anti-conservativeness of the method, a simulation study was performed using $v = 10$, and sample sizes $n = 5, 10, 20, 40$. The data vectors were generated from the multivariate normal distribution with mean zero and identity covariance matrix. Using multivariate normal resampling with *NBOOT* = 500, resampling-based adjusted p-values for testing H_{0i}: $\mu_i = 0$ were computed for all $k = 10$ tests. For each resampled data set, we noted whether the minimum p-value fell below α, for $\alpha = .05, .10$. The analysis was repeated *NSIM* = 4000 times, and the proportion of resampled data sets yielding a false significance was calculated.

Estimates of the false significance rates are plotted in Figure 4.3. Values above the nominal FWE level indicate that the method is anticonservative. The use of estimated

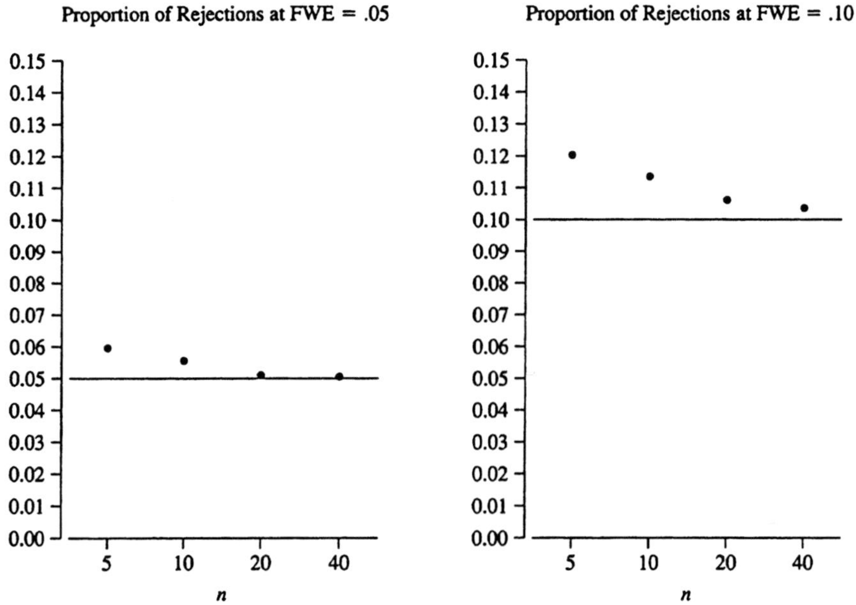

FIGURE 4.3. Rejection proportions with ten independent normal populations, using estimated correlations.

correlations makes the procedure anti-conservative, but to a very small degree. It is also clear that the convergence is extremely rapid for increasing n, corroborating the results of Westfall and LuValle, who proved convergence at a rate of n^{-1} in a related application. Since most large-sample results only claim convergence at a rate of $n^{-1/2}$, this rate of convergence is outstanding, as seen in the graphs of Figure 4.3.

The problem of increased familywise Type I error probabilities associated with using estimated correlation structures is relatively minor. We recommend that correlation structures be incorporated routinely into multiplicity adjustments. If one is very concerned that independence (or at least, approximate independence) may hold for the data, then one may use an independence-based adjustment as the main adjustment technique. Resampling-style adjustments are still useful to estimate the effect of the sample correlation structure upon the adjustments. When sample correlations are very small, and when both methods use the multivariate normal model, there will be little difference between independence-based (Šidák) adjustments and resampling-based adjustments.

Example 4.2 Obesity Biochemical Levels

Data given in Table 4.4 are used in a MANCOVA example discussed by Morrison (1990, pp. 234–236). The response variables are Y_1 = Pigment Creatinine and Y_2 = Chloride of urine samples; these are to be compared for four different obesity groups ("lighter underweight," "heavier underweight," "lighter obese," and "heavier obese") of subjects, adjusting for a single covariate X = Volume (mL).

Table 4.4 Obesity Data*

Category	Creatinine	Pigment Chloride	Volume
Lighter Underweight	17.6	5.15	205
Lighter Underweight	13.4	5.75	160
Lighter Underweight	20.3	4.35	480
Lighter Underweight	22.3	7.55	230
Lighter Underweight	20.5	8.50	235
Lighter Underweight	18.5	10.25	215
Lighter Underweight	12.1	5.95	215
Lighter Underweight	12.0	6.30	190
Lighter Underweight	10.1	5.45	190
Lighter Underweight	14.7	3.75	175
Lighter Underweight	14.8	5.10	145
Lighter Underweight	14.4	4.05	155
Heavier Underweight	18.1	9.00	220
Heavier Underweight	19.7	5.30	300
Heavier Underweight	16.9	9.85	305
Heavier Underweight	23.7	3.60	275
Heavier Underweight	19.2	4.05	405
Heavier Underweight	18.0	4.40	210
Heavier Underweight	14.8	7.15	170
Heavier Underweight	15.6	7.25	235
Heavier Underweight	16.2	5.30	185
Heavier Underweight	14.1	3.10	255
Heavier Underweight	17.5	2.40	265
Heavier Underweight	14.1	4.25	305
Heavier Underweight	19.1	5.80	440
Heavier Underweight	22.5	1.55	430
Lighter Obese	17.0	4.55	350
Lighter Obese	12.5	2.65	475
Lighter Obese	21.5	6.50	195
Lighter Obese	22.2	4.85	375
Lighter Obese	13.0	8.75	160
Lighter Obese	13.0	5.20	240
Lighter Obese	10.9	4.75	205
Lighter Obese	12.0	5.85	270
Lighter Obese	22.8	2.85	475
Lighter Obese	16.5	6.55	430
Lighter Obese	18.4	6.60	490
Heavier Obese	12.5	2.90	105
Heavier Obese	8.7	3.00	115
Heavier Obese	9.4	3.40	97
Heavier Obese	15.0	5.40	325
Heavier Obese	12.9	4.45	310
Heavier Obese	12.1	4.30	245
Heavier Obese	13.2	5.00	170
Heavier Obese	11.5	3.40	220

*Adapted from H. R. Smith, R. Gnanadesikan, and J. B. Hughes, "Multivariate Analysis of Variance (MANOVA)," *Biometrics* **18** 22–41, (1962), with permission of the Biometric Society.

In a multivariate regression model (4.3) for this data, the response matrix **Y** has two columns. The **X** matrix may be written as

$$\mathbf{X} = \begin{pmatrix} 1 & 0 & 0 & 0 & X_{11} \\ 1 & 0 & 0 & 0 & X_{12} \\ \vdots & \vdots & \vdots & \vdots & \vdots \\ 0 & 0 & 0 & 1 & X_{4n_4} \end{pmatrix}$$

and the parameter matrix may be written as

$$\boldsymbol{\beta} = \begin{pmatrix} \beta_{11} & \beta_{12} \\ \beta_{21} & \beta_{22} \\ \beta_{31} & \beta_{32} \\ \beta_{41} & \beta_{42} \\ \beta_{51} & \beta_{52} \end{pmatrix}$$

Of particular interest are comparisons $\theta_{ii'j} = \beta_{ij} - \beta_{i'j}$, $1 \leqslant i < i' \leqslant 4$, and $j = 1, 2$; these comparisons are covariate-adjusted comparisons of the different obesity groups for each of the response variables. There are $6 \times 2 = 12$ identified tests. Each contrast may be written as

$$\theta_{ii'j} = \mathbf{c}'_{ii'}\boldsymbol{\beta}\mathbf{d}_j$$

where $\mathbf{c}'_{12} = (1 \quad -1 \quad 0 \quad 0)$, $\mathbf{c}'_{13} = (1 \quad 0 \quad -1 \quad 0)$, etc.; and where $\mathbf{d}'_1 = (1 \quad 0)$ and $\mathbf{d}'_2 = (0 \quad 1)$. Test statistics are then computable as $t_{ii'j}$ using (4.4).

The raw data for this study were collected by Brown and Beerstecher (1951), and reported by Smith, Gnanadesikan and Hughes (1962) and Morrison (1990). However, for the purposes of multivariate normal resampling-based multiplicity adjustment, only the following summary statistics are required:

$$\mathbf{S} = \begin{pmatrix} 9.65886 & 0.80411 \\ 0.80411 & 3.64817 \end{pmatrix}, \quad \nu = 40$$

$$(\mathbf{X}'\mathbf{X})^{-1} = \begin{pmatrix} .201708 & .156400 & .182383 & .108590 & -.000547 \\ .156400 & .278067 & .240969 & .143472 & -.000723 \\ .182383 & .240969 & .371911 & .167308 & -.000843 \\ .108590 & .143472 & .167308 & .224614 & -.000502 \\ -.000547 & -.000723 & -.000843 & -.000502 & 2.531 \times 10^{-6} \end{pmatrix}$$

$$\begin{pmatrix} t_{12,1} & t_{12,2} \\ t_{13,1} & t_{13,2} \\ t_{14,1} & t_{14,2} \\ t_{23,1} & t_{23,2} \\ t_{24,1} & t_{24,2} \\ t_{34,1} & t_{34,2} \end{pmatrix} = \begin{pmatrix} -0.6051 & 0.6500 \\ 1.0569 & 0.1720 \\ 2.5894 & 2.4111 \\ 1.7814 & -0.4562 \\ 3.0819 & 1.8021 \\ 1.3701 & 2.0010 \end{pmatrix}$$

In this example, single-step multiplicity-adjusted p-values may be obtained using the following algorithm.

Algorithm 4.2 Multivariate Normal Resampling

0. Initialize counting variables: $COUNT_{ii'j} = 0$, $1 \leqslant i < i' \leqslant 4, j = 1, 2$.
1. Generate the vector $\{\hat{\theta}^*_{ii'j};\ 1 \leqslant i < i' \leqslant 4,\ j = 1, 2\}$ from the multivariate normal distribution with mean zero and (12×12) covariance matrix as in (4.5). Generate $\mathbf{S}^*$ independently, distributed as $1/\nu$ times a $\mathcal{W}(\nu, \mathbf{S})$-distributed (Wishart) matrix.
2. Compute the statistics

$$t^*_{ii'j} = \frac{\hat{\theta}^*_{ii'j}}{(\mathbf{d}_j'\mathbf{S}^*\mathbf{d}_j)^{1/2}[\mathbf{c}_{ii'}'(\mathbf{X}'\mathbf{X})^{-1}\mathbf{c}_{ii'}]^{1/2}}$$

3. If $\max_{1 \leqslant i < i' \leqslant 4; j=1,2} |t^*_{ii'j}| \geqslant |t_{ii'j}|$, where the $t_{ii'j}$ are the covariate-adjusted test statistics from the original experiment, then increment the count variable: $COUNT_{ii'j} \leftarrow COUNT_{ii'j} + 1$.
4. Repeat 1–3 N times. The estimated value of $\tilde{p}_{ii'j}$ is $\tilde{p}^{(N)}_{ii'j} = COUNT_{ii'j}/N$.

Software for general implementation of such multiplicity adjustments for the general multivariate regression model are not yet available. Algorithm 4.2 was implemented using a hand-written program in SAS/IML, with $N = 10{,}000$.

A summary of the resulting p-values is given in Table 4.5. Only the $\beta_{21} - \beta_{41}$ comparison achieves significance at the $\alpha = .05$ FWE level, indicating that heavier

Table 4.5 Summary of p-Values for Obesity Data

Contrast	$t_{ii',j}$	Raw_p	Adj_p	Sid_p
y_1 (Creatinine)				
$\beta_{11} - \beta_{21}$	−0.605	0.5485	0.9935	0.9999
$\beta_{11} - \beta_{31}$	1.057	0.2969	0.9116	0.9854
$\beta_{11} - \beta_{41}$	2.589	0.0133*	0.1219	0.1484
$\beta_{21} - \beta_{31}$	1.781	0.0824	0.4984	0.6437
$\beta_{21} - \beta_{41}$	3.082	0.0037*	0.0376*	0.0435*
$\beta_{31} - \beta_{41}$	1.370	0.1783	0.7693	0.9053
y_2 (Chloride)				
$\beta_{12} - \beta_{22}$	0.650	0.5194	0.9913	0.9998
$\beta_{12} - \beta_{32}$	0.172	0.8643	1.0000	1.0000
$\beta_{12} - \beta_{42}$	2.411	0.0206*	0.1770	0.2210
$\beta_{22} - \beta_{32}$	−0.456	0.6507	0.9987	1.0000
$\beta_{22} - \beta_{42}$	1.802	0.0791	0.4847	0.6280
$\beta_{32} - \beta_{42}$	2.001	0.0522	0.3687	0.4745

*p-value < .05.

underweight subjects have significantly higher pigment creatinine than heavier obese subjects, controlling for volume. While this analysis shows a significant effect, a corresponding analysis by Morrison (1990, pp. 234–236) was not able to find significance for this contrast. □

In fairness, Morrison's analysis allows infinitely many covariate-adjusted contrasts, and he identifies a particular linear contrast (not pairwise) of the parameters that is significant. However, it is usually best for both the experimenter and the statistical analyst if specific contrasts of interest are identified. When multiplicity adjustment is performed over infinitely many hypotheses, the tests are likely to lose power for specific tests of interest. Furthermore, when a particular linear contrast among the infinite collection is found significant, the contrast often is uninterpretable from a subject matter standpoint.

A resampling-based solution to the problem of infinite contrasts is presented in Chapter 6 of this book.

Comparing the "Adj_p" column with the "Sid_p" column of Table 4.5, we see that the correlations have little effect on the multiplicity adjustments. While the use of estimated correlations in multiplicity adjustments is questionable, this example shows that such estimated correlations do not make the tests extremely "liberal" in the sense of rejecting hypotheses too often. This observation supports the simulation analysis presented in Figure 4.3.

Contributing to the correlation structure among tests are (1) the use of pairwise tests, where different pairs have common elements, and (2) the correlation across variables. The partial correlation of the Y's, controlling for group and covariate, measures the effect of the latter type of dependence. In this example, the partial correlation was rather small, estimated to be only $.80411/\{9.65886 \times 3.64817\}^{1/2} = .135$ (using the **S** matrix). This leads us to believe that correlations between pairwise comparisons contribute more to the dependence structure among test statistics than do correlations between variables.

4.3 UNIVARIATE REGRESSION ANALYSIS

Many applications are covered by the general multivariate regression model. Simple applications include testing regression coefficients in an ordinary (univariate) regression model, and comparisons of means with repeated-measures data. The following example illustrates the application to ordinary regression.

Example 4.3 Survival of Patients Undergoing Liver Operation

Data given in Table 4.6 are reported by Neter, Wasserman and Kutner (1989), as an example of using regression to predict survival time. There are $n = 54$ patients, with dependent variable Y = Survival Time, and independent variables X_1 = Blood Clotting Score, X_2 = Prognostic Index, X_3 = Enzyme Function Test, and X_4 = Liver Function Test. The dependent variable is log-transformed: $Y' = \ln(Y)$.

Table 4.6 Survival Times of Patients in a Surgical Unit*

Blood Clotting Score	Prognostic Index	Enzyme Function Test	Liver Function Test	Survival Time
6.7	62	81	2.59	200
5.1	59	66	1.70	101
7.4	57	83	2.16	204
6.5	73	41	2.01	101
7.8	65	115	4.30	509
5.8	38	72	1.42	80
5.7	46	63	1.91	80
3.7	68	81	2.57	127
6.0	67	93	2.50	202
3.7	76	94	2.40	203
6.3	84	83	4.13	329
6.7	51	43	1.86	65
5.8	96	114	3.95	830
5.8	83	88	3.95	330
7.7	62	67	3.40	168
7.4	74	68	2.40	217
6.0	85	28	2.98	87
3.7	51	41	1.55	34
7.3	68	74	3.56	215
5.6	57	87	3.02	172
5.2	52	76	2.85	109
3.4	83	53	1.12	136
6.7	26	68	2.10	70
5.8	67	86	3.40	220
6.3	59	100	2.95	276
5.8	61	73	3.50	144
5.2	52	86	2.45	181
11.2	76	90	5.59	574
5.2	54	56	2.71	72
5.8	76	59	2.58	178
3.2	64	65	.74	71
8.7	45	23	2.52	58
5.0	59	73	3.50	116
5.8	72	93	3.30	295
5.4	58	70	2.64	115
5.3	51	99	2.60	184
2.6	74	86	2.05	118
4.3	8	119	2.85	120
4.8	61	76	2.45	151
5.4	52	88	1.81	148
5.2	49	72	1.84	95
3.6	28	99	1.30	75
8.8	86	88	6.40	483
6.5	56	77	2.85	153
3.4	77	93	1.48	191
6.5	40	84	3.00	123
4.5	73	106	3.05	311
4.8	86	101	4.10	398
5.1	67	77	2.86	158
3.9	82	103	4.55	310
6.6	77	46	1.95	124
6.4	85	40	1.21	125
6.4	59	85	2.33	198
8.8	78	72	3.20	313

*From Neter, Wasserman, and Kutner, *Applied Linear Regression Models*, 2nd ed., (1989), pp. 439–440, with permission from Richard Irwin, Inc.

Model-fitting involves investigating the correct functional form of the relationship between dependent and independent variables. To this end, the pool of candidate regressor variables was expanded to allow all interactions and quadratic terms. To simplify the interpretation of the coefficients, all original X's were mean-centered prior to creation of transformations: $X_1' = X_1 - \bar{X}_1, X_2' = X_2 - \bar{X}_2, \dots, X_{12}' = X_1'X_2'$, $X_{13}' = X_1'X_3', \dots, X_{11}' = X_1'^2, X_{22}' = X_2'^2, \dots$. The regression model is then

$$\begin{aligned} Y' = {} & \beta_0 + \beta_1 X_1' + \beta_2 X_2' + \beta_3 X_3' + \beta_4 X_4' \\ & + \beta_{12} X_{12}' + \beta_{13} X_{13}' + \cdots + \beta_{11} X_{11}' + \beta_{22} X_{22}' + \cdots + \epsilon \end{aligned}$$

This model represents a special case of the multivariate regression model, namely, the univariate regression analysis for which $v = 1$.

Consider the problem of testing for nonzero regression parameters. We exclude the intercept from the family of tests, leaving $k = 14$ in this example. As a first step, we seek a simple yes/no answer to the testing problem, rather than a directional determination. The *free* step-down adjustments are useful here.

Such adjustments may not be made more powerful by incorporating logical restrictions: the hypotheses satisfy the *free combinations* condition; therefore, the *restricted* and *free* step-down methods coincide.

A bootstrap analysis was performed on this data. First, the ordinary least squares estimates of the β's were computed, yielding sample residuals

$$\begin{aligned} \hat{\epsilon} = Y' - (& \hat{\beta}_0 + \hat{\beta}_1 X_1' + \hat{\beta}_2 X_2' + \cdots \\ & + \hat{\beta}_{12} X_{12}' + \hat{\beta}_{13} X_{13}' + \cdots + \hat{\beta}_{11} X_{11}' + \hat{\beta}_{22} X_{22}' + \cdots) \end{aligned}$$

The bootstrap data sets were then constructed as

$$Y_i^* = \hat{\epsilon}_i^*, \quad i = 1, \dots, n$$

where $\hat{\epsilon}_i^*$, $i = 1, \dots, n$ is a *with replacement* sample from the sample residuals $\hat{\epsilon}_i$. (As discussed in Section 4.2.2, the distribution of the test statistics depends only on the residuals ϵ_i under H_0^C. Thus we estimate this distribution by re-creating data sets using estimated residuals $\hat{\epsilon}_i^*$.)

All regression test statistics were then re-computed for each bootstrap data set using the same design matrix as in the original sample, and the free step-down adjustments of Algorithm 2.8 were evaluated. (In this problem, "min p" may be replaced with "max $|t|$," leading to definition of successive *maxima* in step 2., and a reversal of the inequality in step 3.)

This analysis was programmed using SAS/IML and 10,000 simulations, and the results are presented in Table 4.7. The results show that only the linear terms for X_1', X_2', and X_3' achieve significance. While centering variables reduces multicollinearity, there is still a strong dependence among the test statistics. This dependence may be analyzed using a principal components decomposition of the correlation matrix of the estimates $\hat{\beta}$: four components explain 78% of thetotal standardized variance, out

Table 4.7 Summary of p-Values for Regression Analysis Survival Data

Parameter	t-Value	Raw_p	Adj_p	Sid_p
β_1	14.188	0.0000*	0.0000*	0.0000*
β_2	20.516	0.0000*	0.0000*	0.0000*
β_3	23.668	0.0000*	0.0000*	0.0000*
β_4	1.266	0.2131	0.4312	0.5128
β_{12}	−1.654	0.1061	0.3966†	0.4973†
β_{13}	−1.550	0.1293	0.3966†	0.4973†
β_{14}	1.917	0.0626	0.3096†	0.4569†
β_{23}	−1.214	0.2319	0.4312†	0.5128†
β_{24}	0.314	0.7552	0.7539	0.7552
β_{34}	1.943	0.0592	0.3096	0.4569
β_{11}	−2.466	0.0182*	0.1320	0.1825
β_{22}	1.585	0.1210	0.3966†	0.4973†
β_{33}	−1.719	0.0936	0.3966	0.4973
β_{44}	−1.934	0.0604	0.3096†	0.4569†

†monotonicity enforced.
*p-value $< .05$.

of a total of 14 estimates. The effect of such correlations is to reduce the p-value adjustments, as can be seen by comparing the "Adjusted p-value" and "Šidák p-value" entries in Table 4.7.

Step-down multiplicity adjustment is not the best methodology for variable selection. Other methods more ideally suited for this purpose include *stepwise* and *all subsets* variable selection, with selection criteria such as the *PRESS* and Mallow's C_p statistics (e.g., Neter, Wasserman, and Kutner, 1989, Chapter 12). However, if the goal is to declare which variables are significant while protecting the FWE in the strong sense, this methodology is an improvement over the Bonferroni method (e.g., Neter, Wasserman, and Kutner, 1989, pp. 159–162), or the modified Bonferroni method (Stoline, 1983). □

4.4 REPEATED MEASURES MULTIVARIATE ANALYSIS

Another special case of the multivariate regression model is the repeated measures model, which is somewhat the reverse of the regression model in that the dimension of the parameters β is $(1 \times v)$ rather than $(b \times 1)$. Usually, such models are not called "regression" models, since the **X** matrix has only a single column consisting entirely of ones. Nevertheless, the general methods described in Section 4.2 apply to this situation.

There are two modes of analysis often used for such data—full multivariate analysis or "block" analysis, where each subject defines a "block." In the latter case, one makes implicit assumptions concerning the covariances among the repeated observations. Resampling-based analysis may be used in either case, with appropriate modifications to account for the assumed form of the model.

4.4.1 Full Multivariate Analysis

The following example illustrates use of the method for multiple comparisons of means from a repeated measures experiment.

Example 4.4 Effects of Anesthetics on Dogs

Table 4.8 displays repeated measures data from $n = 19$ dogs. The four repeated measurements represent a (2×2) factorial application of Halothane ("Present" or "Absent"), and CO_2 pressure ("Low" or "High"). The response data, milliseconds between heartbeats, are displayed graphically in Figure 4.4.

Comparisons of interest include main effects and interactions contrasts. Table 3.11 in Section 3.3 shows linear combinations used to estimate such contrasts. One also might wish to examine all pairwise comparisons. Combining this collection of tests into a single "family" yields the set of $k = 9$ contrasts given by Table 4.9. These 9 contrasts may be viewed as $\mathbf{Y}_2$ variables, in the data analysis paradigm of Section 1.6 in Chapter 1. Here, there are *no* treatment groups, and we are testing simply whether the mean values of our selected $\mathbf{Y}_2$ variables are zero.

Repeated measures data may be modelled using the multivariate regression model:

$$\begin{pmatrix} Y_{11} & \cdots & Y_{14} \\ Y_{21} & \cdots & Y_{24} \\ \vdots & \ddots & \vdots \\ Y_{n1} & \cdots & Y_{n4} \end{pmatrix} = \begin{pmatrix} 1 \\ 1 \\ \vdots \\ 1 \end{pmatrix} (\mu_1 \quad \mu_2 \quad \mu_3 \quad \mu_4) + \begin{pmatrix} \epsilon_{11} & \cdots & \epsilon_{14} \\ \epsilon_{21} & \cdots & \epsilon_{24} \\ \vdots & \ddots & \vdots \\ \epsilon_{n1} & \cdots & \epsilon_{n4} \end{pmatrix}$$

Table 4.8 Dog Anesthesia Data,* Milliseconds between Heartbeats

	Halothane Absent		Halothane Present	
Dog	CO_2 High	CO_2 Low	CO_2 High	CO_2 Low
1	426	609	556	600
2	253	236	392	395
3	359	433	349	357
4	432	431	522	600
5	405	426	513	513
6	324	438	507	539
7	310	312	410	456
8	326	326	350	504
9	375	447	547	548
10	286	286	403	422
11	349	382	473	497
12	429	410	488	547
13	348	377	447	514
14	412	473	472	446
15	347	326	455	468
16	434	458	637	524
17	364	367	432	469
18	420	395	508	531
19	397	556	645	625

*Data used with permission of Dr. J. Atlee.

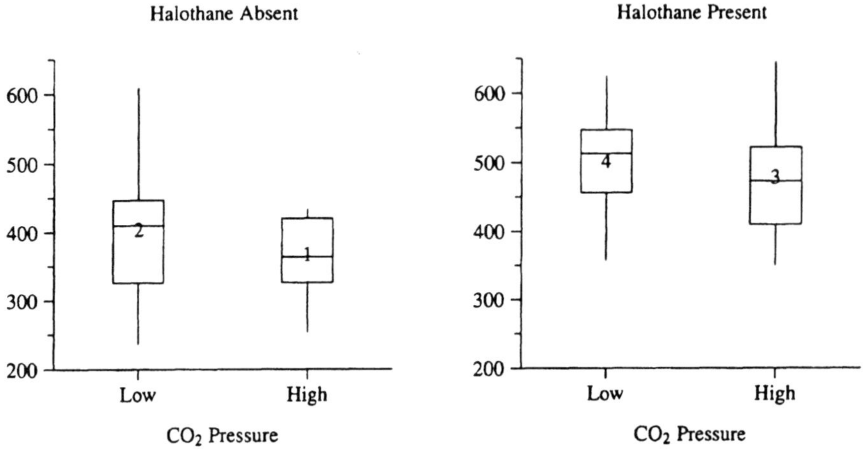

FIGURE 4.4. Box plots for dog anesthesia data. Mean values are indicated by number of treatment group.

The mean vector $(\mu_1 \quad \mu_2 \quad \mu_3 \quad \mu_4)$ replaces the parameter matrix $\boldsymbol{\beta}$, and the least squares estimate of $\boldsymbol{\beta}$ becomes the ordinary sample mean vector. Estimated parameters for these data are

$$\hat{\boldsymbol{\mu}} = (\hat{\mu}_1 \quad \hat{\mu}_2 \quad \hat{\mu}_3 \quad \hat{\mu}_4) = (368.21 \quad 404.63 \quad 479.26 \quad 502.89)$$

and

$$\mathbf{S} = \begin{pmatrix} 2819.29 & 3568.42 & 2943.49 & 2295.35 \\ 3568.42 & 7963.14 & 5303.98 & 4065.44 \\ 2943.49 & 5303.98 & 6851.32 & 4499.63 \\ 2295.35 & 4065.44 & 4499.63 & 4878.99 \end{pmatrix} \qquad (n = 19, \ \nu = 18)$$

Using these estimates, test statistics (4.4) are computed using the contrasts found in Table 4.9: $\mathbf{d}_1' = (+1 \quad -1 \quad 0 \quad 0)$, $\mathbf{d}_2' = (+1 \quad 0 \quad -1 \quad 0)$, and so on; $\mathbf{c}_i \equiv 1$, $f_i \equiv 0$, and $(\mathbf{X}'\mathbf{X})^{-1} = 1/n$. Thus the test statistics become

$$T_i = \frac{\hat{\boldsymbol{\mu}}\mathbf{d}_i}{(\mathbf{d}_i'\mathbf{S}\mathbf{d}_i/n)^{1/2}}$$

and the unadjusted two-sided p-values are computed as $p_i = \Pr(|T_\nu| > |t_i|)$.

To perform resampling-based adjustments using the multivariate normal model, we require only the statistics t_i, the matrix $\mathbf{D} = (\mathbf{d}_1 : \cdots : \mathbf{d}_9)$, the matrix $\mathbf{S}$, and the degrees of freedom $\nu = 18$. We consider a *single-step* adjustment to allow directional decisions. The following algorithm summarizes the method.

Algorithm 4.3 Multivariate Normal Resampling for Repeated Measures

0. Initialize counting variables: $COUNT_i = 0, i = 1, \ldots, 9$.
1. Generate $(\hat{\theta}_1^*, \ldots, \hat{\theta}_9^*)$ from the multivariate normal distribution with mean 0 and covariance matrix $\mathbf{D}'\mathbf{S}\mathbf{D}/\mathbf{n}$. (This expression is derived from the covariance

Table 4.9 Contrast Tests for Dog Anesthesia Data

	Contrast Coefficients				p-Values[1]			p-Values[2]		
Effects	μ_1	μ_2	μ_3	μ_4	Raw_p	Adj_p	Sid_p	Raw_p	Adj_p	Sid_p
Pairwise										
1 vs. 2	+1	−1	0	0	.0170(−)	.1001	.1431	.0117	.0631	.1002
1 vs. 3	+1	0	−1	0	.0000(−)	.0000	.0000	.0000	.0000	.0000
1 vs. 4	+1	0	0	−1	.0000(−)	.0000	.0000	.0000	.0000	.0000
2 vs. 3	0	+1	−1	0	.0001(−)	.0008	.0008	.0000	.0000	.0000
2 vs. 4	0	+1	0	−1	.0000(−)	.0000	.0001	.0000	.0000	.0000
3 vs. 4	0	0	+1	−1	.0643(−)	.2907	.4502	.0960	.3644	.5969
ANOVA										
Halothane	−1	−1	+1	+1	.0000(+)	.0000	.0000	.0000	.0000	.0000
CO_2	+1	−1	+1	−1	.0019(−)	.0138	.0170	.0036	.0214	.0320
Interaction	+1	−1	−1	+1	.5294(−)	.9297	.9989	.5196	.9261	.9986

*Sign of estimated effect in parentheses.
[1]Multivariate normal model.
[2]Randomized complete block (compound symmetry) model.

form of equation (4.6).) Generate $\mathbf{S}^*$, distributed independently as $1/\nu$ times a $\mathcal{W}(\nu, \mathbf{S})$ random variable.

2. Compute the statistics

$$t_i^* = \frac{\hat{\theta}_i^*}{(\mathbf{d}_i'\mathbf{S}^*\mathbf{d}_i/n)^{1/2}}, \qquad i = 1, \ldots, 9$$

4. If $\max_{1 \leq j \leq 9} |t_j^*| \geq |t_i|$, where the t_i are the t-statistics from the original data, then increment the count variable: $COUNT_i \leftarrow COUNT_i + 1$.
5. Repeat 1–3 N times. Compute $\tilde{p}_i^{(N)} = COUNT_i/N$.

The results of this analysis are presented in Table 4.9. All pairs of treatments are significantly different, with the exception of the "3 vs. 4" and possibly the "1 vs. 2" pair. The responses are significantly higher when Halothane is present, and the responses are significantly higher for low CO_2 pressure. There is no significant interaction: the mean difference between low and high CO_2 pressure does not appear to depend upon the presence or absence of Halothane. □

It should be noted that the pairwise contrasts (1 vs. 2, 1 vs. 3, etc.) are not as meaningful as the main effects and interaction contrasts in this example. Including many contrasts makes the multiplicity problem worse, and limits our ability to conclude significance for meaningful tests. Alternatively, one might (1) minimize the importance of the pairwise tests through "weighting" (Chapter 6, Section 6.1), or (2) drop the pairwise tests altogether.

4.4.2 "Block" Design and Analysis

As an alternative analysis, we might consider "dog" to be a "block" factor having 19 levels, and perform all such comparisons in the context of a randomized complete block design. Because there is no replication in this design, the error variance is estimated using the interaction sum of squares. This model has only one variable ($v = 1$), rather than $v = 4$. The error degrees of freedom are larger, $\nu = 54$ rather than $\nu = 18$. When this model is used, tests will usually be more powerful than tests which use the repeated measures model, at the expense of adding the assumption that the true covariance has the "compound symmetry" structure. Under compound symmetry, we assume that the variances of the four measurements are identical, and that the correlations are equal for all pairs of measurements. (See Ott, 1988, p. 795, for example.) On the other hand, the repeated measures multivariate analysis of Example 4.4 makes no assumptions concerning the form of the covariance matrix.

Table 4.9 includes the unadjusted and resampling-based p-values for all contrasts, when the randomized complete block model

$$Y_{ij} = \mu_i + \beta_j + \epsilon_{ij}$$

is used. In this example, $i = 1, \ldots, 4$ (treatments) and $j = 1, \ldots, 19$ (dogs as blocks). The model assumes that the μ_i are fixed constants, that the β_j are independent and normally distributed random variables with mean zero and variance σ_b^2, that the ϵ_{ij} are independent of the β_j, and are also independent and normally distributed random variables with mean zero and variance σ_e^2.

A given contrast $\mathbf{d}_i'\boldsymbol{\mu} = d_{i1}\mu_1 + \cdots + d_{i4}\mu_4$ may be estimated using $\mathbf{d}_i'\bar{\mathbf{y}}_i$, with the corresponding t-statistic

$$t_i = \frac{\mathbf{d}_i'\bar{\mathbf{y}}_i}{(\hat{\sigma}_e^2\mathbf{d}_i'\mathbf{d}_i/19)^{1/2}}$$

The residual mean square estimate is

$$\hat{\sigma}_e^2 = \sum\sum(y_{ij} - \bar{y}_{i.} - \bar{y}_{.j} + \bar{y}_{..})^2/54$$

and the unadjusted p-values are computed from the t-distribution with $\nu = 54$ degrees of freedom.

Let

$$\mathbf{D} = \begin{pmatrix} \mathbf{d}_1' \\ \mathbf{d}_2' \\ \vdots \\ \mathbf{d}_9' \end{pmatrix}$$

Then the joint null distribution of the nine test statistics is multivariate t with dispersion matrix

$$\rho_{\mathbf{D}} = [\text{diag}(\mathbf{DD}')]^{-1/2}\mathbf{DD}'[\text{diag}(\mathbf{DD}')]^{-1/2}$$

and degrees of freedom $\nu = 54$. It is easy to simulate values from this distribution. The *adjusted p-value* entries in Table 4.9 which appear under the compound symmetry analysis were computed using 20,000 simulations.

At the FWE = .10 level, an additional contrast (1 vs. 2) is significant when the compound symmetry model is used. This may be attributable to the increase in power. However, some p-values are *larger* using the compound symmetry model than they are for the multivariate normal model. This phenomenon may be attributed to the different estimates of standard deviations used by the two methods.

4.5 COMPARING ALTERNATIVE METHODS

So far we have presented three different methods of resampling: permutation resampling (rerandomization), bootstrap resampling, and normal resampling (sampling from a normal distribution). A choice regarding selection of resampling method must be made in any given application. Compounding the problem of selecting a resampling method is the additional decision concerning which *marginal* test procedures to resample. Should the tests be based on normal theory (as in most of our examples so far), should they be nonparametric (rank- or permutation-based), or should they be based upon robust estimates, such as trimmed means or other statistics that down-weight outliers? No particular combination of resampling method and base tests can be uniformly best. In the next two sections, we offer comments and suggestions to help in making these choices.

4.5.1 Comparing Normal and Bootstrap Resampling

Among the three resampling methods presented for continuous data applications, the bootstrap and normal-based resampling methods are closest in spirit and in resulting adjustments. Both methods resample mean-zero residuals. If the empirical residual distributions are fairly normal-looking (e.g., as suggested by quantile–quantile plots), then the bootstrap adjustments should be very similar to the normal-based adjustments. The results are expected to be similar in large sample sizes even when the residual distribution is nonnormal, since the base tests used for testing means are asymptotically distribution-free. However, as discussed in Section 2.5, the approximations can be poor even with moderate to large sample sizes. The adequacy of the approximation depends on the degree of nonnormality (particularly skewness) of the true distributions, and on the type of test considered (e.g., one-sample test or multi-sample contrast test).

The dog anesthesia example in Section 4.2.2 is a moderately sized data set exhibiting nonnormal characteristics. For the validity of the unadjusted tests presented in Table 4.9, the distributions of the indicated linear contrasts should be approximately normal. The normal quantile-quantile plots of Figure 4.5 show serious skewness and kurtosis problems, particularly for the contrasts $Y_1 - Y_2, Y_2 - Y_3, Y_2 - Y_4, Y_3 - Y_4$, and $-Y_1 - Y_2 + Y_3 + Y_4$. The effect of such nonnormalities upon the resampling-based inference is illustrated in Figure 4.6, which displays the histograms of the

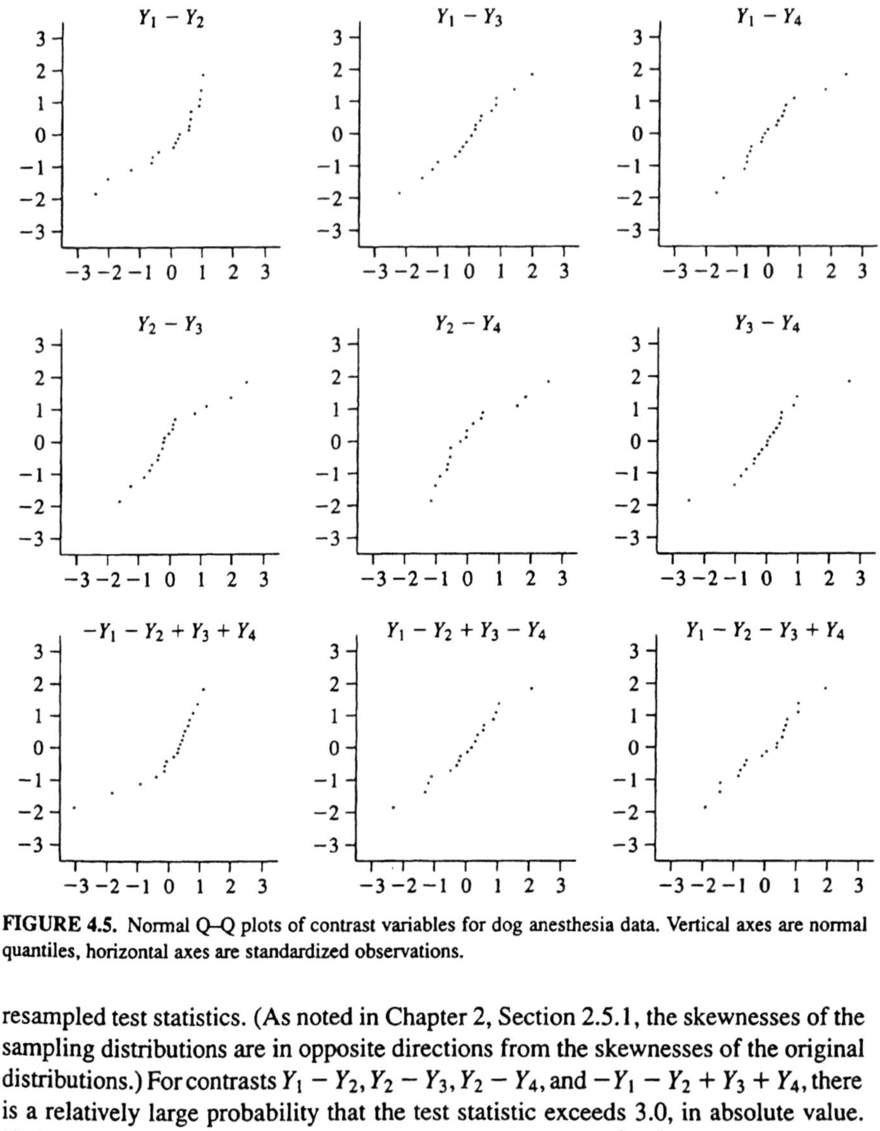

FIGURE 4.5. Normal Q–Q plots of contrast variables for dog anesthesia data. Vertical axes are normal quantiles, horizontal axes are standardized observations.

resampled test statistics. (As noted in Chapter 2, Section 2.5.1, the skewnesses of the sampling distributions are in opposite directions from the skewnesses of the original distributions.) For contrasts $Y_1 - Y_2, Y_2 - Y_3, Y_2 - Y_4$, and $-Y_1 - Y_2 + Y_3 + Y_4$, there is a relatively large probability that the test statistic exceeds 3.0, in absolute value. Under normal resampling, this probability is very small, $\Pr(|T_{19}| > 3.0) = .0074$.

The distributions of Figure 4.6 suggest that the bootstrapped t-statistics can be more extreme than is expected under normal sampling. Such extreme values can cause the resampling-based multiplicity-adjusted p-values to be larger: the maximum absolute t-statistic will exceed a critical value more frequently under bootstrap sampling than under normal sampling. Table 4.10 shows that some bootstrap adjustments are larger in this application; in particular, the CO_2 ANOVA effect is no longer significant at the $\alpha = .05$ FWE level. Accounting for skewness, the unadjusted bootstrap p-values also tend to be larger than their normal counterparts.

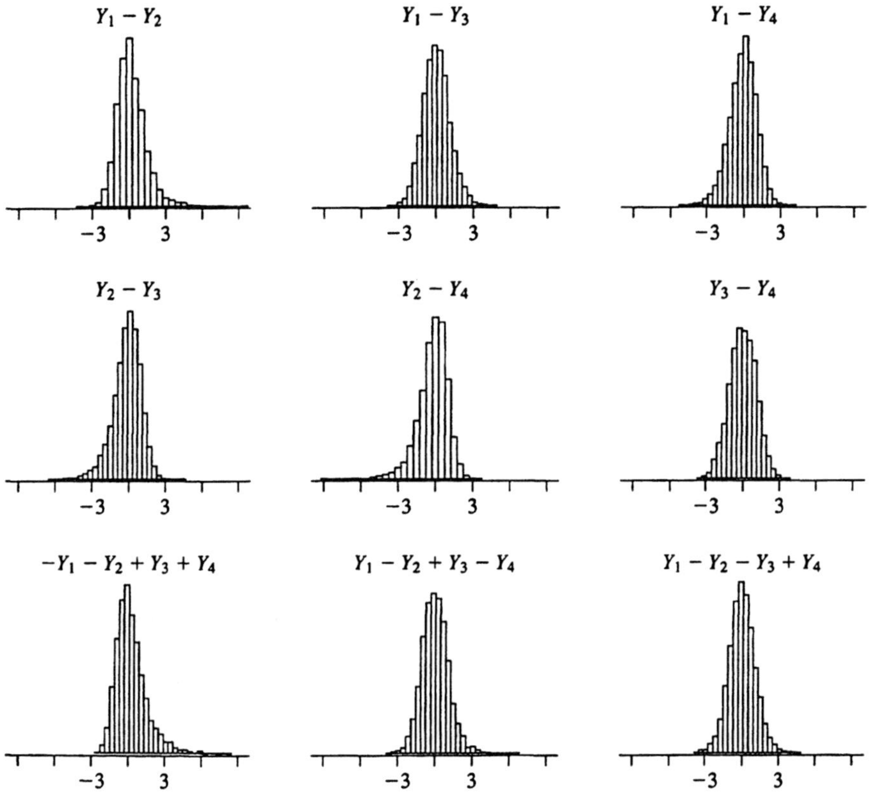

FIGURE 4.6. Histograms (normalized to common area) of resampled t-statistics using dog anesthesia data.

Given the discrepancy between the bootstrap and normal-based analyses, it might be prudent to choose the more conservative analysis (i.e., the one with the larger p-values.) This way, claimed "significant" results are more likely to be real. On the other hand, if one is trying to demonstrate "insignificance," (as happens frequently, for example, in toxicology studies performed by product manufacturers), then it might be wise to present both forms of evidence. If the test is "insignificant" using both measures, then one might be more confident in this conclusion; however, if the test is "significant" using one method and "insignificant" using another, then one must explain the discrepancy, perhaps using quantile-quantile plots and simulation analyses.

Our experience has shown that (1) the resampling-based bootstrap adjustments are quite similar to normal-theory adjustments when the data are approximately normal, (2) the bootstrap adjustments may differ substantially from the normal adjustments when the data are skewed, and (3) the bootstrap adjustments outperform the normal adjustments in terms of level and/or power when the data are skewed. Considering

Table 4.10 Bootstrap Tests for Dog Anesthesia Data in Block Model

	Bootstrap *p*-Values	
Effects	Unadjusted	Adjusted
Pairwise		
1 vs. 2	0.0393	0.1488
1 vs. 3	0.0000	0.0013
1 vs. 4	0.0000	0.0001
2 vs. 3	0.0013	0.0166
2 vs. 4	0.0013	0.0043
3 vs. 4	0.0621	0.3213
ANOVA		
Halothane	0.0001	0.0004
CO_2	0.0020	0.0540
Interaction	0.5171	0.9301

these observations, we recommend using bootstrap adjustments over normal-theory adjustments.

4.5.2 Comparing Permutation Resampling with Normal and Bootstrap Resampling

The bootstrap and normal resampling models are quite similar, since the normal model is a special case of the location shift model assumed for bootstrap resampling. These models are usually justified by assuming the data are randomly sampled from a real or conceptual large population, where the parameters of the models are population statistics. On the other hand, the randomization model requires no assumption of a "population"—the data generating process is completely determined by the random assignment of units to treatment groups. Thus, permutation-based inference is exact from this viewpoint, provided one can evaluate the permutation distributions with sufficient accuracy. This exactness is quite appealing to statisticians: permutational inference has been called the "Gold Standard" by Strawderman and Mehta (1990), and the "Platinum Standard" by Tukey (1988). In multiple testing applications where the subset pivotality condition holds *exactly* (such as the clinical study of Section 4.1), the rerandomization method of inference is even more attractive.

There are two fundamental distinctions between permutation and bootstrap resampling algorithms: First, the bootstrap resampling method, as presented here for continuous location-shift involves resampled *centered* data; in permutation analysis one resamples the *raw* data. Second, the permutation-based resampling method is a *without replacement* sampling method; the bootstrap samples are made *with replacement*.

Which method is preferred? While the permutation-based approach is preferable because of fewer assumptions (and because of metallurgy, according to some statisticians!), the bootstrap method can accommodate more complex models. Permutation-based inference in the presence of multiple continuous covariates is extremely complicated even in univariate testing, but this case poses no new problems for the bootstrap. Even in the pairwise comparisons application of Chapter 3, with balanced data and no covariates, the subset pivotality condition does not hold for permutational inference. However, this condition does hold for bootstrap inference in simple and more complex models, which also favors the use of bootstrap resampling.

When the two methods can be applied easily, they tend to have similar properties. Romano (1989) shows that they are asymptotically equivalent in certain applications, which favors permutational inference because of its "exact" finite-sample interpretability.

The issue of whether to perform resampling in the permutation or bootstrap framework is somewhat controversial. We return to this controversy in Chapter 5, pointing out some of the key distinctions when analyzing binary data.

4.5.3 Choice of Test Statistic

For most analyses, we have assumed that the choice of a "base" test is the t-statistic. While resampling can help attain nominal coverage levels using these statistics, it does not guarantee good power. Miller (1986) comments,

> ... whereas the t test is somewhat *robust for validity*, it is not *robust for efficiency*.*

This applies to the bootstrapped t test as well, though the bootstrapped test is more *robust for validity*, meaning that the actual and nominal levels more nearly coincide.

When the data exhibit severely nonnormal characteristics, a more efficient test procedure should be used. For example, with the percentage increase data of Example 4.1, p-values were reduced when rank-transformed tests were used. Log-transformation is also effective in making tests more powerful (e.g., Berry, 1987). Tests based on "trimmed" data, or based on data whose outliers are downweighted via M-estimation also may improve the power (e.g., Staudte and Sheather, 1990). Shorack (1982) describes bootstrap models for robust regression that might be adapted for our multiple testing applications in multivariate regression models.

One must be careful when applying a bootstrap resampling plan to nonstandard test statistics. It is possible that the hypothesis tested using an alternative test statistic is *false* in the bootstrap population. For example, if a test statistic tests for differences in *medians*, and the resampling distribution forces equal *means*, but not equal distributions otherwise, then the bootstrapped p-values may be incorrect. When resampling distributions with unequal medians, a p-value for testing medians generated within a bootstrap simulation tends to be *too small*. Thus the bootstrap estimate,

*With permission of Barbara B. Miller.

$\Pr(P^*_{\text{bootstrap}} \leqslant p_{\text{observed}})$ will tend to be *too large*. One must take care to insure the null hypotheses tested hold in the resampling population.

Example 4.5 Multiple Sign Tests—Dog Anesthesia Data
Some of the distributions shown in Figure 4.5 appear nonnormal. Rather than test *means*, one might prefer to test whether *medians* of the original distributions differ from zero. The *sign test* can be used for this purpose. There are $k = 9$ such tests, six pairwise contrasts, two main effects, and one interaction. We are treating this problem as a single-sample multivariate analysis, with $n = 19$ observations and $k = 9$ variables (the $\mathbf{Y}_2$ matrix of Chapter 1, Section 1.6).

To perform the two-tailed sign test for a zero median, count the number of values above and below zero, say Y_+ and Y_-, respectively. Let $Y_m = \max(Y_+, Y_-)$. The unadjusted p-value is

$$p = 2\Pr(Y \geqslant Y_m)$$

where Y has a binomial distribution with probability $\pi = .5$. The sample size used for the binomial distribution is $n_{\neq 0} = n - Y_+ - Y_-$, the number of observations that differ from zero.

To create a bootstrap "population" having zero medians, one may *median-center* the variables. Then these centered data may be resampled, and multiplicity adjustments computed. It is important that the data be *median-centered,* rather than *mean-centered,* to insure that all null hypotheses are true in the bootstrap population.

Note that the correlation matrix of the median-centered data is identical to that of the raw data, so correlation is incorporated in the multiplicity adjustments.

Table 4.11 Multiple Sign Tests for Dog Anesthesia Data

Effects	$\max(Y_+, Y_-)$	Number $\neq 0$	Raw_p*	Bootstrap Raw_p	Bootstrap Adj_p	Perturbed Bootstrap Raw_p	Perturbed Bootstrap Adj_p
Pairwise							
1 vs. 2	12(−)	17	.1435	.0928	.4599	.0622	.3474
1 vs. 3	18(−)	19	.0001	.0001	.0007	.0001	.0011
1 vs. 4	18(−)	19	.0001	.0000	.0007	.0000	.0011
2 vs. 3	16(−)	19	.0044	.0031	.0203	.0048	.0320
2 vs. 4	16(−)	19	.0044	.0026	.0203	.0041	.0320
3 vs. 4	15(−)	18	.0075	.0051	.0374	.0041	.0320
ANOVA							
Halothane	18(+)	19	.0001	.0002	.0007	.0002	.0011
CO_2	15(−)	19	.0192	.0139	.0804	.0205	.1271
Interaction	10(+)	19	1.0000	.9238	1.0000	1.0000	1.0000

*Two-tailed binomial sign test.

Note also that the odd sample size $n = 19$ creates a possible problem: the median is in fact one of 19 the data values. Thus, there will be at least one zero for every variable to be resampled, creating a positive probability that one observation will equal the median in the resampling population. To assess the impact of this problem, we try two resampling schemes: in one scheme, we sample the ordinary median-centered data, and in the other scheme, we sample and *perturb* the median-centered data. In the perturbed data, data values are adjusted by adding ϵ to every value, where the ϵ's are mean zero independent normal random variables, with standard deviations equal to .00001. Thus, in the resampled and perturbed data, no values will actually equal zero, yet the median of the distribution remains zero. Efron (1981) discusses use of perturbations, although his main goal is to make the bootstrap distribution continuous.

Table 4.11 displays the results of resampling-based analysis using 10,000 resampled data sets, for each resampling method. The adjusted p-values are single-step.

Bootstrap unadjusted p-values for variables with all 19 observations different from zero are identical to the binomial sign test p-values, up to Monte Carlo error. However, the ordinary bootstrap unadjusted p-values differ from the binomial sign test p-values, because of the problem with resampling zero values. For the variables with one or more differences equal to zero, the unperturbed bootstrap p-values are slightly closer.

Most raw p-values of the multiple sign tests are somewhat larger than the parametric p-values of Table 4.9. However, there is one sign test with a *smaller* p-value. The sign test flags the "3 vs. 4" comparison as significant, even with *adjusted* p-values. This comparison is insignificant with the t-test, even with *unadjusted* p-values! The discrepancy is explained by Figure 4.5, which shows distinct outliers in the upper and lower tails of the distribution of the $Y_3 - Y_4$ variable. Such outliers make parametric methods less powerful, but have little or no affect on nonparametric methods.

□

CHAPTER 5

Binary Data Applications

Binary data are frequent in statistical applications. Whenever the data are coded as one of two mutually exclusive categories, such as won/lost, lived/died, bankrupt/non-bankrupt, etc., the data are binary. Categorical data having more than two possible outcomes may be coded as *multivariate* binary: if a survey response takes one of the three categories A, B, or C, then the responses may be completely determined using the three binary codes "A/not A," "B/not B," and "C/not C." There is a perfect dependence structure among these responses, since any two responses completely determines the third. Fortunately, resampling methods incorporate these and other dependence structures automatically.

Binary data are obviously nonnormal because of discreteness. When the counts are very sparse for some categories, the sampling distributions of the test statistics may be very discrete, and also nonnormal. In such cases multiple testing procedures based on asymptotically normal distributions can be very inaccurate, often having actual FWE rates far below nominal, with an attendant loss of power. Resampling-based methods for multiplicity adjustment are particularly attractive in such applications, since the effects of discreteness are incorporated, thereby improving FWE level-robustness and power.

When comparing binary responses over multiple treatment groups, the hypotheses are often phrased as comparisons of proportions, or of linear combinations of proportions. It is useful to visualize such hypotheses using the "structured data analysis" paradigm of Section 1.6, where the variables in the $\mathbf{Y}_2$ matrix are binary. It is also very convenient to frame questions about the treatment groups in terms of linear contrasts. There are a variety of methods for testing such hypotheses, including the Cochran-Armitage test (Cochran, 1954; Armitage, 1955), of which the usual two-sample binomial Z test is a special case; the Freeman-Tukey arcsine transformation of the Cochran-Armitage test (Freeman and Tukey, 1950); as well as exact rerandomization tests (Heyse and Rom, 1988; Farrar and Crump, 1988), of which the "Fisher exact test" is a special case (e.g., Davis, 1986).

While binomial testing methods usually result in asymptotically normal or chi-squared test statistics, the true sampling distributions can be extremely nonnormal for sparse data. The degree of nonnormality of the sampling distributions is related more to *sparseness* than sample size: if $n = 1000$ observations are sampled and the underlying binomial strike rate is .1%, then the data will be quite sparse, with only one "strike" expected out of 1000 sampled observations. If the strike rate is π, it is the smallness of $n\pi$ or $n(1-\pi)$ that is the issue. Frequently $\min\{n\pi, n(1-\pi)\} < 5$ is used to identify "small" strike rates.

In this chapter we present some resampling methods for analyzing multiple tests with multivariate binary data, demonstrating how the effects of such sparseness may be handled. The multivariate binomial distribution is presented and discussed in Section 5.1. In Section 5.2, we review and compare methods for testing univariate hypotheses with binary data, including the chi-square, Cochran-Armitage, Freeman-Tukey, and randomization tests. When several univariate tests are considered, there is a multiplicity problem, and in Section 5.3 we present bootstrap- and permutation-based multiplicity adjustments. In Section 5.4 we compare the bootstrap and permutation methods for binary data, for both univariate and multiple testing applications. In Section 5.5 we present an application using single-sample multivariate binary data, and in Section 5.6 we show how historical information may be incorporated via resampling.

5.1 PRELIMINARY DEFINITIONS

This section follows Westfall and Young (1989a) rather closely, where we assume that the observable data vectors **Y** may be modeled as having come from multivariate binomial distributions. Define the notation

$$\mathbf{Y} \sim \mathcal{MVB}_v(\boldsymbol{\pi}, n, \mathbf{D})$$

for a v-component multivariate binomial random vector

$$\mathbf{Y}' = (Y_1 \quad \dots \quad Y_v)$$

with binomial marginal distributions $Y_j \sim \mathcal{BIN}(\pi_j, n)$ and dependence structure specified by **D**.

When $n = 1$, the distribution $\mathcal{MVB}_v(\boldsymbol{\pi}, 1, \mathbf{D})$ of **Y** is called *multivariate Bernoulli.* In this case **D** denotes a particular probability distribution on the 2^v binary v-tuples subject to the constraint that $E(\mathbf{Y}) = \boldsymbol{\pi}$.

If $\mathbf{B}_1, \dots, \mathbf{B}_n$ denote independent and identically distributed (i.i.d.) multivariate Bernoulli vectors ($\mathbf{B}_i \sim \mathcal{MVB}_v(\boldsymbol{\pi}, 1, \mathbf{D})$), then the vector of summates has the multivariate binomial distribution

$$\mathbf{Y} = \sum \mathbf{B}_i \sim \mathcal{MVB}_v(\boldsymbol{\pi}, n, \mathbf{D})$$

Situations where this model applies include surveys, where all responses are coded using binary variables; adverse event clinical studies, in which the components of **Y** represent frequencies of different adverse events (or side effects); and animal carcinogenicity studies, in which the components of $\mathbf{Y}$ may represent frequencies of tumors at any of several lesion/site combinations.

Since the multinomial model is a special case of the multivariate binomial model, the scope of applications includes multinomial data. When the vectors are multinomial, the $\mathcal{MVB}_v(\boldsymbol{\pi}, 1, \mathbf{D})$ distribution places probability π_j on the vector

$$\mathbf{b}_j' = (0 \quad \cdots \quad 0 \quad 1 \quad 0 \quad \cdots \quad 0)$$

having a 1 in position j and 0's elsewhere. For multinomial data, $\sum \pi_j = 1$, so there is no possibility of multiple occurrences of 1's in a vector.

If there are $i = 1, \ldots, g$ treatment groups with n_i experimental units per group, the available data are $\mathbf{B}_{11}, \ldots, \mathbf{B}_{1n_1}, \ldots, \mathbf{B}_{g1}, \ldots, \mathbf{B}_{gn_g}$, where the vectors $\mathbf{B}_{i1}, \ldots, \mathbf{B}_{in_i}$ within each group are independent and identically distributed as $\mathcal{MVB}_v(\boldsymbol{\pi}_i, 1, \mathbf{D}_i)$, and where all $n = \sum n_i$ vectors are independent. Now $\boldsymbol{\pi}_i' = (\pi_{i1} \quad \cdots \quad \pi_{iv})$: π_{ij} denotes the probability of a "1" for variable j in group i. For each $i = 1, \ldots, g$, the frequency vectors $\mathbf{Y}_i = \sum_j \mathbf{B}_{ij}$ are distributed as $\mathcal{MVB}_v(\boldsymbol{\pi}_i, n_i, \mathbf{D}_i)$.

The parameter matrix

$$\Pi = \begin{pmatrix} \pi_{11} & \pi_{12} & \cdots & \pi_{1v} \\ \pi_{21} & \pi_{22} & \cdots & \pi_{2v} \\ \vdots & \vdots & \ddots & \vdots \\ \pi_{g1} & \pi_{g2} & \cdots & \pi_{gv} \end{pmatrix}$$

corresponds to the summary matrix $\mathbf{S}$ of the structured data analysis (Figure 1.2 of Section 1.6 in Chapter 1). Neither the columns nor the rows need sum to 1.0. The rows might sum to 1.0, for example when the variables are multinomial.

To illustrate this structure, consider the following survey:

Q1: Do you have extra-sensory perception?

A1: 1. Yes 2. No 3. Not Sure

Q2: Have you ever seen a UFO?

A2: 1. Yes 2. No

Suppose further that these questions are asked of three distinct groups, say college graduates, high school (non-college) graduates, and those who did not graduate from high school. There are five response variables: each person chooses one of A11, A12, A13 (from question one), and one of A21, A22 (from question 2). The two choice variables are coded as "1," and the remaining three variables are coded as "0." Then

$$\Pi = \begin{pmatrix} \pi_{11} & \pi_{12} & \pi_{13} & \pi_{14} & \pi_{15} \\ \pi_{21} & \pi_{22} & \pi_{23} & \pi_{24} & \pi_{25} \\ \pi_{31} & \pi_{32} & \pi_{33} & \pi_{34} & \pi_{35} \end{pmatrix}$$

where π_{ij} is the proportion of responses that fall in category j among individuals in group i. For example, π_{24} is the proportion of high school (non-college) graduates who say they have seen a UFO. Here $\pi_{i1} + \pi_{i2} + \pi_{i3} = 1$ and $\pi_{i4} + \pi_{i5} = 1$, so the row totals sum to 2.0. Including redundant categories such as column 5 is no problem—the dependence structures are incorporated via resampling.

5.2 UNIVARIATE TESTING METHODS

Testing in the multiple-group multivariate binomial setting is analogous to testing in multiple-group continuous-data applications such as MANOVA. Omnibus tests for equality of proportions across groups within a column are denoted H_j: $\pi_{1j} = \pi_{2j} = \cdots = \pi_{gj}$, and contrasts which compare specific groups within a column are denoted H_{lj}: $c_{l1}\pi_{1j} + c_{l2}\pi_{2j} + \cdots + c_{lg}\pi_{gj} = 0$. In the MANOVA case, the same types of hypotheses are considered, with the π's replaced by μ's. If there are v variables and c tests per variable, we have $k = c \times v$ total hypothesis tests.

In this section we present standard univariate methods for testing the indicated hypotheses. Since each hypothesis involves a single variable, the j subscript is unnecessary, and will be dropped for the remainder of this section. In Section 5.3 we return to multiple testing applications in which several variables are considered, each indexed by j.

5.2.1 Chi-Squared Test

With binary data, the hypothesis of equality of proportions for all groups is H: $\pi_1 = \cdots = \pi_g$. This is the analog of the ANOVA hypothesis H: $\mu_1 = \cdots = \mu_g$ for continuous data. The hypothesis of equal proportions is tested using the chi-squared statistic, derived from the following $(2 \times g)$ contingency table:

	Occurrence	Nonoccurrence	Total
	Y_1	$n_1 - Y_1$	n_1
	Y_2	$n_2 - Y_2$	n_2
	$\vdots$	$\vdots$	$\vdots$
	Y_g	$n_g - Y_g$	n_g
Total	Y_+	$n - Y_+$	n

The expected values under the null hypothesis are

$$E_i = n_i Y_+ / n$$

for the "Occurrence" column, and

$$E_i' = n_i - E_i$$

for the "Nonoccurrence" column. The chi-squared test statistic is then

$$T = \sum_i \left\{ \frac{(Y_i - E_i)^2}{E_i} + \frac{(Y_i' - E_i')^2}{E_i'} \right\}$$

where $Y_i' = n_i - Y_i$. The null distribution of T is approximately χ^2_{g-1}, and the null hypothesis is usually rejected when T exceeds $\chi^2_{g-1}(\alpha)$, the $1 - \alpha$ quantile of the chi-squared distribution with $g - 1$ degrees of freedom.

5.2.2 Cochran-Armitage Test

Like the ANOVA F-test, the chi-squared test is unspecific. An analyst may wish to know *which* proportions differ. In some instances, the treatment groups form *ordered* categories, such as control, low, and high doses of a treatment. With ordered categories, one might expect an increasing (or decreasing) *trend* in the "Occurrence" column. If the increase is very gradual, the chi-squared test might not detect it. However, if the test statistic is chosen to reflect such trends, it may be possible to detect significant effects which are missed by the chi-squared test. The Cochran-Armitage test (Cochran, 1954; Armitage, 1955; Agresti, 1990) is useful in such situations.

Let $d_1, \ldots, d_g$ denote user-specified *scores* for the different groups. In an experiment where groups represent varying doses of a compound, the d_i may represent the dose values themselves, or perhaps a monotonic transformation of the dosage values. For example, if the groups are ordered as group 0 = Control, group 1 = Low Dose, and group 2 = High Dose, then the scores $d_0 = 0$, $d_1 = 1$, $d_2 = 2$ are commonly used. In general, $d_0, d_1, d_2, \ldots$, are numerical values representing the relative strengths of the categories.

The Cochran-Armitage procedure may also be used when the categories are unordered. In such cases the scores d_i may be chosen to reflect specific contrasts among treatment proportions. Let $\hat{\pi} = Y_+/n$ denote the pooled probability estimate under the complete null hypothesis $H\colon \pi_1 = \cdots = \pi_g$. Then the numerator of the Cochran-Armitage trend test is

$$\text{Trend} = \sum_{i=1}^{g} d_i(Y_i - n_i\hat{\pi}) \tag{5.1}$$

Again under the complete null hypothesis, the variance is estimated as

$$\hat{V} = \hat{\pi}(1 - \hat{\pi}) \sum_{i=1}^{g} n_i(d_i - \bar{d})^2 \tag{5.2}$$

where $\bar{d} = \sum_{i=1}^{g} n_i d_i / n$. Equation (5.2) may be derived assuming binomial distributions for the Y_i. Putting these together yields the Cochran-Armitage test statistic

$$T = \frac{\text{Trend}}{\hat{V}^{1/2}} \tag{5.3}$$

Under the null hypothesis of equal proportions, the distribution of T is approximately standard normal.

A benefit of the Cochran-Armitage test is that it is more powerful than the chi-squared test when the scores d_i reflect the true underlying alternative hypothesis configuration. The Cochran-Armitage test also allows single-tail testing to detect increasing or decreasing dose effects. On the other hand, the analyst needs prior information concerning anticipated patterns to choose the d_i. If the d_i are chosen badly, then the chi-square test may outperform the Cochran-Armitage test.

The null hypotheses tested by the Cochran-Armitage and chi-square procedures are not identical: the null hypothesis tested by the chi-squared test is a *special case* of the null hypothesis tested by the Cochran-Armitage test. To see how, note that

$$\mathrm{E(Trend)} = \sum_{i=1}^{g} n_i(d_i - \bar{d})\pi_i$$

which is zero whenever the hypothesis

$$H: \sum_{i=1}^{g} n_i(d_i - \bar{d})\pi_i = 0$$

is true. The value $\bar{d}$ is the weighted average used in (5.2). The hypothesis H is certainly true under the global null configuration $\pi_i \equiv \pi$, but is also true for other configurations of the π where not all are equal. In particular, when the scoring coefficients d_i are chosen to be monotonically increasing in the dose groups, the Cochran-Armitage test tends not to detect certain alternative configurations, such as when there is an initial *rise* in the probability level for small dose groups, followed by a *decline* in the probability level for larger dose groups, lending an "umbrella" appearance to the probability profile. Alternative scorings, for example $d_1 = -1$, $d_2 = +2$, $d_3 = -1$ in a three group experiment, are more powerful for detecting such effects. If the dose response plateaus, then coefficients that mimic the plateau would be more powerful, e.g., $-2, 1, 1$. If there is a threshold, then the coefficients $-1, -1, 2$ would be more powerful. Note that the test procedure is invariant to shifts in the scoring coefficients: the scores $-1, -1, 2$ produce the same test statistic as the scores $0, 0, 3$. Further discussions and guidelines concerning choice of d_i may be found in Agresti (1990), and Kimeldorf, Sampson, and Whitaker (1992).

When there are only two groups, the Cochran-Armitage test statistic reduces to the usual two-sample Z statistic

$$Z = \frac{\hat{\pi}_1 - \hat{\pi}_2}{\sqrt{\hat{\pi}(1 - \hat{\pi})(1/n_1 + 1/n_2)}}$$

However, with more than two groups, the Cochran-Armitage and two-sample Z tests differ, even when the d_j are chosen to reflect a two-sample comparison. For example, suppose the scores were $d_1 = -1, d_2 = 0$, and $d_3 = +1$ in a three-group experiment.

While these scores produce a test that is most sensitive to differences in groups 1 and 3, the denominator uses a pooled probability estimate from *all three* groups, making the test statistic different from the usual two-sample Z-test that compares groups 1 and 3.

This discrepancy between the test procedures can lead to drastically different inferences. For example, suppose there are three groups, with $n_1 = n_2 = n_3 = 500$, $\hat{\pi}_1 = .01$, $\hat{\pi}_2 = .42$, and $\hat{\pi}_3 = .05$. When testing $H\colon \pi_3 - \pi_1 = 0$ using the two-sample Z procedure, the pooled probability estimate is $\hat{\pi} = .03$ and the test statistic a highly significant

$$Z = \frac{.05 - .01}{\sqrt{.03(1 - .03)(1/500 + 1/500)}} = 3.71$$

However, if the Cochran-Armitage procedure is used with $d_1 = -1, d_2 = 0$, and $d_3 = +1$, the pooled probability estimate is $\hat{\pi} = .16$, and the test statistic is insignificant, with

$$T = \frac{-1(5 - 80) + 0(210 - 80) + 1(25 - 80)}{\sqrt{.16(1 - .16)[500(-1)^2 + 500(0)^2 + 500(1)^2]}} = 1.73$$

The difference in pooled probability estimates causes the problem: had the pooled probability estimate for the Cochran-Armitage test also been .03, the results of the two procedures would have been identical.

The trend statistic can also mislead by declaring statistical significance too frequently. Again suppose there are three groups, with $n_1 = n_2 = n_3 = 500$, but $\hat{\pi}_1 = .14$, $\hat{\pi}_2 = .01$, and $\hat{\pi}_3 = .18$. When testing $H\colon \pi_3 - \pi_1 = 0$ using the two-sample Z procedure, the pooled probability estimate is $\hat{\pi} = .16$ and the test statistic is insignificant at the two-sided $\alpha = .05$ level:

$$Z = \frac{.18 - .14}{\sqrt{.16(1 - .16)(1/500 + 1/500)}} = 1.73$$

The Cochran-Armitage procedure is used with $d_1 = -1$, $d_2 = 0$, and $d_3 = +1$, but the pooled probability estimate is $\hat{\pi} = .11$. In this case the test statistic is *significant*, with

$$T = \frac{-1(70 - 55) + 0(5 - 55) + 1(90 - 55)}{\sqrt{.11(1 - .11)[500(-1)^2 + 500(0)^2 + 500(1)^2]}} = 2.02$$

Again, the difference in pooled probability estimates causes the problem. Had there been many more groups with small probability estimates, the pooled estimate $\hat{\pi}$ would have been even smaller, and the difference between the two statistics would be further exagerrated.

Even though specific contrasts or trends may be specified using the Cochran-Armitage method, it is safest to consider the resulting tests as tests of the complete null hypothesis, since the variance is estimated using data from all treatment groups. The same holds true for multiple-group rerandomization procedures, discussed below. This problem may be alleviated using variance-stabilizing transformations, such as the arcsine transformation.

5.2.3 Arcsine-Transformed Procedures

The problems with the Cochran-Armitage method stem from the fact that variance of the sample proportion $\hat{\pi}_i$ depends on the value of π_i. To avoid these problems one might transform the data, making variances of the estimates $\hat{\pi}_i$ free of the underlying parameters π_i. The *arcsine transformation*

$$f_{AS}(p) = \arcsin(\sqrt{p}) \tag{5.4}$$

can be used for this purpose. A variant and slight improvement of this transformation is the *Freeman-Tukey double-arcsine transformation*, also called the *average angular transformation*, discussed by Freeman and Tukey (1950). If a proportion π is estimated using the fraction of occurrences y/n, then the double arcsine transformation is defined as

$$f(y,n) = \arcsin\{[y/(n+1)]^{1/2}\} + \arcsin\{[(y+1)/(n+1)]^{1/2}\} \tag{5.5}$$

Miller (1978) gives the inverse of this function, from which the sample proportion y/n may be recovered from a given value of the transformation $f(y,n)$:

$$y/n = 1/2\{1 - \text{sgn}(\cos f)[1 - (\sin f + (\sin f - 1/\sin f)/n)^2]^{1/2}\}$$

where

$$\text{sgn}(x) = \begin{cases} -1, & \text{for } x < 0 \\ 0, & \text{for } x = 0 \\ 1, & \text{for } x > 0 \end{cases}$$

For large n, the transformation $f(y,n)$ is essentially equivalent to $2 \times f_{AS}(y/n)$, which estimates $2f_{AS}(\pi)$. Freeman and Tukey (1950) show that if Y has the binomial distribution $\mathcal{BIN}(\pi, n)$, then the variance of $f(Y,n)$ is approximately

$$\text{Var}(f(Y,n)) \approx \frac{1}{n+.5}$$

which does not depend on π. Thus, when comparing several treatment groups using a particular contrast hypothesis $H: \sum_i c_i f_{AS}(\pi_i) = 0$, where $\sum c_i = 0$, the following test statistic may be used:

$$T = \frac{\sum_i c_i f(Y_i, n_i)}{\sqrt{\sum_i c_i^2/(n_i + .5)}}$$

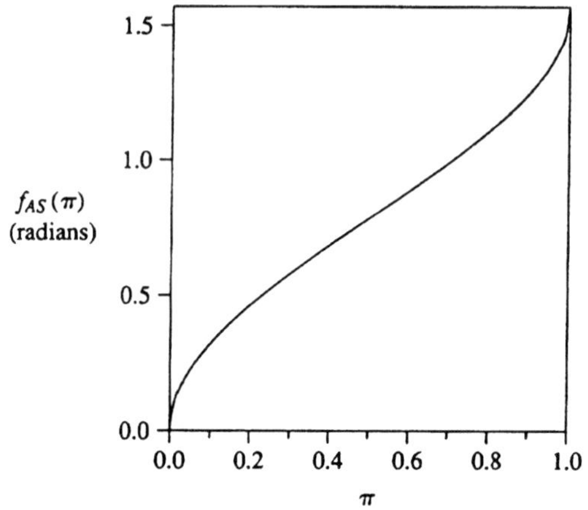

FIGURE 5.1. Arcsine transformation.

These test statistics are approximately standard normal under the hypotheses $H: \sum_i c_i f_{AS}(\pi_i) = 0$. When the contrast coefficients c_i are chosen to compare two groups only, the hypotheses $H: \sum_i c_i f_{AS}(\pi_i) = 0$ and $H: \sum_i c_i \pi_i = 0$ are equivalent. Otherwise, the hypotheses are different. Figure 5.1 illustrates why: the arcsine function emphasizes differences at the extremes of the [0, 1] interval range. For example, $f_{AS}(.02) - f_{AS}(.01) = .142 - .100 = .042$, more than four times as large as $f_{AS}(.52) - f_{AS}(.51) = .805 - .795 = .010$.

As is the case with the Cochran-Armitage statistic, the global null hypothesis is a special case of a specific contrast hypothesis, arcsine transformed or not.

If the contrasts $c_i = n_i(d_i - \bar{d})$ are chosen, the variance formula becomes

$$\operatorname{Var}\left(\sum_i c_i f(Y_i, n_i)\right) \approx \sum_i \frac{n_i^2(d_i - \bar{d})^2}{n_i + .5}$$

which is very similar to the variance of the Cochran-Armitage trend statistic, excluding the $\hat{\pi}(1 - \hat{\pi})$ term.

Unlike the Cochran-Armitage statistic, the Freeman-Tukey trend statistic is not sensitive to different pooled probability estimates. Consider, as in Section 5.2.2, three groups, and the contrast defined by $c_1 = -1$, $c_2 = 0$, and $c_3 = +1$. The resulting test statistic using the Freeman-Tukey statistic remains the same whether or not data from group 2 is included.

When data are sparse, the distribution of the Freeman-Tukey arcsine-transformed statistic is more normal than the distribution of the Cochran-Armitage statistic. However, both methods break down with extremely sparse data, so extra caution is urged in such situations. The typical "rule of thumb" given for sparseness is $\min[n\pi, n(1 - \pi)] < 5$, but this rule is probably too restrictive. To better understand the operating characteristics of the procedure, one should simulate the performance of the method for a variety of underlying strike rates, using sample sizes similar to those under study. It turns out that methods such as the Cochran-Armitage test, or its arcsine-transformed version, tend to be conservative when strike rates are very low.

However, on occasion, they may be anticonservative, and the utility of simulation analysis is to identify the problem and its potential magnitude, so one may better interpret the results.

5.2.4 Randomization Tests

While normal approximations for p-values of the Cochran-Armitage and Freeman-Tukey tests can be inaccurate for sparse cells, the rerandomized p-values are *exact* under the randomization null hypothesis that all permutations of the binary observations are equally likely. The general ideas for randomization and rerandomization analysis given in Chapter 3 and Chapter 4 apply equally here—binary data are just a special case of the more general structure. The main difference is that rerandomization p-values and their associated distributions may be computed more easily in the binary case.

Corresponding to the Cochran-Armitage statistic, the test statistic

$$T = \sum d_i Y_i$$

is frequently used in randomization analysis. (It is not necessary to center or scale the statistic as indicated in equations (5.1)–(5.3) since the rerandomized p-value is invariant under such transformations, Davis, 1986).

If the d_i are chosen to test for an increasing trend, then the null hypothesis is rejected when T is large. The rerandomized p-value is

$$p = \Pr(T \geqslant t \,|\, \text{all permutations are equally likely})$$

Here, T denotes the random trend statistic, whose distribution is determined by the randomization of experimental units to the different treatment groups, and t denotes the value actually observed in the experiment. Replacing $T \geqslant t$ with $T \leqslant t$ results in the lower-tailed p-value, used to test for a decreasing trend.

The alternatives may be two-tailed as well. In this case there is some controversy over whether to double the tail probability or to use the tail area of the "reflected score." To define the reflected score, note that the expectation of the trend T is given by

$$\mathrm{E}(T) = \sum d_i n_i \hat{\pi}$$

where $\hat{\pi} = \sum y_i / \sum n_i$, the pooled probability estimate. If the observed value t_{obs} is greater than $\mathrm{E}(T)$, then the distance $t_{\text{obs}} - \mathrm{E}(T)$ measures how "extreme" the value is, in the upper tail of the distribution. It seems natural to assign the same level of "extremity" to those values t in the lower tail, for which $t \leqslant \mathrm{E}(T) - [t_{\text{obs}} - \mathrm{E}(T)]$, i.e., when $t \leqslant 2\mathrm{E}(T) - t_{\text{obs}}$. The reflected two-sided permutational p-value is calculated as $p = \Pr(T \geqslant t_{\text{obs}}) + \Pr(T \leqslant 2\mathrm{E}(T) - t_{\text{obs}})$. If t_{obs} is less than $\mathrm{E}(T)$, the reflected two-sided permutational p-value would be $p = \Pr(T \leqslant t_{\text{obs}}) + \Pr(T \geqslant 2\mathrm{E}(T) - t_{\text{obs}})$, and if $t_{\text{obs}} = \mathrm{E}(T)$, then the two-sided p-value is simply 1.0. Currently, PROC MULTTEST computes rerandomized two-tailed tests using the "reflected score" approach.

Yates (1984) discusses these concepts, and recommends doubling the tail probability instead. In other words, he recommends $p = 2\Pr(T \geqslant t_{\text{obs}})$ when $t_{\text{obs}} \geqslant \text{E}(T)$, and $p = 2\Pr(T \leqslant t_{\text{obs}})$ when $t_{\text{obs}} \leqslant \text{E}(T)$. Such p-values are critized since they do not necessarily refer to the probability of *any* set of experimental outcomes. See Yates (1984) for further discussion.

The rerandomized p-value can be computed analytically in many cases, without requiring resampling. When the n binary data points are permuted, the total Y_+ remains fixed, but the totals in each cell vary. Let Y_i^* be the number of occurrences in the ith cell following a rerandomization. The probability of observing any particular configuration $y_1^*, \ldots, y_g^*$ (where $0 \leqslant y_i^* \leqslant n_i$ and $\sum y_i^* = Y_+$) is calculated from the multivariate hypergeometric distribution as

$$\Pr(Y_1^* = y_1^*, Y_2^* = y_2^*, \ldots, Y_g^* = y_g^*) = \frac{\binom{n_1}{y_1^*}\binom{n_2}{y_2^*}\cdots\binom{n_g}{y_g^*}}{\binom{n}{Y_+}} \tag{5.6}$$

The p-value is the probability that the rerandomized trend T^* exceeds the observed trend t, and is given by

$$p = \sum_{\{\mathbf{y}^* \mid \sum d_i y_i^* \geqslant t\}} \Pr(Y_1^* = y_1^*, Y_2^* = y_2^*, \ldots, Y_g^* = y_g^*) \tag{5.7}$$

Fast and efficient methods for calculating (5.7) analytically have been developed and made available in various software packages. Cyrus Mehta and others have developed network algorithms for the computation of such p-values. These results have been published in research papers by Mehta and Patel (1983); Mehta, Patel, and Tsiatis (1984); Mehta, Patel, and Senchaudhuri (1988); and Mehta and Wei (1988). Some methods are commercially available in the *StatXact* software (CyTel Software Corporation, 137 Erie Street, Cambridge, MA 02139). PROC MULTTEST of SAS/STAT also provides exact rerandomized p-values of the form (5.7).

When there are only two groups, the randomization test reduces to *Fisher's exact test* (e.g., Yates, 1984) for comparing binomial proportions in two samples. When there are more than two groups, the randomization test does not reduce to the two-sample test for particular scoring coefficients; for example, use of the scorings $d_1 = -1, d_2 = 0, d_3 = +1$ in a three-group experiment does not produce the same result as the Fisher exact test for comparing groups one and three, since the variance of the test statistic depends on data from all three groups. Like the Cochran-Armitage test of Section 5.2.2, it is safest to assume that the complete null hypothesis is being tested.

5.3 MULTIPLE TESTING WITH BINARY DATA

Binary data can pose special problems in the estimation of a complete null distribution. In the continuous data applications of Chapters 3 and 4, such null distributions were estimated by resampling residuals from a regression or ANOVA model. This approach fails with binary data since the resampled data sets will contain positive and negative residual values, which are not 0–1 binary values. What is needed is a bootstrap resampling method which produces binary data sets.

Following Westfall's (1985) bootstrap method for constructing simultaneous confidence intervals for multivariate binomial data, Westfall and Young (1989a) introduce the bootstrap resampling method for multiple testing in multivariate binary applications. This method turns out to be very similar to permutation-based approaches given by Brown and Fears (1981), Heyse and Rom (1988), and Farrar and Crump (1988).

The permutation methods treat continuous and binary data similarly—the data are not centered in either case. However, the bootstrap method requires centering prior to pooling in the case of continuous data. For binary data, bootstrap resampling is done directly, without pooling.

Another special problem encountered with binary data is the failure of the subset pivotality condition when multiple comparisons are performed within a particular variable. Unlike the continuous data applications, in which subset pivotality fails only for randomization methods, the condition fails both for bootstrap and randomization methods when binary data are analyzed using Cochran-Armitage tests. The examples given in Section 5.2 demonstrate this problem. Suppose there are only three groups with equal sample sizes n, with occurrence probabilities π_1, π_2, and π_3. The pairwise hypotheses $H_1: \pi_1 = \pi_2$, $H_2: \pi_1 = \pi_3$, and $H_3: \pi_2 = \pi_3$ are to be tested using the appropriate Cochran-Armitage scoring coefficients. Under the complete null hypothesis, all test statistics are asymptotically (as n becomes large) distributed as $\mathcal{N}(0, 1)$.

On the other hand, suppose $\pi_1 = \pi_2 = \pi$, but $\pi_3 \neq \pi$. The Cochran-Armitage test for $H_1: \pi_1 = \pi_2$ is asymptotically distributed as

$$T_1 \underset{\text{(approx.)}}{\sim} \mathcal{N}\left(0, \tfrac{\pi(1-\pi)}{\pi'(1-\pi')}\right)$$

where $\pi' = \pi + (\pi_3 - \pi)/3$. Clearly, the distribution of the test statistic is not the same under the complete and partial null hypotheses.

The randomization method suffers as well: pairwise comparison p-values calculated under the complete randomization null hypothesis are not valid for subset randomization hypotheses. (This problem was noted also for randomization analysis of continuous ANOVA data in Chapter 3.)

On the other hand, the Freeman-Tukey arcsine method described in Section 5.2 avoids this defect. Thus, when performing pairwise comparisons using resampling-based multiplicity adjustment, the Freeman-Tukey tests are preferred over the Cochran-Armitage tests.

If the multiple tests concern *variables only* (not multiple comparisons or contrasts within variables), then the subset pivotality holds for the Cochran-Armitage, the Freeman-Tukey, and the randomization methods.

5.3.1 Bootstrap Resampling

When the treatment effects are identical, so are the underlying multivariate Bernoulli distributions: $\boldsymbol{\pi}_1 = \cdots = \boldsymbol{\pi}_g = \boldsymbol{\pi}$ and $\mathbf{D}_1 = \cdots = \mathbf{D}_g = \mathbf{D}$. We have $n = \sum n_i$

binary data vectors $\mathbf{B}_{ij}$, independent and identically distributed as $\mathcal{MVB}_v(\boldsymbol{\pi}, 1, \mathbf{D})$. These data vectors may be resampled using the ordinary bootstrap method: generate $\mathbf{B}_{11}^*, \ldots, \mathbf{B}_{gn_g}^*$ *with replacement* from the combined data $\mathbf{B}_{11}, \ldots, \mathbf{B}_{gn_g}$.

The resulting bootstrap data vectors are guaranteed to be multivariate binary. The probability distribution of the $\mathbf{B}_{ij}^*$'s is $\mathcal{MVB}_v(\hat{\boldsymbol{\pi}}, 1, \hat{\mathbf{D}})$, where $\hat{\boldsymbol{\pi}} = \sum_{i,j} \mathbf{B}_{ij}/n$, and where $\hat{\mathbf{D}}$ is the estimated dependence structure. Letting $\mathbf{b} = (b_1, \ldots, b_v)$ denote an arbitrary binary v-tuple, we have

$$\Pr(\mathbf{B}_{ij}^* = \mathbf{b}) = \frac{\#\text{ of } \{\mathbf{B}_{ij}^* = \mathbf{b}\}}{n}$$

Our multiplicity adjustments may be carried out within the framework of this resampling distribution. Univariate tests, such as Cochran-Armitage or Freeman-Tukey, may be specified for each variable. If c such trends (or contrasts) are considered, one for each variable, then there are $c \times v$ tests in total. As in Chapter 1, Section 1.6, these test statistics and their associated p-values may be arranged in a $(c \times v)$ summary test statistic matrix $\mathbf{T}$. The p-values may be adjusted according to the distribution of the minimum p-value as computed via bootstrapping. Algorithm 2.5 or 2.8 of Chapter 2 may be applied for single-step or step-down adjustments, respectively; both algorithms are available in PROC MULTTEST.

When performing bootstrap resampling, the unadjusted p-values are usually computed using the large-sample tests such as the two-sample Z, the Cochran-Armitage, or the Freeman-Tukey arcsine test. Permutation-style tests also may be used as base tests for the bootstrap adjustments, but it is more logical and consistent to use permutation resampling if permutation tests are used as base tests.

Example 5.1 Adverse Events in Clinical Trials

New drugs are commonly tested on human subjects for evidence of safety and efficacy. From the standpoint of safety, the detection of possible *adverse events*, sometimes called *side effects*, is of interest. It is not unusual for drugs to have side effects, but it is sometimes difficult to distinguish side effects from chance phenomena because of the number of side effects usually examined and the variability of the process. It is particularly difficult to detect unexpected side effects.

Table 5.1 is reproduced from Westfall and Young (1989a), and gives the incidences of 28 different side effects, denoted as S_1–S_{28}, measured on four treatment groups of about 40 patients each in an actual clinical trial. For the purposes of confidentiality, the actual types of adverse events are not given, but the list includes standard items such as headaches and nausea. Some patients reported side effects that were not anticipated; these unanticipated side effects are also included in the list.

The four treatments are low and high dose levels of two drugs: A_1, A_2, and B_1, B_2, with actual names of compounds disguised again to insure confidentiality. The questions of interest were the following: First, are the two drugs different with respect to any side effects, and second, for each drug, is there a difference between dose levels? These questions can be formulated by the three linear contrasts $-A_1 - A_2 + B_1 + B_2$ (A vs. B), $-A_1 + A_2$ (A dosage effect), and $-B_1 + B_2$ (B dosage effect).

Table 5.1 Adverse Events in a 4 Group Clinical Trial*

Adverse Event	A_1	A_2	B_1	B_2	Total Incidence
S_1	4	4	12	13	33
S_2	5	4	4	4	17
S_3	3	2	3	3	11
S_4	1	4	1	3	9
S_5	0	2	3	2	7
S_6	0	0	2	2	4
S_7	1	2	2	0	5
S_8	0	0	2	3	5
S_9	1	2	1	0	4
S_{10}	0	0	2	0	2
S_{11}	1	0	0	0	1
S_{12}	1	1	0	0	2
S_{13}	0	1	0	0	1
S_{14}	1	1	0	0	2
S_{15}	0	0	2	0	2
S_{16}	0	1	0	1	2
S_{17}	1	0	0	0	1
S_{18}	1	0	0	0	1
S_{19}	1	1	0	0	2
S_{20}	0	0	0	1	1
S_{21}	0	1	1	0	2
S_{22}	1	0	0	0	1
S_{23}	0	0	1	0	1
S_{24}	0	1	0	0	1
S_{25}	0	1	0	0	1
S_{26}	0	1	0	0	1
S_{27}	1	0	0	0	1
S_{28}	14	20	19	17	70
Total Patients	41	39	40	40	160

*From Westfall and Young (1989a), with permission of the American Statistical Association.

Dosing and previous treatment experience suggest that event S_1 differs for the treatment groups; the unadjusted Freeman-Tukey test for the A vs. B comparison supports this claim with $p = .001$. The remaining adverse events, S_2 through S_{27}, are a standard list of other possible side effects, and variable S_{28} is a composite variable noting whether an individual had *any* side effect. Given the *a priori* knowledge concerning variable S_1, it would be reasonable to leave this variable out of the family of tests over which multiplicity adjustment is to be performed. We will perform the analysis two ways, leaving S_1 in, for a total of $28 \times 3 = 84$ tests, and removing S_1, leaving $27 \times 3 = 81$ tests.

For each test, the arcsine-transformation procedure of Section 5.2.3 was done using the scoring coefficients $(-1, -1, 1, 1)$, $(-1, 1, 0, 0)$, and $(0, 0, 1, -1)$. The alternatives

Table 5.2 Two-Tailed *p*-Values for Adverse Events Data

Adverse Event	*A* vs. *B*			Dose Effect, *A*			Dose Effect, *B*		
	Raw_p	Adj_p	Sid_p	Raw_p	Adj_p	Sid_p	Raw_p	Adj_p	Sid_p
S_1	.001*	.016*	.073	.939	1.000	1.000	.813	1.000	1.000
S_2	.813	1.000	1.000	.798	1.000	1.000	1.000	1.000	1.000
S_3	.760	1.000	1.000	.714	1.000	1.000	1.000	1.000	1.000
S_4	.782	1.000	1.000	.166	1.000	1.000	.340	1.000	1.000
S_5	.171	1.000	1.000	.117	0.987	1.000	.669	1.000	1.000
S_6	.031*	.567	.922	.986	1.000	1.000	1.000	1.000	1.000
S_7	.475	1.000	1.000	.565	1.000	1.000	.126	0.998	1.000
S_8	.014*	.271	.683	.986	1.000	1.000	.669	1.000	1.000
S_9	.277	1.000	1.000	.565	1.000	1.000	.316	1.000	1.000
S_{10}	.280	1.000	1.000	.986	1.000	1.000	.126	0.998	1.000
S_{11}	.484	1.000	1.000	.331	1.000	1.000	1.000	1.000	1.000
S_{12}	.156	1.000	1.000	.966	1.000	1.000	1.000	1.000	1.000
S_{13}	.473	1.000	1.000	.302	1.000	1.000	1.000	1.000	1.000
S_{14}	.156	1.000	1.000	.966	1.000	1.000	1.000	1.000	1.000
S_{15}	.280	1.000	1.000	.986	1.000	1.000	.126	0.998	1.000
S_{16}	.993	1.000	1.000	.302	1.000	1.000	.316	1.000	1.000
S_{17}	.484	1.000	1.000	.331	1.000	1.000	1.000	1.000	1.000
S_{18}	.484	1.000	1.000	.331	1.000	1.000	1.000	1.000	1.000
S_{19}	.156	1.000	1.000	.966	1.000	1.000	1.000	1.000	1.000
S_{20}	.479	1.000	1.000	.986	1.000	1.000	.316	1.000	1.000
S_{21}	.993	1.000	1.000	.302	1.000	1.000	.316	1.000	1.000
S_{22}	.484	1.000	1.000	.331	1.000	1.000	1.000	1.000	1.000
S_{23}	.479	1.000	1.000	.986	1.000	1.000	.316	1.000	1.000
S_{24}	.473	1.000	1.000	.302	1.000	1.000	1.000	1.000	1.000
S_{25}	.473	1.000	1.000	.302	1.000	1.000	1.000	1.000	1.000
S_{26}	.473	1.000	1.000	.302	1.000	1.000	1.000	1.000	1.000
S_{27}	.484	1.000	1.000	.331	1.000	1.000	1.000	1.000	1.000
S_{28}	.765	1.000	1.000	.126	0.998	1.000	.659	1.000	1.000

*p-value $< .05$.

were two-sided in all cases, and the standard normal approximation was used to compute the unadjusted p-values. Bootstrapping from the pooled binary data vectors, 10,000 resampled data sets were generated, and the arcsine-transformed test statistics were re-computed. Table 5.2 gives the resulting p-values and *free* step-down adjustments, all computed using PROC MULTTEST. The headings "Raw_p," "Adj_p," and "Sid_p" refer to the unadjusted Freeman-Tukey p-value, the resampling-based step-down adjustment, and the Šidák step-down adjustment, respectively.

We conclude from Table 5.2 that the suspected effect, S_1, remains significant even after multiplicity adjustment. Otherwise, there are no surprising adverse events. The raw p-values .014 (A vs. B, S_8) and .031 (A vs. B, S_6) are statistically significant, but such small p-values are not surprising when considered in the context of 84 tests. The adjusted p-values, .271 and .567, show that p-values as small as .014 occur frequently when 84 tests are performed.

If one wished to be less conservative, the suspected variable S_1 could be excluded from the analysis, making a total of $27 \times 3 = 81$ tests. When these tests are excluded and the data re-analyzed, all effects are insignificant. The tests have only slightly less conservative adjusted p-values—.257 instead of .271 for A vs. B, S_8, and .562 instead of .567 for A vs. B, S_6.

Rather than perform multiplicity adjustment over a collection of tests, one may prefer a *meta-analysis* test (or *omnibus* test) to *combine*, rather than *isolate* individual tests. The unadjusted test for variable S_{28}, a combined variable, is such a meta-analytic test. Although the suspected (and highly significant) variable S_1 affects the combined variable S_{28}, the tests for the combined variable are insignificant *even before* multiplicity adjustment.

This analysis essentially duplicates the analysis of Westfall and Young (1989a), with the exceptions that (1) the Freeman-Tukey tests are used, rather than the Cochran-Armitage tests, and (2) step-down tests are used, rather than single-step tests. The Freeman-Tukey test statistics are preferable both for identifying specific pairwise contrasts and from the pivotality standpoint, and the step-down method produces more powerful tests. Despite the different methods used, the conclusions are the same for this data. □

The composite variable, S_{28}, is statistically redundant; however, it addresses a relevant question. One beauty of the vector resampling approach is that the dependence structure induced by the redundancy is incorporated in the adjusted p-values.

Another consideration in the analysis of such data is that the extreme sparseness of side effects for many variables makes it difficult to detect significances. Following a decision made *a priori*, the data could more reasonably be analyzed by consolidating variables by type or by composite rare events.

In this example, there were many such sites with rare frequencies. Perhaps the major benefit of performing resampling-based adjustment on multivariate binary data is that sites with rare frequencies contribute less (or not at all) to the adjusted p-values. This is very noticeable when one compares the resampling-based adjustments with the Šidák adjustments. For the most significant effect, A vs. B in variable S_1, the Šidák adjustment is insignificant at $\tilde{p} = .073$, whereas the resampling-based adjustment is significant at $\tilde{p} = .016$. The Šidák adjustment is based upon the assumptions that p-values are independent and uniformly distributed; however, the actual p-value distribution differs markedly from the uniform distribution when data are sparse.

To illustrate the lack of uniformity, consider the resampling distributions of the Freeman-Tukey p-values in the previous example. Variables S_{27} and S_{21} are sparse, with one and two occurrences, respectively. The pooled probability estimates are $\hat{\pi}_{27} = 1/160 = .00625$ and $\hat{\pi}_{21} = 2/160 = .0125$. Variable S_8 is only moderately sparse, with five occurrences and a pooled probability estimate $\hat{\pi}_8 = 5/160 = .03125$; and variable S_1 has a moderate probability of occurrence with $\hat{\pi}_1 = 33/160 = .20625$. The marginal bootstrap distributions for the A vs. B test statistics may be simulated by generating data for each of the four groups as binomial random variables with the given sample sizes and estimated probabilities.

The cumulative distribution functions of the p-values were estimated using 40,000 simulated data sets, and the results are displayed in Figure 5.2. All p-values were

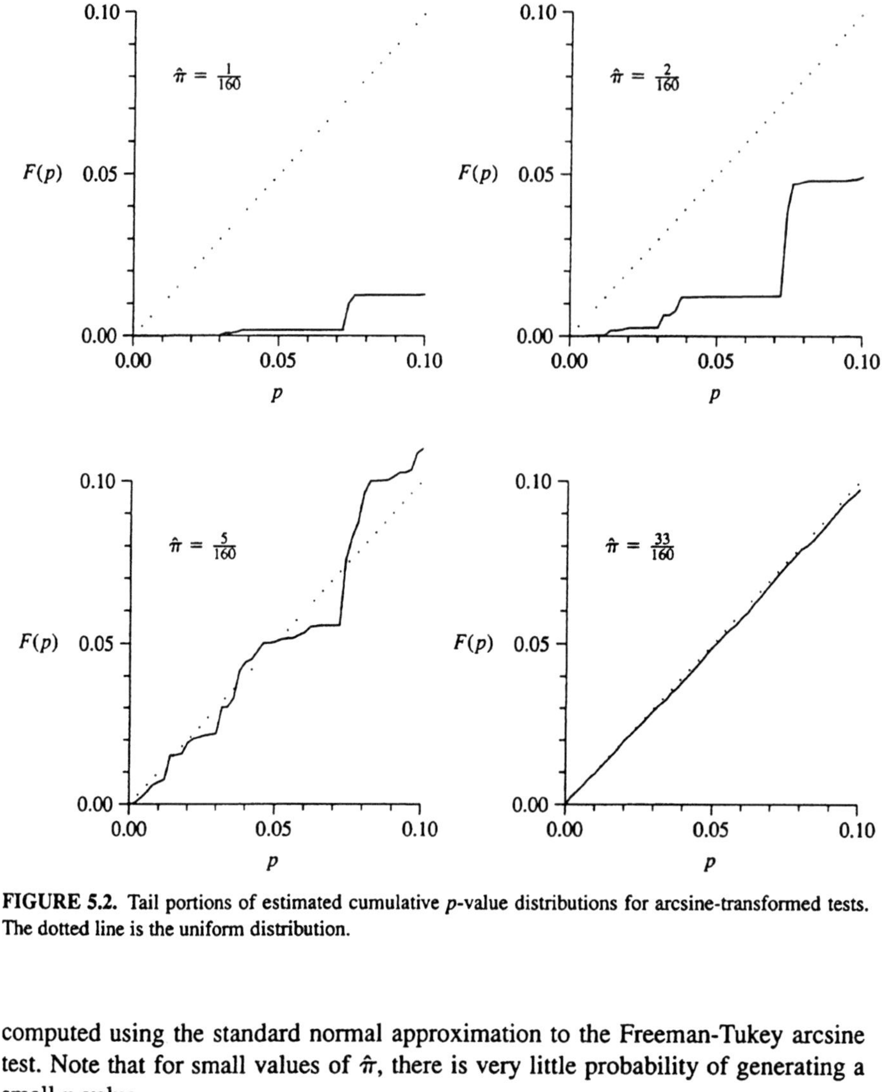

FIGURE 5.2. Tail portions of estimated cumulative p-value distributions for arcsine-transformed tests. The dotted line is the uniform distribution.

computed using the standard normal approximation to the Freeman-Tukey arcsine test. Note that for small values of $\hat{\pi}$, there is very little probability of generating a small p-value.

The multiplicity adjustments use the distribution of the minimum p-value. As Figure 5.2 demonstrates, sites with small probability estimates are unlikely to generate small p-values. Therefore, such sites will have little impact upon the multiplicity adjustment of small p-values. When there are zero occurrences for a particular variable, then $\hat{\pi} = 0$, and the variable will not contribute at all to the multiplicity adjustment in the current experiment.

That sites with small probabilities contribute less to multiplicity adjustments has been recognized as a desirable property. Tarone (1990) suggests modifying the Bonferroni method to account for rare marginal frequencies. A nice feature of resampling methods is that such modifications are automatic.

5.3.2 Permutation Resampling

Using notation similar to that in Section 4.1 where we discussed continuous data, the complete randomization null hypothesis for binary data may be stated as

> H: All random permutations of the observed vectors $\mathbf{B}_{11}, \mathbf{B}_{12}, \ldots, \mathbf{B}_{gn_g}$ into the g groups, with sample sizes n_i, are equally likely.

Thus, resampling-based analysis may be performed by *randomly shuffling* the data vectors. In other words, the rerandomized binary data vectors $\mathbf{B}_{ij}^*$ are obtained by sampling the observed pooled data vectors $\mathbf{B}_{11}, \ldots, \mathbf{B}_{gn_g}$ *without replacement.* Brown and Fears (1981), Heyse and Rom (1988), and Farrar and Crump (1988) use "exact" permutation tests as base tests, and consider resampling-based adjustments for these tests.

In permutation-based resampling we generate vectors $\mathbf{B}_{11}^*, \ldots, \mathbf{B}_{gn_g}^*$ *without replacement* from the combined data $\mathbf{B}_{11}, \ldots, \mathbf{B}_{gn_g}$. The marginal totals are fixed for all resampled data sets:

$$\sum_{i=1}^{g}\sum_{j=1}^{n_i} \mathbf{B}_{ij}^* = \sum_{i=1}^{g}\sum_{j=1}^{n_i} \mathbf{B}_{ij}$$

Therefore, the permutation distribution of all the trend tests is the same for all resampled data sets. Calculation of p-values for randomized trend tests can be time-consuming, but the fact that all permutation distributions are identical allows some savings: the distributions need only be tabulated for the original data set, and the p-values for all resampled data sets can be determined from such tables.

For example, consider the multivariate binary data in Table 5.3. There are three groups and three variables. For each variable, the trend statistic is

$$\text{Trend} = 0 \times Y_1 + 1 \times Y_2 + 2 \times Y_3$$

where Y_j denotes the number of occurrences for treatment group j. Using the exact method given in Section 5.2.4, upper-tailed p-values are computed for each variable; the resulting probabilities are given in Table 5.4. Given a particular data set containing binary vectors, sampled without replacement, the totals can be computed easily, and the resampled p-values can be obtained from the given table. One random permutation of the data is given in Table 5.4, with p-values calculated from Table 5.5. It is the occurrence of small p-values such as the .0567 for variable B_1 that brings home to investigators the fact that Type I errors can and do happen.

Adjusted p-values are calculated by generating many data sets as in Table 5.4, and by using either Algorithm 2.5 or 2.8 for single-step or step-down adjustments, respectively. For this data, 20,000 data sets were generated, and the step-down adjusted p-values were computed to be .5555, .0456, and .6216 for variables B_1, B_2, and B_3, respectively. The smallest unadjusted p-value, $p = .01942$ for B_2, remains significant after adjustment.

Table 5.3 Binary Vectors for Permutation Analysis

Treatment	B_1	B_2	B_3
I	1	0	0
	1	0	0
	0	0	1
	1	0	0
	0	1	0
	0	1	1
II	1	0	0
	1	0	0
	1	0	1
	1	1	0
	1	1	0
	1	1	1
III	1	1	0
	1	1	0
	0	1	1
	1	1	0
	1	1	0
	0	1	1
Total	13	11	6
Trend	14	15	6
p-value	.3789	.0194	.6163

Table 5.4 A Random Permutation of Binary Vectors

Treatment	B_1	B_2	B_3
I	1	1	0
	1	0	1
	0	1	1
	0	1	0
	0	1	1
	1	1	0
II	1	1	0
	0	0	1
	1	0	0
	1	1	1
	0	1	1
	1	0	0
III	1	0	0
	1	0	0
	1	0	0
	1	1	0
	1	1	0
	1	1	0
Total	13	11	6
Trend	16	9	3
p-value	.0567	.9259	.9840

Table 5.5 Permutation Distributions for Trend Statistics

S_1		S_2		S_3	
Trend	p-Value	Trend	p-Value	Trend	p-Value
8	1.00000	5	1.00000	0	1.00000
9	.99930	6	.99981	1	.99995
10	.98880	7	.99679	2	.99801
11	.94328	8	.98058	3	.98395
12	.82423	9	.92590	4	.93331
13	.62115	10	.80524	5	.81211
14*	.37885	11	.61180	6*	.61625
15	.17577	12	.38820	7	.38375
16	.05672	13	.19476	8	.18789
17	.01120	14	.07410	9	.06669
18	.00070	15*	.01942	10	.01605
		16	.00321	11	.00199
		17	.00019	12	.00005

*Observed trend.

While resampling of vectors preserves the dependence structure, the adjusted p-values can be approximated by assuming independence. Heyse and Rom (1988) demonstrate this phenomenon using animal carcinogenicity data. Using the data and distributions of Tables 5.3 and 5.5, the independence-based adjusted p-value for the trend test in the B_2 variable is

$$\begin{aligned}\tilde{p} &= \Pr(\min P_j^* \leqslant .01942)\\ &= 1 - \Pr(P_1^* > .01942)\Pr(P_2^* > .01942)\Pr(P_3^* > .01942)\\ &= 1 - (1 - .01120)(1 - .01942)(1 - .01605)\\ &= .0460\end{aligned}$$

which is virtually identical to the resampling result, .0456. In this example, dependence structure has little effect. However, accounting for discreteness is *very* important. An ordinary Šidák adjustment results in $\tilde{p} = 1 - (1 - .01942)^3 = .0571$, which is too large because it does not account for discreteness.

Use of the independence assumption is appealing, as it provides an exact answer, with an additional assumption. Use of resampling provides an approximate answer (approximate because of Monte Carlo error), but without the independence assumption. Comparison of the two methods will ordinarily show that the resampling-based method gives smaller p-values, but often the difference is minor.

One must note that roundoff error can easily result in incorrect probability calculations of probabilities such as $\Pr(\min P_j^* \leqslant .01942)$ with discrete data. Here, the actual p-value is not .01942, but an infinitely repeating decimal, $p_{\min} = \Pr(T \geqslant 15 \mid \text{Total} = 11)$, that has been rounded to .01942. Further rounding to .0194 illustrates the problem: mathematically, $\Pr(\min P_j^* \leqslant .0194)$ will not include the discrete event $\min P_j^* = p_{\min}$ in its determination.

Caution in rounding off is extremely crucial in discrete applications. This problem can persist in spite of the high degree of precision afforded by modern software. Values which are mathematically equal may differ by a minuscule amount when calculated in slightly different ways, and the computer may call them "different." To alleviate the problem, algorithms should use a "fuzz" operator. Many software packages have a "fuzz" comparison operator, for which the binary relation "$x \leqslant y$" is equivalent to "$x \leqslant y + \delta$" where δ is an extremely small value, but larger than the machine's internal rounding error. Calculation of resampling-based adjusted p-values should be defined using such a "fuzz" comparison, particularly when dealing with discrete data. PROC MULTTEST of SAS/STAT performs all comparisons using "fuzz" comparisons. RESAMPLING STATS contains a FUZZ $< \delta >$ command, where δ is the fuzz value. GAUSS386 also has fuzzy comparisons functions, such as $d = FLE(x, y)$, for which

$$d = \begin{cases} 1, & \text{if } x < y + \delta \\ 0, & \text{if } x \geqslant y + \delta \end{cases}$$

The default is $\delta = 10^{-15}$, but this parameter may be changed.

Example 5.2 Animal Carcinogenicity Study
Westfall and Young (1989a) analyze data on tumor incidence for 38 organs of male rats. This data is taken from the National Toxicology Program's technical report 225 on Red 9 (1982). The control sample size is $n_1 = 50$; low dose sample size $n_2 = 50$; and the high dose group contains $n_3 = 49$ male rats. The data are summarized in Table 5.6, with $(0, 1, 2)$ scores for trend statistics. For each tumor site, a permutational upper-tailed p-value is reported, as well as a multiplicity-adjusted value obtained using 20,000 rerandomized data sets, all computed using PROC MULTTEST. Suspect tumor sites are #4–the spleen, and #5–the liver. There is a highly significant dose effect for tumors in the spleen. The effect is less pronounced for liver tumors, but is still significant even after multiplicity adjustment. □

Several comments concerning the preceding example are in order. First, in the original analysis of Westfall and Young (1989a), 38 tumor sites were included in the analysis. In this re-analysis of the data, all sites having zero observed tumors were discarded, leaving a total of 26. Deletion of these sites with zero observed tumors cannot affect the adjusted p-values since it is impossible to generate any permutational p-value other than 1.0 from such a site. Further reduction in variables is possible by eliminating sites with small marginal totals. In this application, when the total number of observed tumors is two, the minimum permutational p-value occurs when all occurrences are in the high dose group, Trend = 4. The permutational probability of this outcome is computed from the multivariate hypergeometric distribution as

$$\begin{aligned} &\Pr(\text{Trend} = 4 \,|\, \text{Total} = 2) \\ &= \frac{\binom{50}{0}\binom{50}{0}\binom{49}{2}}{\binom{149}{2}} \\ &= .1067 \end{aligned}$$

Table 5.6 NTP Tumor Data[1]

Tumor Site	Control $n_1 = 50$	Low Dose $n_2 = 50$	High Dose $n_3 = 49$	Trend	Raw p-Value	Adjusted p-Value
1	2	1	2	5	.5995	.9999
2	5	2	1	4	.9778	1.0000
3	2	0	1	2	.8504	1.0000
4	0	1	28	57	.0000*	.0000*
5	1	6	9	24	.0068*	.0413*
6	0	0	1	2	.3289	.9991
7	0	0	1	2	.3289	.9991
8	1	0	1	2	.6622	1.0000
9	0	0	1	2	.3289	.9991
10	8	8	5	18	.8432	1.0000
11	18	15	15	45	.7530	1.0000
12	6	7	5	17	.6649	1.0000
13	1	2	0	2	.8504	1.0000
14	0	4	1	6	.3858	.9991
15	2	1	0	2	.9637	1.0000
16	49	48	47	142	.7993	1.0000
17	0	1	0	1	.6644	1.0000
18	7	2	0	2	.9996	1.0000
19	1	1	1	3	.6259	.9999
20	0	0	1	2	.3289	.9991
21	1	0	0	0	1.0000	1.0000
22	1	0	0	0	1.0000	1.0000
23	2	0	1	2	.8504	1.0000
24	0	1	0	1	.6644	1.0000
25	0	1	0	1	.6644	1.0000
26	13	4	16	36	.2552	.9567

[1]From Westfall and Young (1989a), with permission of the American Statistical Association.
*p-value < .05.

Since no significant $\alpha = .05$ results may be found when there are two or fewer occurrences, these sites also may be excluded from the analysis without affecting the multiplicity adjustment. In the preceding example, there were ten such sites, and eliminating these will reduce the total to 16 sites.

Clearly, unthinking application of the Bonferroni or Šidák adjustments is not appropriate. For example, if one applied a step-down Bonferroni adjustment to the p-value for the liver tumors (tumor site #5 from Table 5.6), when all 38 sites were considered, the adjusted p-value would be $37 \times .0068 = .2516$. (Since this is the second-smallest p-value, the multiplier is 37 instead of 38 in a step-down adjustment.) Recognizing that sites with zero tumor frequencies cannot possibly contribute to the adjusted p-value, one may use $26 - 1$ as a multiplier, yielding the adjusted p-value $25 \times .0068 = .1700$, which is still too large. In fact, in this example, only sites with total tumors above 5 (and less than $149 - 5 = 144$) may generate a p-value less than or equal to .0068; therefore, only 11 possible sites may affect the adjusted p-value for liver tumors. Applying the correction (11 − 1) yields a Bonferroni adjustment of .0680, or a Šidák adjustment of .0660, much closer to the resampled value .0413, but still insignificant at $\alpha = .05$. Further discussion of modifying the Bonferroni correction is given by Tarone (1990). Rather than modify the Bonferroni correction, we prefer to perform multiplicity adjustment using resampling, which incorporates discreteness and dependence structures automatically.

Multiple comparisons within a tumor type are also of interest, and may be tested using Fisher exact tests. It is natural to consider upper-tailed comparisons of the Control groups vs. the various treated groups. Brown and Fears (1981), Farrar and Crump (1988), and Westfall and Young (1989a) considered methods for multiplicity adjustment in such cases, using rerandomization and/or bootstrap methods. Since the rerandomization distribution used by these authors was the complete null distribution, assuming no difference between any of the treatment groups for variable j, the resulting adjusted p-values are best seen as measures of evidence for the complete null hypothesis of no treatment effect in variable j. This is because pairwise comparison p-values calculated under the complete randomization null hypothesis are not valid under subset randomization hypotheses.

Nevertheless, the resulting p-values are measures of evidence against the complete null hypothesis in variable j which *are sensitive to* pairwise deviations from the complete null, and therefore are of interest. If the NTP data presented in Table 5.6 is analyzed using upper-tailed Fisher exact tests rather than randomization trend tests, the number of tests increases from 26 to 52. There is also much more dependence among the test statistics, since both comparisons are against the same control group. The adjusted p-values for this analysis were computed in step-down fashion, using PROC MULTTEST with 10,000 rerandomized data sets, and the results for the most extreme p-values are given in Table 5.7. We are still able to claim a statistical significance for the S_5 tumor site—liver tumors—even though the size of the family of tests has doubled. This demonstrates the importance of incorporating dependence structures into the multiplicity adjustment: without incorporating such dependence structures the adjusted p-value would likely be insignificant at the .05 FWE level.

On a more philosophical note, the common use of only upper tailed testing in these situations is questionable. There is no *a priori* basis for thinking that treatments can *only* cause an increase in tumor rates. Young and Gries (1984) noted simultaneous dose-related decreases and increases in tumor rates in rodent carcinogenicity studies. Weinberg and Storer (1985) note the simultaneous decrease and increase in

Table 5.7 Pairwise Comparisons in NTP Tumor Data

Comparison	Raw_p	Permutation Adj_p	Bootstrap Adj_p
C vs. H, S_4	.0000	.0000	.0000
C vs. H, S_5	.0071	.0451	.0434
C vs. L, S_5	.0559	.3529	.3712
C vs. L, S_{14}	.0587	.4408	.4620

tumor incidence in rodent studies and call compounds associated with such effects "ambiguous carcinogens." In this study the decrease at tumor site 18 is statistically more dramatic than the increase at tumor site 5, with a lower-tailed permutational trend p-value of .0023, compared to the permutational upper-tailed trend p-value of .0068 listed in Table 5.6 for S_5. In the case examined by Young and Gries, it appeared that the effects were linked within the animal: animals exhibiting one type of tumor were much less likely to exhibit the other. The analysis of multiple variables using only upper-tailed tests can obscure the fact that several variables may be related to the same underlying phenomenon. While the adjusted p-values are correct in the sense that they adjust for such correlations, restriction to upper-tailed significances may lead to an incomplete interpretation.

These comments echo Bailar (1991), who takes a more extreme point of view: "Use two-tail tests in statistical hypothesis testing. *Always*.*" This position may be somewhat extreme in carcinogenicity studies, where the lower tail is of little concern when the safety of a compound is on trial. One defense of one-tailed tests in the case of rodent long-term tests is that investigators are interested *only* in increases in tumors. Certainly the consequences of a declared increase or decrease are different. Nevertheless, a scientist would certainly like to know if a compound *reduces* tumors, and if these determinations are planned to be made from a rodent carcinogenicity study *in addition to* determinations of increasing tumors, then two-sided tests are indicated. At the very least, one might consider *allocating* the error rate differently to upper-tailed and lower-tailed alternatives, so it is more easy to detect increases. Methods for allocating error rates using resampling are discussed in Chapter 6.

5.4 COMPARISON OF BOOTSTRAP AND PERMUTATION METHODS

Table 5.7 shows the results of resampling the Fisher exact test using the bootstrap as well as rerandomization. The distinction is very minor in this example, as in many applications. However, in some situations the methods give different results. We consider these two methods in more detail, comparing them in univariate and multiple testing applications, and identifying similarities and discrepancies. We also compare the philosophical underpinnings of each method.

5.4.1 Univariate Testing Comparisons

How does the bootstrapped binomial test perform in the simplest of situations? Consider the univariate comparison of proportions from two independent samples, testing $H: \pi_1 = \pi_2$ versus $H': \pi_1 < \pi_2$. The data is $Y_1 \sim \mathcal{BIN}(\pi_1, n_1)$ and $Y_2 \sim \mathcal{BIN}(\pi_2, n_2)$, with Y_1 distributed independently of Y_2. If the data are pooled and resampled, then the bootstrapped observations will be generated as $Y_1^* \sim \mathcal{BIN}(\hat{\pi}, n_2)$ and $Y_2^* \sim \mathcal{BIN}(\hat{\pi}, n_2)$, where $\hat{\pi}$ is the pooled probability estimate, $\hat{\pi} = (Y_1 + Y_2)/(n_1 + n_2)$. Suppose a test statistic T is used for testing H, such as

*Reprinted from *Chance* 4(2) 27–38 (1991), with permission of Springer-Verlag and J. C. Bailar.

the two-sample Z-statistic or the arcsine-transformed test, with larger values of T suggesting the alternative hypothesis. Then the bootstrapped p-value is

$$p^* = \Pr(T^* \geqslant t \mid \hat{\pi})$$

where T^* is the resampled test statistic, and t is the observed value in the original experiment. In small sample situations this probability can be computed easily without resampling. Realizing that the test statistic is a function of the values y_1 and y_2, and that there are only $(n_1 + 1)(n_2 + 1)$ possible (y_1, y_2) outcomes, one simply identifies the set of (y_1, y_2) pairs for which $T^*(y_1, y_2) \geqslant t$, and computes the probability of the set. The probability of an individual (y_1, y_2) belonging to the set of such outcomes is

$$\begin{aligned} &\Pr\big((Y_1^*, Y_2^*) = (y_1, y_2) \mid \hat{\pi}\big) \\ &= \binom{n_1}{y_1} \hat{\pi}^{y_1}(1 - \hat{\pi})^{n_1 - y_1} \times \binom{n_2}{y_2} \hat{\pi}^{y_2}(1 - \hat{\pi})^{n_2 - y_2} \end{aligned} \tag{5.8}$$

For example, suppose $n_1 = 7, n_2 = 5$, and that the observed values are $y_1 = 1$ and $y_2 = 4$. Using the two-sample Z-test, we have $T = 2.2764$, with $p = .0114$ based on the normal approximation. The set of outcomes producing test statistics at least as extreme as 2.2764 are $(0, 3)$, $(0, 4)$, $(0, 5)$, $(1, 4)$, $(1, 5)$, and $(2, 5)$ (see Table 5.8). Using the binomial probability $\hat{\pi} = 5/12$, the probabilities of these outcomes total $p^* = .0226$, using (5.8) and summing over the six individual (y_1, y_2) outcomes shown in Table 5.8.

The bootstrap p-value differs, depending on the base statistic T chosen. If this statistic was chosen to be the difference of proportions $T = Y_2/n_2 - Y_1/n_1$, then a different p-value will usually result. When $n_1 = 7$, $n_2 = 5$, and the observed values are $y_1 = 1$ and $y_2 = 4$, the outcome $(0, 3)$ does *not* produce a more extreme T, unlike the case where T represents the two-sample Z-statistic. In this case, the bootstrap p-value would be .0169 instead of .0226.

On the other hand, rerandomization produces the Fisher exact test, whether the base statistic T is the two-sample Z-statistic, the difference of proportions, or any of

Table 5.8 Binomial Two-Sample Z-Statistics

	y_2					
y_1	0	1	2	3	4	5
0	—	1.2358	1.8330	2.3664	2.8983	3.4641
1	−0.8827	0.2619	1.0142	1.6562	2.2764	2.9277
2	−1.3093	−0.3381	0.4140	1.0887	1.7566	2.4744
3	−1.6903	−0.8281	−0.0990	0.5855	1.2867	2.0702
4	−2.0702	−1.2867	−0.5855	0.0990	0.8281	1.6903
5	−2.4744	−1.7566	−1.0887	−0.4140	0.3381	1.3093
6	−2.9277	−2.2764	−1.6562	−1.0142	−0.2619	0.8827
7	−3.4641	−2.8983	−2.3664	−1.8330	−1.2358	—

several test statistics discussed in Davis (1986). Using the given n_i and y_i data, the total $y_1 + y_2 = 5$ is fixed for all rerandomized data sets, and we wish to compute the probability that Y_2^* is at least 4, given that $Y_1 + Y_2 = 5$. We have

$$\begin{aligned} \Pr(Y_2^* \geqslant 4 \mid Y_1 + Y_2 = 5) \\ &= \frac{\binom{7}{1}\binom{5}{4} + \binom{7}{0}\binom{5}{5}}{\binom{12}{5}} \\ &= .0455 \end{aligned}$$

In this example, the Fisher exact test provides a larger p-value than the Z-test or the bootstrapped versions. In fact, p-values for rerandomization procedures with binary data are usually larger than those of parametric procedures such as the Z-test. Using the binomial model, the Fisher exact test is *always* conservative: assuming $H: \pi_1 = \pi_2 = \pi$ is true, the probability of observing a Fisher exact p-value smaller than α is always less than or equal to α, for all π. This problem has motivated the use of "mid p-values" when reporting the results of randomization tests (e.g., Lancaster, 1961). In this method, randomization p-values are made smaller by averaging the actual observed p-value with the next smaller p-value that could possibly be observed from the randomization distribution. Using the unconditional binomial model, the probability of observing a mid-p-value less than α should better approximate the nominal level α.

Figure 5.3 displays the degree of conservativeness of the Fisher exact test as a function of the underlying parameter π, for different sample size configurations. If the probability of observing a p-value less than .05 is less than .05, then the test is conservative. The probabilities

$$\Pr(P_{\text{Fisher}} \leqslant .05 \mid \pi)$$

are graphed assuming the binomial model. These are calculated by identifying all (y_1, y_2) locations for which the Fisher exact p-value is less than or equal to α, then calculating the probability of this set using (5.8), with $\hat{\pi}$ replaced by the actual assumed value of π. Surprisingly, the method remains quite conservative even for moderately large sample sizes. For example, with $n_1 = n_2 = 50$, the maximum rejection probability is only slightly larger than .03.

Bootstrap p-values can be computed using the Fisher exact test as a base test, just as they can be computed using the two-sample Z-test or any other statistic. Using the Fisher exact p-values, a bootstrap p-value may be computed analytically as

$$p^* = \Pr(P_{\text{Fisher}}^* \leqslant p_{\text{Fisher}} \mid \hat{\pi})$$

In words, this p-value is the sum of the binomial products (5.8) over all (y_1, y_2) which yield a Fisher exact p-value smaller than or equal to the observed value. In the example with $n_1 = 7$, $n_2 = 5$, $y_1 = 1$, and $y_2 = 4$, the set (y_1, y_2) yielding Fisher exact p-values smaller than or equal to the observed value .0455 is identical to that

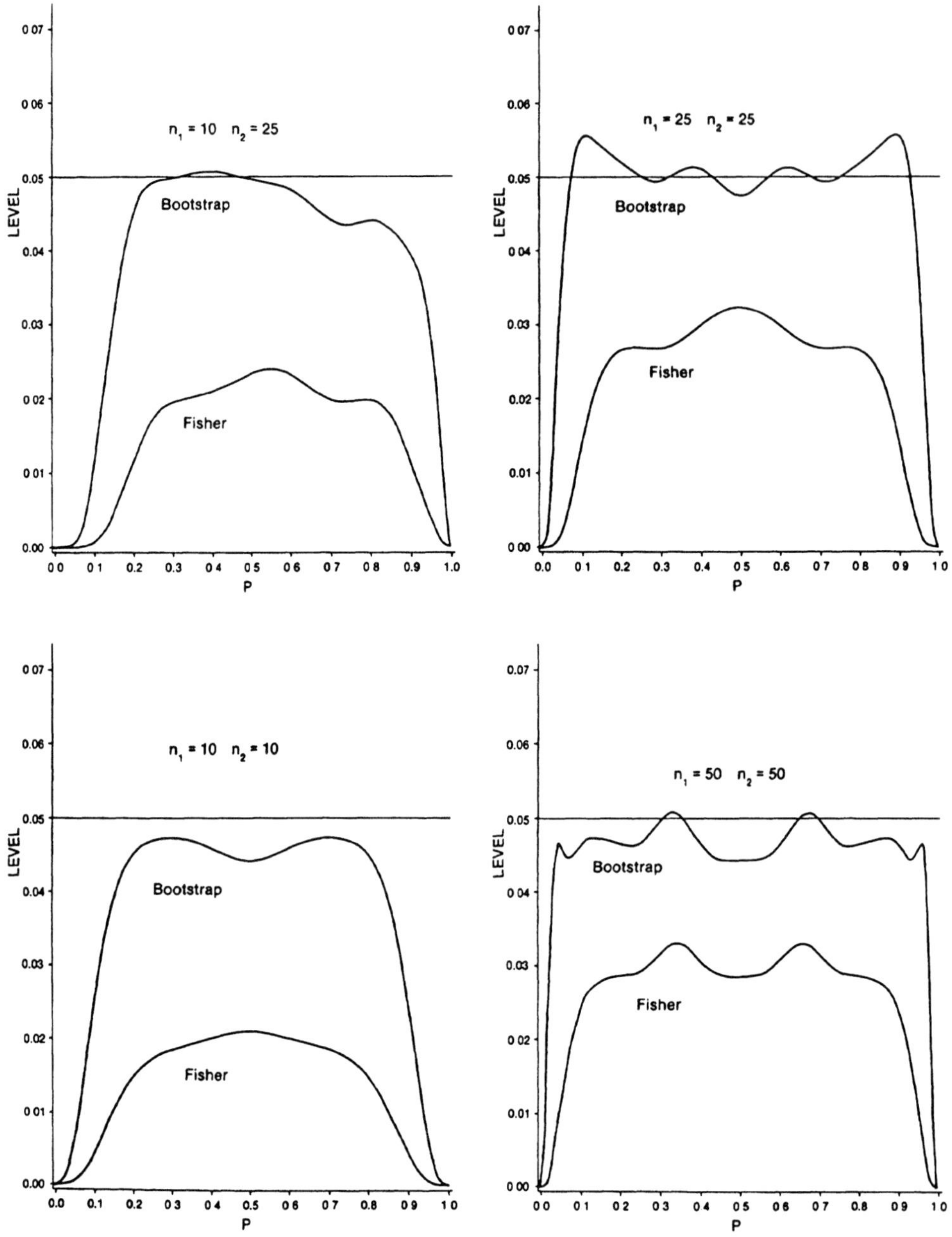

FIGURE 5.3. Actual significance levels (as functions of the true binomial probability, p) for nominal $\alpha = .05$ tests using upper-tailed Fisher exact tests and bootstrapped Fisher exact tests. Four different sample-size scenarios are given, with n_1 and n_2 denoting control and treatment sample sizes. Although it is sometimes biased, the actual size of the bootstrapped test better approximates the nominal $\alpha = .05$ level. From Westfall and Young (1989a), with permission of the American Statistical Association.

of the two-sample Z-test, namely (0,3), (0,4), (0,5), (1,4), (1,5), and (2,5). This bootstrapped p-value is equal to the corresponding bootstrapped Z-test p-value in this example; however, these p-values are often different.

Under product binomial sampling, the probability of observing a bootstrapped version of the Fisher exact p-value that is less than .05 may be computed as before with the Fisher exact p-value: first locate all (y_1, y_2) locations for which the bootstrapped p-value is less than or equal to α, then calculate the probability of this set using (5.8), with $\hat{\pi}$ replaced by the actual assumed value of π. These probabilities are graphed in Figure 5.3, showing that the actual rejection proportions are much closer to the nominal .05 level. Figure 5.3 also implies that the exact test has less power than the bootstrapped version: there are fewer observed values (y_1, y_2) yielding a significant $p \leqslant .05$ conclusion.

5.4.2 Multiple Testing Comparisons

It is more difficult to analytically compare the two methods in the multiple testing situation. The methods are easy to study via simulation, although this analysis is computer-intensive.

We consider a simulation experiment under the complete null hypothesis for which H: $\pi_{ij} \equiv \pi$ for $i = 1, 2$ groups and $j = 1, \ldots, 6$ variables. There are six binary variables and two treatment groups, so a total of six upper-tailed tests were performed for each simulated data set. All variables are generated independently—there is no dependence structure, but adjusted p-values are computed using vector resampling. In the multivariate normal case, use of vector resampling makes the adjustments somewhat anti-conservative when there is no dependence structure in the population.

Figure 5.4 displays the proportion of simulations for which *at least* one adjusted p-value was less than .05. Since the six null hypotheses are all true, this is the proportion of simulated data sets yielding *at least one* Type I error. The raw p-values were chosen to be the Fisher exact p-values in all situations, and the bootstrap and permutation resampling methods were compared. There were 1,000 simulated data sets, and all adjusted p-values were computed using 1,000 resampled data sets for each simulated data set. The "Null Hypothesis" graph in Figure 5.4 shows that the true FWE rate for both permutation and bootstrap resampling depends upon the underlying value of π. However, the bootstrap method rejects hypotheses more frequently, and its FWE more closely approximates the nominal level. This conclusion is similar to the univariate testing result shown in Figure 5.3: the true FWE of the bootstrap method more closely approximates the nominal FWE = .05 level.

Under the alternative hypothesis, the probabilities are specified to be $\pi_{1j} \equiv .01$ and $\pi_{2j} \equiv \pi_2$. The "Alternative Hypothesis" graph of Figure 5.4 displays proportions of samples for which at least one hypothesis H_j: $\pi_{1j} = \pi_{2j}$ was correctly rejected, for $\pi_2 > .01$. Again the bootstrap method rejects null hypotheses more frequently, and therefore has higher power.

In both the "Null Hypothesis" and the "Alternative Hypothesis" graphs of Figure 5.4, the bootstrap method rejects more frequently than the permutation method. This property is desirable in both cases. In the null case, the permutation method is too

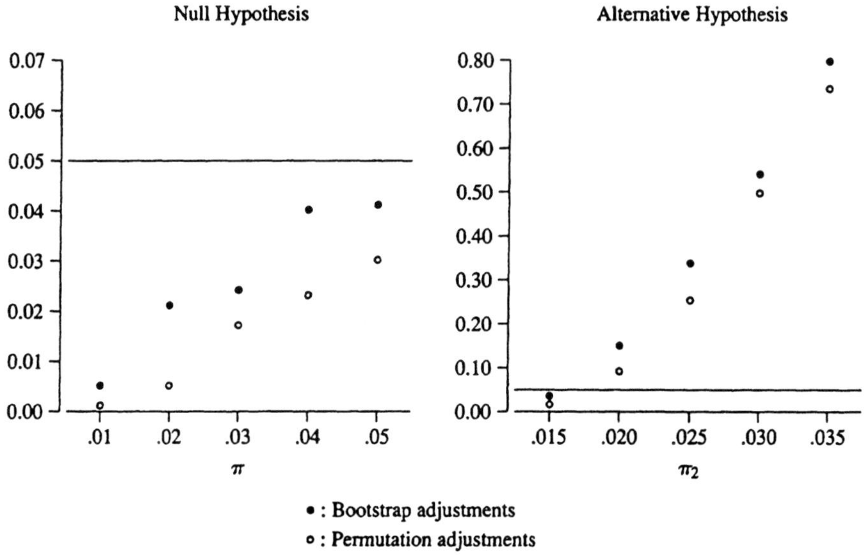

FIGURE 5.4. Rejection proportions with multiple binomial tests.

conservative, so rejecting hypotheses more often brings the true FWE closer in line with the nominal FWE. In the alternative hypothesis case, more rejections mean more correct decisions, again a desirable property.

The binomial distributions used in these simulations generate sparse data, particularly for the complete null configuration. In this case one is likely to find the largest difference between the bootstrap and permutation approach. As suggested by Table 5.7, the discrepancy between the methods is small in some applications. Our experience suggests that the methods are closer when the distribution of $\min_{1 \leq j \leq k} P_j$ is more continuous. As the sample size increases, there is a greater chance that a location will have a large total frequency, in turn causing the $\min_{1 \leq j \leq k} P_j$ distribution to be more continuous. This observation is supported by Romano (1989), who demonstrates asymptotic equivalence of bootstrap and permutation resampling in a related application.

5.4.3 Philosophical Comparisons

The conservativeness of the Fisher exact test, and of randomization procedures in general, is well known. Whether this is an important issue is the subject of considerable controversy. Yates (1984) strongly favors the use of the Fisher exact test because of its exact interpretation in the randomization model. Under the randomization null hypothesis that the treatments identically affect the responses, the Fisher exact p-value is exactly the probability of observing a result as large or larger than the observed result. In this case, the probability is assumed to come *only* from the initial randomization of experimental units to treatment groups. Within this context, the rerandomization p-values are exact, and p-values derived from the Z-test and bootstrap resampling are too small.

However, suppose that the experimental units are randomly selected from a larger population and then randomized to treatment groups. We may want to consider the outcomes if other samples were chosen. While the permutation model forces constant marginal totals for all rerandomizations, the bootstrap model allows these to vary, as they naturally would in repeated samples from a larger population. A reasonable model for such an experiment is the binomial model, and, as Figure 5.3 demonstrates, the Fisher exact p-values are certainly not "exact" within this context.

Whether to choose a permutation model, conditional on the marginal totals, or a bootstrap model, not conditioned on the marginal totals, has been discussed at length by Upton (1982), Yates (1984), Rice (1988), D'Agostino, Chase, and Belanger (1988), and Little (1989). The key point of contention in these articles is a question as old as statistics: "Where does the probability come from?" If it is assumed to come from repeated sampling of a larger population, then the binomial model seems most appropriate, implicitly favoring bootstrap resampling. However, if the probability is assumed to come from the randomization of experimental units to treatment groups, then the randomization model is most appropriate, favoring permutational resampling.

Tied to the question "Where does the probability come from?" is the question, "What is the scope of inference?" from a particular experiment. In the case of a permutation test, conditioned on the margins, the probabilities are argued to be correct even if the experimental units are not a random sample from some super population. When there is a statistically significant result, it is fair to say that treatment A is better than treatment B for these units. In a strict sense, one can not say that A is expected to be better than B in other situations. If on the other hand, the units are randomly selected from some super population, then a significant result implies A is better than B for that population.

For example, in the case of rodent carcinogenicity studies, it is common to use inbred lines of animals. Every rat is genetically virtually identical to every other rat. In a strict sense, there is very little scope of inference: although treatments are assigned to animals at random so permutational p-values are "correct," they are correct only in the sense that B is more carcinogenic than A within this fixed genotype of rats. Obviously, it would be a stronger statement to say that B is more carcinogenic than A for all rats, as that would make the generalization to humans more plausible.

If in rodent testing, the animals were selected from a large pool of animals representing diverse genetics and were raised in different environments, then the scope of inference would be much wider. In this case, sampling with replacement, or bootstrap sampling, would mimic the original animal selection in its scope. Furthermore, it would be appropriate that the margins be allowed to vary: a bootstrap sample that had more or fewer tumors than found in the original experiment would mimic a new experiment where more or fewer of a particular tumor would be expected. So bootstrap sampling mimics sampling from a super population and the desire to infer to that population; permutation sampling is consonant with the idea that the experimental material is limited in its scope of inference. Using bootstrap sampling should introduce more variation into the resampling process, but it is not expected to extend the scope beyond the population from which the units were sampled: inbred rats will not become outbred rats just because bootstrap sampling is used!

In judging between bootstrap and permutation resampling, the first question is how the experimental material is obtained. On the philosophical side, if the experimental material is unique in some important sense, and if the experiment is truly a randomized comparative trial, then permutational resampling and inference within the scope of the experiment are appropriate. If the experimental material is truly a random sample from a larger population, and if we desire inference to extend to that population, then bootstrap resampling seems more appropriate. On the technical side, there are conflicting considerations: as indicated in Figures 5.3 and 5.4, the bootstrap may achieve the nominal α levels more accurately and have higher power than the permutational approach. On the other hand, the bootstrap tests are not unbiased, which can cause error rates to exceed nominal levels on occasion.

Little (1989) comments that, even from the standpoint of the binomial super population model, there are good arguments for conditioning on the marginal totals. However, as noted by Little, arguments that favor conditioning upon marginal totals (permutational analysis) also naturally lead to conditioning upon *all* the data. In this case probabilistic conclusions must be obtained using Bayesian methods. These methods require the specification of a prior distribution for the unknown model parameters, then, after the data have been observed, the prior distribution is updated via Bayes theorem to produce a posterior distribution. Probabilistic inference for the parameters is then completely determined by the posterior distribution.

There are many inferential methods for binomial parameters. The bootstrap and permutational approaches are quite similar, as they both consider other data sets that could have happened, assuming a complete null hypothesis. The Bayesian approach is very different from both of these methods, in that other possible data sets are not considered when defining a measure of evidence for or against the null hypothesis: one conditions on *all* the observed data, then calculates the *probability* that the null hypothesis is either true or false (Berger and Sellke, 1987, Casella and Berger, 1987). It is our view that the distinction between bootstrap and permutation resampling approaches is relatively minor when one considers them in the same context as the Bayesian approach: both resampling approaches consider data sets that could have been observed, and are therefore in equal violation of the "likelihood principle" (Berger and Wolpert, 1984). If one accepts the likelihood principle, the given resampling approaches are neither appropriate nor relevant. Figure 5.5 depicts the

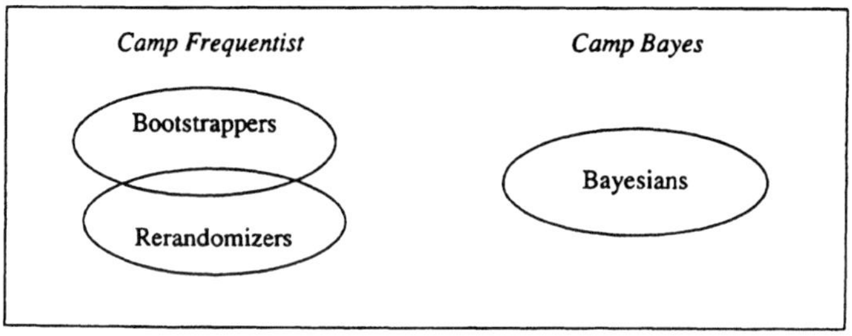

FIGURE 5.5. Frequentist and Bayesian camps.

relationships of the three methods: both resampling approaches are in the frequentist "camp," with the permutation approach only slightly closer to the Bayesian "camp" since inference is conditional upon at least *some* of the data.

Because we adopt a frequentist point of view throughout this book, we are not overly concerned with distinctions between bootstrap and permutation resampling methods. However, we *are* concerned that the Bonferroni-type adjusted p-values are excessively conservative, mainly resulting from sparse data. Bonferroni adjustments appear to be equally bad, whether the probability is assumed to come from the binomial distribution or from the permutation distribution. On the other hand, the effect of discreteness is appropriately accounted for via resampling.

The circumstances surrounding the selection of experimental units, and the intended scope of inference, are factors that sometimes determine whether bootstrap or rerandomization should be used for resampling. In other cases, the bootstrap might be used because of pragmatism: in complex models such as those involving covariates, permutational models are very difficult to specify, yet bootstrap models are relatively simple. When both methods apply equally well, the decision is a matter of individual preference.

5.5 MULTIVARIATE BINARY SINGLE-SAMPLE ANALYSIS

Till now we have considered binary data vectors for several treatment groups. It is also possible to test multiple hypotheses with single-sample data. Experimental frequencies may be compared with established control frequencies, or in the case of genetic experiments, with theoretical frequencies based on hypothesized distributions.

With a single sample, permutation resampling provides no insight into sampling variation, since the sample proportions are identical for every resampled data set. However, bootstrap sampling is ideally suited for evaluating sample-to-sample variability with such data.

As in the multi-sample case, denote the observable binary vectors by $\mathbf{B}_1, \ldots, \mathbf{B}_n$, assumed i.i.d. $\mathcal{MVB}_v(\boldsymbol{\pi}, 1, \mathbf{D})$. The vector of probabilities is $\boldsymbol{\pi} = (\pi_1 \ldots \pi_v)$, and $\mathbf{D}$ represents the dependence structure, again an unknown probability distribution on all 2^v binary v-tuples, subject to the constraint that $\mathrm{E}(B_i) = \boldsymbol{\pi}$. The vector of binomial frequencies $\mathbf{Y} = \sum \mathbf{B}_i$ is distributed as $\mathcal{MVB}_v(\boldsymbol{\pi}, n, \mathbf{D})$, and its individual elements Y_j are distributed as $\mathcal{BIN}(\pi_j, n), j = 1, \ldots, v$.

Hypotheses of interest are H_{0j}: $\pi_j = \pi_{0j}, j = 1, \ldots, v$, for specified constants π_{0j}. Letting $\hat{\pi}_j = Y_j/n$, the common Z-statistic is

$$Z_j = \frac{\hat{\pi}_j - \pi_{0j}}{\sqrt{\pi_{0j}(1 - \pi_{0j})/n}}$$

For reasonably large sample sizes, significance may be determined from the standard normal approximation for Z_j. There are many possible variants on this method—one may use arcsine-transformed tests, or small-sample methods which use the binomial distribution directly.

One problem with multiple binomial tests using single-sample data is how to estimate a complete null distribution. This problem may be circumvented by attacking the problem somewhat differently, obtaining instead simultaneous confidence intervals for all parameters π_j. The multiple hypothesis testing question may be answered by noting whether the hypothesized values π_{0j} lie within their respective simultaneous intervals. The usual marginal $100(1-\alpha)\%$ confidence intervals for the parameters π_j are

$$\hat{\pi}_j \pm z_{\alpha/2}\sqrt{\frac{\hat{\pi}_j(1-\hat{\pi}_j)}{n}} \tag{5.9}$$

where $z_{\alpha/2}$ denotes the $1-\alpha/2$ quantile of the standard normal distribution. Much like the applications given in Chapters 3 and 4 concerning simultaneous confidence intervals in univariate and multivariate continuous models, a goal is to define an appropriate critical value Q_α to replace the value $z_{\alpha/2}$, making the *simultaneous* coverage level approximately $1-\alpha$.

Westfall (1985) considers this problem in a comparison of HLA-Antigen proportions to population baselines. Rather than use the intervals (5.9), he uses simultaneous intervals of the "Clopper-Pearson" type (Clopper and Pearson, 1934). Although often called "exact," these intervals are not really exact because of the discrete nature of the binomial distribution. On the other hand, they are always *conservative* (actual confidence level $\geqslant \alpha$), achieving the nominal $100(1-\alpha)\%$ level only in rare situations (Angus and Schafer, 1984).

Two good reasons for choosing the "exact" Clopper-Pearson intervals over (5.9) are: (1) symmetric intervals are undesirable for proportions near zero, and (2) the resampled pivotal statistics behave poorly when the marginal frequencies are close to zero.

The Clopper-Pearson intervals are now defined. Drop the "j" subscript for now, and note that the random count Y is distributed as $\mathcal{BIN}(\pi, n)$. If the observed value is y, then the Clopper-Pearson upper limit is the value $\pi_U(y)$ such that $\Pr(Y \leqslant y \mid \pi_U(y)) = \alpha/2$ (if $y = n$ then define $\pi_U(y) = 1$). The Clopper-Pearson lower limit is defined similarly as the value $\pi_L(y)$ such that $\Pr(Y \geqslant y \mid \pi_L(y)) = \alpha/2$ (if $y = 0$ then define $\pi_L(y) = 0$). The marginal (not simultaneous) $100(1-\alpha)\%$ Clopper-Pearson confidence interval is then the set of values $I_\alpha(y)$, defined as follows:

$$I_\alpha(y) = \begin{cases} [0, \pi_U(y)), & \text{for } y = 0 \\ (\pi_L(y), \pi_U(y)), & \text{for } 0 < y < n \\ (\pi_L(y), 1], & \text{for } y = n \end{cases}$$

As an example, suppose $n = 52$ and we observe $y = 2$. Then the 95% Clopper-Pearson upper limit π_U is given by

$$\sum_{i=0}^{2}\binom{52}{i}\pi_U^i(1-\pi_U)^{(52-i)} = .025$$

and the 95% Clopper-Pearson lower limit π_L is given by

$$\sum_{i=2}^{52} \binom{52}{i} \pi_L^i (1 - \pi_L)^{(52-i)} = .025$$

These equations must be solved numerically, yielding $\pi_L = .00469$ and $\pi_U = .13213$. The confidence interval for π is then $I_{.05} = (.00469, .13213)$.

If data values $y_1, \ldots, y_v$ are observed, intervals $I_{j\alpha} = I_\alpha(y_j)$ can be computed for every j. The marginal coverage level of each interval is approximately $100(1 - \alpha)\%$, but for simultaneous inference, the interesting quantity is

$$C_\alpha = \Pr(\pi_j \in I_\alpha(Y_j), \text{ for all } j)$$

This quantity is analytically intractable, but is easily estimated via bootstrap resampling. Let $\mathbf{B}_1^*, \ldots, \mathbf{B}_n^*$ be a bootstrap sample from the original vectors $\mathbf{B}_1, \ldots, \mathbf{B}_n$, and denote the bootstrap frequencies computed from these resampled vectors Y_j^*. In the bootstrap "population," the Y_j^* are distributed as $\mathcal{BIN}(\hat{\pi}_j, n)$, where $\hat{\pi}_j = Y_j/n$. Thus, the simultaneous coverage probability C_α may be estimated by

$$\hat{C}_\alpha = \Pr(\hat{\pi}_j \in I_\alpha(Y_j^*), \text{ for all } j) \tag{5.10}$$

using bootstrap sampling.

For fixed α, the simultaneous coverage estimate (5.10) is easily simulated. Compute Clopper-Pearson intervals for every $i = 0, \ldots, n$ *before* resampling, and then identify the intervals $I_\alpha(Y_j^*)$ from this precomputed list for each bootstrap sample. Note that since Y_j^* may only take values in $0, \ldots, n$, there is no need to recompute Clopper-Pearson intervals for every resampled data set. If all original sample proportions $\hat{\pi}_j$ are contained in the resampling-based intervals $I_\alpha(Y_j^*)$, then increment the count variable. The bootstrap estimate of $\hat{C}_\alpha$ is the proportion of simulations for which all $\hat{\pi}_j$ are contained in the resampling-based intervals $I_\alpha(Y_j^*)$.

One is typically interested in making $\hat{C}_\alpha$ equal to a prespecified value, such as 95%. To do this using the bootstrap analysis presented here, we must consider several marginal α's. Table 5.9 is adapted from Westfall (1985), and shows simultaneous intervals for HLA antigen frequencies from a sample of $n = 52$ alcoholics. In this example it would be of great interest to identify a "marker gene" associated with alcoholism. This identification would support a psychological basis for the disease and help guide researchers to possible cures. On the other hand, it would be unfortunate to start on a very expensive wild goose chase based on a statistical false positive result caused by multiple testing.

There were $v = 27$ separate antigen types identified. Population reference frequencies are also provided, and the asterisk indicates those intervals that do not include the population reference value. The simultaneous confidence estimate at the bottom of Table 5.9 gives the proportion of resampled data sets having all population proportions within the individual Clopper-Pearson limits.

Table 5.9 Confidence Intervals for Antigen Frequencies[1] from n = 52 Alcoholics

		Marginal Confidence Level				
Antigen	y_j	90%	95%	99%	99.5%	Population Reference
A1	13	(.155, .368)	(.140, .389)	(.115, .432)	(.106, .449)	.291
A2	28	(.416, .658)	(.395, .678)	(.354, .715)	(.340, .729)	.494
A3	18	(.237, .469)	(.220, .491)	(.188, .534)	(.176, .549)	.253
A11	11	(.123, .326)*	(.111, .347)	(.088, .390)	(.080, .406)	.118
A23(9)	7	(.065, .238)*	(.056, .258)*	(.041, .298)*	(.036, .314)	.039
A24(9)	3	(.016, .142)*	(.012, .159)*	(.007, .196)	(.005, .210)	.159
A25(10)	2	(.007, .116)	(.005, .132)	(.002, .166)	(.001, .180)	.038
A29(w19)	4	(.027, .167)	(.021, .185)	(.013, .223)	(.011, .238)	.078
A30(w19)–A31(w19)[2]	6	(.052, .215)	(.044, .234)	(.031, .274)	(.027, .290)	.099
A32(w19)	1	(.000, .088)	(.000, .103)	(.000, .134)	(.000, .147)	.069
B7	21	(.289, .527)*	(.270, .549)*	(.235, .591)	(.222, .606)	.241
B8	11	(.123, .326)	(.111, .347)	(.088, .390)	(.080, .406)	.204
B13	4	(.027, .167)	(.021, .185)	(.013, .223)	(.011, .238)	.047
B14	1	(.000, .088)	(.000, .103)	(.000, .134)	(.000, .147)	.083
B15	4	(.027, .167)	(.021, .185)	(.013, .223)	(.011, .238)	.137
B17	4	(.027, .167)	(.021, .185)	(.013, .223)	(.011, .238)	.096
B18	5	(.039, .192)	(.032, .210)	(.021, .249)	(.018, .264)	.083
Bw22	6	(.052, .215)*	(.044, .234)	(.031, .274)	(.027, .290)	.045
B27	2	(.007, .116)	(.005, .132)	(.002, .166)	(.001, .180)	.074
B35	7	(.065, .238)	(.056, .258)	(.041, .298)	(.036, .314)	.167
B40	10	(.108, .304)	(.096, .325)	(.075, .368)	(.068, .384)	.148
B44(12)	10	(.108, .304)	(.096, .325)	(.075, .368)	(.068, .384)	.275
B49(12)	2	(.007, .116)	(.005, .132)	(.002, .166)	(.001, .180)	.030
B51(5)	9	(.093, .283)	(.082, .303)	(.063, .345)	(.057, .361)	.099
Cw2	3	(.016, .142)	(.012, .159)	(.007, .196)	(.005, .210)	.050
Cw3	13	(.155, .368)	(.140, .389)	(.115, .432)	(.106, .449)	.170
Cw4	10	(.108, .304)	(.096, .325)	(.075, .368)	(.068, .384)	.130
Simultaneous confidence estimate		25.3%	49.6%	88.8%	94.6%	

[1] Adapted from P. H. Westfall, "Simultaneous Small-Sample Multivariate Bernoulli Confidence Intervals," *Biometrics* **41** 1001–1013, with permission of the Biometrics Society.
[2] Antigens pooled because of difficulty in identification.
*Denotes significant difference from population reference.

The intervals show that the A23(9) antigen is significantly different from the population reference at an approximate 88.8% simultaneous confidence level, but no comparisons are significant at the approximate 94.6% simultaneous confidence level. In other words, the adjusted p-value for the A23(9) comparison is between .112 and .054. Thus, no antigen frequency is significantly different from the corresponding population reference value at the .05 familywise significance level.

5.6 INCORPORATING EXTERNAL INFORMATION

The preceding example highlights a possible problem with the confidence interval approach. The confidence interval for the antigen A23(9) was based on a sample proportion $\hat{\pi} = \frac{7}{52} = .135$, resulting in a much wider interval than would be obtained had the sample proportion been near the population reference value of .039. A simple Z-statistic which incorporates the reference value is $\sqrt{52}(.135 - .039)/\sqrt{.039(1 - .039)} = 3.576$, but if one uses the sample value .135 rather than .039 in the variance calculation, the Z-value is 2.026. Clearly, results can differ depending upon how historical and/or other reference information is used.

When testing proportions coming from two or more samples, this issue arises in the use of "pooled" probability estimates from all samples. If, in the absence of treatment, there is essentially zero probability of observing a particular tumor, side effect, or other phenomenon, then *any* occurrence of such an outcome in a treatment group should be cause for concern. However, using standard methods for multi-sample binary tests, whether conditional (permutational) or unconditional, the probability estimate for occurrence of that outcome is taken from the pooled data. This can result in an excessively large estimated probability, which may in turn result in a statistically insignificant test, when the result really ought to be considered "significant."

A possible solution to this problem is to resample the *control data only*. In this way, all probability estimates are taken from the control group, and the estimated null hypothesis probabilities may not be biased in an upward direction because of pooling with the treatment data. This is in the spirit of the first rule of Hall and Wilson (1991): resampling should be performed in a way that reflects the null hypothesis. However, there are two problems with resampling control data only. First, if there are more identified binary categories in the treated groups than in the control group (a common occurrence), then the resampled data sets will produce no data for comparison with these "extra" categories. From a bootstrap resampling point of view, the proportions for these categories are estimated to be zero; therefore, such categories will not appear in any bootstrap data sets. A second, and related problem, is that we would like the bootstrap distribution to be as accurate as possible. If data are resampled from a single group, then the statistical accuracy will tend not to be as high as if data are sampled from all groups. There is "borrowed strength" from sampling data in all groups together: while there is a distinct possibility of increased bias when data are pooled, this bias may be offset by a reduction in variance.

Resampling control data only results in inaccurately estimated null distributions when the control group is small. For control groups with very large sample sizes, and with all categories of interest represented, resampling only from the control data may prove to be more accurate and powerful than resampling the pooled data. For example, if binary vectors $\mathbf{B}_{11}, \ldots, \mathbf{B}_{1n_1}$ are the observed control sample, and if the $\mathbf{B}_{21}, \ldots, \mathbf{B}_{2n_2}$ are the observed treated sample, then create bootstrap data sets $\mathbf{B}^*_{11}, \ldots, \mathbf{B}^*_{1n_1}$ and $\mathbf{B}^*_{21}, \ldots, \mathbf{B}^*_{2n_2}$ as independently generated *with replacement* samples from the control data $\mathbf{B}_{11}, \ldots, \mathbf{B}_{1n_1}$ only. Then recompute all test statistics from the resampled data, for example, arcsine-transformed tests if these were used to analyze the original data, and adjust p-values as usual.

When sample sizes are not large, this type of resampling is inappropriate because of potential instability and possible exclusion of categories. In such cases it may be desirable to include historical control information for resampling, should such data exist. Without claiming to provide a definitive solution to the problem of incorporating historical controls, the following method provides a framework within which the data from current experiments may be evaluated against historical controls.

There are important reasons for including a control group with any experiment. Without a control, one would not know whether certain conditions unique to the given time, laboratory, sample, or experimenters' technique might have contributed to an "unusual" outcome for a given variable in a given treatment group. Any method which incorporates historical control information should recognize this fact and therefore give more weight to the current control data.

Let the vectors $\mathbf{B}_{11}^C, \ldots, \mathbf{B}_{1n_{C1}}^C$ denote current control vectors, with n_{C1} observations. Let $\mathbf{B}_{21}^C, \ldots, \mathbf{B}_{2n_{C2}}^C, \ldots, \mathbf{B}_{g1}^C, \ldots, \mathbf{B}_{gn_{Cg}}^C$ denote current active treatment groups, assuming $g - 1$ current active treatment groups, and one current control, for a total of g current groups. There are n_{Cj} observations in current group j. Let $\mathbf{B}_{11}^H, \ldots, \mathbf{B}_{1n_{H1}}^H, \ldots, \mathbf{B}_{h1}^H, \ldots, \mathbf{B}_{hn_{Hh}}^H$ denote historical control data sets, with sample sizes n_{Hj} per group. If binary categories appear in some vectors but not others, simply expand all vectors using zeros for the additional categories, so all vectors have the same dimension, v.

The following generalized resampling plan recreates resampled data sets using all available data. Define resampling-based weights $w_{C1}, w_{C2}, \ldots, w_{Cg}, w_{H1}, \ldots, w_{Hh}$, all of which are nonnegative and sum to one. Then recreate the current experiment by resampling the given data using the weights to define the probability that an observation vector is selected from a given group:

$$\mathbf{B}_{ij}^{C*} = \begin{cases} \mathbf{B}_{kl}^C, & \text{with probability } \frac{w_{Ck}}{n_{Ck}}, \text{ for } k = 1, \ldots, g \text{ and } l = 1, \ldots, n_{Ck} \\ \mathbf{B}_{kl}^H, & \text{with probability } \frac{w_{Hk}}{n_{Hk}}, \text{ for } k = 1, \ldots, h \text{ and } l = 1, \ldots, n_{Hk} \end{cases}$$

In essence, individual data vectors are sampled from group $C1$ with probability w_{C1}, from group $C2$ with probability w_{C2}, and so on. This generalized resampling method reduces to the bootstrap method where the vectors are pooled over all current groups and then sampled with replacement, discussed in Section 5.3.1. To make this correspondence, choose the historical weights $w_{Hk} \equiv 0$, and the current weights proportional to the group sizes, $w_{Ck} = n_{Ck}/\sum_j n_{Cj}$.

Selection of the weights w is subjective. If categories from all current active treatment groups are represented in at least one of the control groups, then it seems reasonable to set the weights for all current treatment groups to zero. Among the remaining weights, it also seems reasonable to assign a large proportion, say 50% or more, to the current control. The weights for the historical control groups might be evenly divided among the remaining groups. One might examine the effect of allowing the weight for the current control, w_{C1}, to be variable, in order to assess the effect of incorporating historical control information. For example, if all historical controls are equally weighted, a comparison of analyses with $w_{C1} = .50$ and $w_{C1} = .90$ will highlight the difference between incorporating a large amount versus a small amount of historical control information.

Incorporation of historical control data may improve the estimation of null frequencies. This can make tests more sensitive, particularly when the historical relative frequency of a particular outcome is very small. While univariate tests may be made more stringent, the corresponding effect from a multiple testing perspective can be to make tests *less stringent*, particularly if there are many outcomes observed in historical data that are not observed in the current experiment. For example, there may be 100 tumor types that have been identified in a collection of historical data, but only 50 of those are identified in a current experiment. When resampled data sets are created from such historical data, the number of tests, k, is 100 rather than 50: the minimum p-value distribution is computed by considering the minimum of 100 resampled p-values. If the historical frequencies of the 50 types not observed in the current data are all small, as they should be, then the effect of incorporating these tests into the multiplicity adjusted p-values will not be great, since sparse binary data have little effect on the false positive rate. On the other hand, if there are large differences between such frequencies in historical data and current data, it suggests that the current experiment is substantially different from historical experiments, and that the historical data should be appropriately downweighted in the resampling scheme.

The fact that multiplicity adjustment includes some categories, e.g., tumor types or side effects, that appear in historical data but not in the current data is sensible from the standpoint of bootstrap multiplicity adjustment. With bootstrap resampling, we recreate data sets that are likely to appear in replications of the experiment. The same categories will not appear in every replication of an experiment in real life; use of bootstrap resampling with historical data mimics this fact. Performing multiplicity adjustment over all categories that *could have* occurred accounts for the *selection effect* that is present in a given multivariate binary experiment where the number of categories is variable: in effect, many tests of hypotheses have *already been performed* when one excludes a collection of variables not appearing in the current experiment. In other words, if there are no occurrences of a particular binary variable, in either the control or the treatment groups, then the experimenter implicitly claims that the difference between groups is insignificant with respect to this variable. Bootstrap sampling using historical data allows the investigator to account for such selection effects explicitly.

CHAPTER 6

Further Topics

In this chapter we present additional developments of the resampling-based multiple testing method. Since this field of study is new, there will undoubtedly be many additions, refinements, and improvements to current knowledge as time progresses. In this chapter we cover a variety of topics. Section 6.1 introduces the *weighted p-value*, which allows the researcher to place more emphasis on tests considered more important. In Sections 6.2 and 6.3 we apply the general method to multiple tests for variances and correlations, respectively. In Section 6.4, the resampling-based method is generalized, allowing infinitely many tests in the family. Section 6.5 discusses power of the resampling-based methods, both for comparing alternative methods and for determining sample size. Some asymptotic results are presented in Section 6.6. Finally, Section 6.7 discusses the use of resampling-based multiple testing in more complex models, such as logistic regression and censored multivariate survival analysis.

6.1 ALLOCATING ERROR: WEIGHTED p-VALUES

So far we have assumed that all hypotheses H_i are equally important in our multiplicity adjustments. This is implicit in the standard Bonferroni correction: all p-values are multiplied by the same constant, k, making it equally difficult to obtain a significant result for any of the k tests. However, in many circumstances, the various hypotheses are not equally important. For example, in a two-way ANOVA model, main effects contrasts might be considered more important than interaction contrasts. If so, it is reasonable to allocate a larger proportion of the FWE to the tests of primary importance, making it easier to obtain a significant result.

For example, when considering growth curves for rats in Section 1.6, we considered an FWE of .05, and allocated .02 for each of the two *Gain* contrasts and .005 for each of the *Shape* contrasts. Thus, 40% of the FWE was allocated to each of the *Gain* contrasts, while only 10% was allocated to each of the *Shape* contrasts. If the p-value for either *Gain* contrast would have been less than $.40 \times .05 = .02$, we would have declared that particular test "statistically significant"; if the p-value for either *Shape* contrast would have been less than $.10 \times .05 = .005$, we would have declared that particular test "statistically significant." In keeping with our thesis that results of multiple testing should be reported using adjusted p-values, this allocation of errors suggests the following adjustments: $\tilde{p} = \frac{1}{.4}p = (2.5)p$ for the *Gain* contrasts, and $\tilde{p} = \frac{1}{.1}p = (10)p$ for the *Shape* contrasts.

In general, suppose that the proportion of the FWE allocated for hypothesis H_i is a_i, where $\sum a_i = 1$. In the methods used previously, the a_i were all equal, i.e. $a_i \equiv \frac{1}{k}$. If the Bonferroni method is used, then the p-values should be adjusted as

$$\tilde{p}_i = w_i p_i, \text{ where } w_i = \frac{1}{a_i}$$

If $a_i \equiv \frac{1}{k}$, then $w_i \equiv k$, and the weighted adjustments reduce to the simple Bonferroni adjustments. The weighted Šidák adjustments are analogously defined:

$$\tilde{p}_i = 1 - (1 - p_i)^{w_i}$$

The FWE is controlled using these adjustments. For the weighted Bonferroni adjustments:

$$\begin{aligned}
&\Pr(\text{Reject at least one } H_i \mid H_0^C) \\
&= \Pr\left(\min_{1\leqslant i\leqslant k} P_i \leqslant a_i\alpha \mid H_0^C\right) \\
&\leqslant \sum_{i=1}^{k} \Pr(P_i \leqslant a_i\alpha \mid H_0^C) \\
&= \sum a_i\alpha \qquad (\text{assuming } P_i \sim \mathcal{U}[0,1] \text{ under } H_{0i}) \\
&= \alpha \sum a_i = \alpha \quad \text{since} \sum a_i = 1
\end{aligned}$$

For the weighted Šidák adjustments, the actual FWE is exactly equal to α under independent and uniformly distributed p-values:

$$\begin{aligned}
\Pr(\text{Reject at least one } H_i \mid H_0^C) &= \Pr\left(\min_{1\leqslant i\leqslant k} 1 - (1 - P_i)^{w_i} \leqslant \alpha \mid H_0^C\right) \\
&= 1 - \Pr(1 - (1 - P_i)^{w_i} > \alpha, \text{ for all } i \mid H_0^C) \\
&= 1 - \prod (1 - \alpha)^{1/w_i} = \alpha
\end{aligned}$$

Like the unweighted Šidák method, we believe that the weighted Šidák method is conservative in many applications where the p-values are not independent.

The weighted p-value method may be improved by resampling. Rather than use the raw p-value p_i as the base test statistic for testing H_i, use the *weighted* p-value $p_{iw} = w_i p_i$. (Note that when *higher* weights w_i are given, the hypothesis H_i is considered *less important*.) Now, define the resampling-based single-step adjusted p-value as before, with the raw p-values replaced by the weighted p-values:

$$\tilde{p}_{iw} = \Pr\left(\min_{1\leqslant j\leqslant k} P_{jw} \leqslant p_{iw} \mid H_0^C\right)$$

As an example of this method, suppose there are just two hypotheses, H_1 and H_2, with $a_1 = .20$ and $a_2 = .80$, implying weights $w_1 = 5.0$ and $w_2 = 1.25$. Further, suppose the p-values (P_1, P_2) are independent. Then the adjusted p-values are

$$\begin{aligned}\tilde{p}_{1w} &= 1 - \Pr(5P_1 > 5p_1)\Pr(1.25P_2 > 5p_1) \\ &= \begin{cases} 1 - (1 - p_1)(1 - 4p_1), & \text{for } p_1 < .25 \\ 1, & \text{for } p_1 \geqslant .25 \end{cases}\end{aligned}$$

and

$$\begin{aligned}\tilde{p}_{2w} &= 1 - \Pr(5P_1 > 1.25p_2)\Pr(1.25P_2 > 1.25p_2) \\ &= 1 - (1 - p_2/4)(1 - p_2)\end{aligned}$$

To declare significance of H_i at the FWE = .05 level, it is required that $\tilde{P}_{iw} \leqslant .05$, or $1 - (1 - p_1)(1 - 4p_1) \leqslant .05$ for H_1 and $1 - (1 - p_2/4)(1 - p_2) \leqslant .05$ for H_2. Using unadjusted p-values, these conditions are $p_1 \leqslant .01008$ and $p_2 \leqslant .04033$, respectively. Thus, these adjustments are slightly less conservative than the strict Bonferroni method, for which $p_1 \leqslant .01$ and $p_2 \leqslant .04$ are required.

The analytic calculation of multiplicity adjustments using weighted p-values is much more difficult than the corresponding calculations for equally weighted adjustments. Take, for example, the single-step multiplicity adjustments for the pairwise comparison p-values in a balanced homoscedastic normal ANOVA: since the p-values are monotonically related to the absolute t-statistics, the minimum p-value corresponds to the maximum absolute t-statistic, and multiplicity adjustments can be computed directly from the studentized range distribution. Use of weighted p-values complicates this situation, making analytic evaluation of adjusted p-values difficult if not impossible. However, the resampling method provides a simple solution. One simply replaces the raw p-values with the weighted p-values in the resampling algorithm.

Example 6.1 Re-Analysis of Rat Growth Curves with Weighted p-Values
We demonstrate the calculation of weighted p-values with a re-analysis of the rat growth curve data of Table 1.5. Considering the actual body weights at each of the five time points (the $\mathbf{Y}_1$ matrix of Table 1.5), the data are reasonably modeled as a MANOVA with five variables and three treatment groups. Consider the model

$$\mathbf{Y} = \mathbf{X}\boldsymbol{\beta} + \boldsymbol{\epsilon}$$

where $\mathbf{Y}$ is a $(n \times 5)$ matrix of body weights at the different times, $\mathbf{X}$ is the matrix whose columns are $\{0, 1\}$ indicators of group membership, and $\boldsymbol{\beta}$ is the matrix of mean body weights:

$$\boldsymbol{\beta} = \begin{pmatrix} \mu_{10} & \mu_{11} & \mu_{12} & \mu_{13} & \mu_{14} \\ \mu_{20} & \mu_{21} & \mu_{22} & \mu_{23} & \mu_{24} \\ \mu_{30} & \mu_{31} & \mu_{32} & \mu_{33} & \mu_{34} \end{pmatrix}$$

Here, μ_{it} refers to the mean weight in treatment group i, $i = 1, 2, 3$, at time t, $t = 0, 1, 2, 3, 4$. For any given treatment i, we measure *Gain* as $\mu_{i4} - \mu_{i0}$, as defined in Section 1.6. The *Shape* variable is defined as $\mu_{i2} - (\mu_{i0} + \mu_{i4})/2$. Comparison of *Gain* for different treatment groups is then expressed in the general form $\mathbf{c}'\boldsymbol{\beta}\mathbf{d}$ of Chapter 4; for example, the comparison of *Gain* for the Control vs. Thyroxin animals is expressed as

$$(-1 \quad 1 \quad 0)\begin{pmatrix} \mu_{10} & \mu_{11} & \mu_{12} & \mu_{13} & \mu_{14} \\ \mu_{20} & \mu_{21} & \mu_{22} & \mu_{23} & \mu_{24} \\ \mu_{30} & \mu_{31} & \mu_{32} & \mu_{33} & \mu_{34} \end{pmatrix}\begin{pmatrix} -1 \\ 0 \\ 0 \\ 0 \\ 1 \end{pmatrix}$$

We will consider a large number of tests falling into four classes: (1) treatment vs. control comparisons of *Gain* (two comparisons), (2) treatment vs. control comparisons of *Shape* (two comparisons), (3) treatment vs. control comparisons of mean weight at each time point ($2 \times 5 = 10$ comparisons), and (4) nonlinearity of the growth curve for each treatment group (three comparisons). Hypotheses in group (4) are not contrasts of treatment groups, they are tests of whether the individual shape parameters $\mu_{i2} - (\mu_{i0} + \mu_{i4})/2$ are equal to zero, for $i = 1, 2, 3$. There are very high correlations among these tests, arising from the high correlations between observations at successive time points, and from the many comparison tests against the common control.

Although the assignment of weights to the various tests is subjective, the weights allow the experimenter to include economic and ethical considerations in the data analysis. The tests in class (1) are considered the most important here, and the largest proportion of the FWE will be allocated to these tests. For example, one might allocate .2 for each test in group (1) and .04 for all remaining tests. The corresponding weights are then 5 and 25, respectively, for group (1) and non-group (1) tests.

Table 6.1 displays the results of the 17 tests. It includes raw p-values, weighted Bonferroni p-values, and resampling-based adjustments to the weighted Bonferroni p-values. The resampling-based adjustments used 10,000 data sets simulated from the estimated multivariate normal model, as discussed in Chapter 4. While the parameter estimates are not displayed in this table, the reader is referred to Figure 1.3 for insight into the nature of the hypotheses tested. The effect of incorporating dependence structure is apparent in Table 6.1: the resampling-based weighted adjusted p-values are smaller than the weighted Bonferroni counterparts. For the comparison of *Shape* in Control vs. Thiouracil, the weighted Bonferroni adjustment is insignificant at the FWE = .05 level, but the resampling-based weighted adjustment is significant at this level. (Note that the weighted Bonferroni calculations are performed using raw p-values that have more decimals than given by the table, and that numerical accuracy of p-value calculation is important.) In the test for linearity of the Thiouracil rat growth curve, the weighted Bonferroni adjustment is insignificant at the FWE = .01 level. However, when correlations are incorporated via resampling, this test becomes significant at the FWE = .01 level. □

Table 6.1 Raw and Weighted p-Values for Rat Growth Data

	Test	Raw p-Value	Weight	Weighted Bonferroni p-Value	Adjusted Weighted p-Value
Gain	Control vs Thyroxin	.7845	5	1.0000	.9753
	Control vs Thiouracil	.0000*	5	.0001*	.0000*
Shape	Control vs Thyroxin	.1399	25	1.0000	.9506
	Control vs Thiouracil	.0020*	25	.0512	.0350*
W_0	Control vs Thyroxin	.4990	25	1.0000	1.0000
	Control vs Thiouracil	.7391	25	1.0000	1.0000
W_1	Control vs Thyroxin	.5238	25	1.0000	1.0000
	Control vs Thiouracil	.5585	25	1.0000	1.0000
W_2	Control vs Thyroxin	.8137	25	1.0000	1.0000
	Control vs Thiouracil	.0278*	25	.6939	.3566
W_3	Control vs Thyroxin	.5483	25	1.0000	1.0000
	Control vs Thiouracil	.0013*	25	.0330*	.0234*
W_4	Control vs Thyroxin	.6521	25	1.0000	1.0000
	Control vs Thiouracil	.0000*	25	.0011*	.0006*
Linearity	Control	.4202	25	1.0000	1.0000
	Thyroxin	.0132*	25	.3297	.1891
	Thiouracil	.0004*	25	.0111*	.0079*

*p-value $< .05$.

6.2 COMPARING VARIANCES

Analysis of variability is becoming increasingly important, particularly in the quality control arena. Resampling methods may be adapted to make multiple comparisons of variances as well as means. We will describe parametric and nonparametric multiple testing methods for comparing variances. Parametric methods may differ dramatically from nonparametric methods when there are outliers or other nonnormal characteristics: inference for variances is much more sensitive to departures from normality than is inference for means.

6.2.1 One-Way ANOVA: Normal Model

Consider the one-way ANOVA model

$$Y_{ij} = \mu_i + \epsilon_{ij} \qquad i = 1, \ldots, g;\ j = 1, \ldots, n_i \tag{6.1}$$

where the μ_i are fixed constants, and the ϵ_{ij} are independently distributed as $\mathcal{N}(0, \sigma_i^2)$. Consider hypotheses of the form $H_{ii'}$: $\sigma_i^2 = \sigma_{i'}^2$. These hypotheses may include all pairwise comparisons or a selected collection of comparisons, such as all treatments vs. the same control. Alternative hypotheses are generally two-tailed, but upper- and lower-tailed tests are allowed.

The F-statistic

$$F_{ii'} = \frac{s_i^2}{s_{i'}^2}$$

may be used to test $H_{ii'}$, where s_i^2 is the within-group variance estimate for group i. For a two-tailed test, the p-value is

$$p_{ii'} = 2\min\{\Pr(F_{n_i-1,n_{i'}-1} > F_{ii'}),\ \Pr(F_{n_i-1,n_{i'}-1} < F_{ii'})\}$$

where F_{ν_1,ν_2} denotes a random variable having the F distribution with ν_1 and ν_2 numerator and denominator degrees of freedom, respectively. For the upper-tailed alternative $H'_{ii'}: \sigma_i^2 > \sigma_{i'}^2$ the p-value is $p_{ii'} = \Pr(F_{n_i-1,n_{i'}-1} > F_{ii'})$, and for the lower-tailed alternative the p-value is $p_{ii'} = \Pr(F_{n_i-1,n_{i'}-1} < F_{ii'})$.

Denote the p-values $P_1, \ldots, P_k$. Note that $k = \binom{g}{2}$ for all pairwise comparisons, and $k = g - 1$ when all treatments are compared to a control. The resampling-based multiplicity adjustments are then easily defined using the general methods of Chapter 2. The subset pivotality condition holds exactly as well: the distributions of all test statistics are free of the parameters σ_i^2. This is an ideal situation for resampling-based testing. We need only simulate data Y_{ij}^*, independent and identically distributed as $\mathcal{N}(0, 1)$, recompute all p-values P_j^* from this simulated data, then compare them with the observed p-values p_j. Testing may be done using step-down methods or single-step methods.

This technique reduces to the analytical method described by Hartley (1950) when the sample sizes are equal. The advantages of using the parametric resampling-based approach include (1) the results can be conveniently displayed using adjusted p-values, (2) there is no requirement that sample sizes be equal, and (3) step-down testing is easily accommodated. As always, the negative side of the resampling-based approach is simulation error.

Example 6.2 Comparing Variances in Litter Weight Data

Table 3.11 displays litter weight data for mice. Previous analyses failed to detect differences in mean values for the various dosage groups. We now attempt to detect differences in the variances.

Let the criterion variable Y be the average non-birth weight for the litter, as in Example 3.4. The data are plotted in Figure 6.1, suggesting larger variability in the dosage groups than in the control. Summary statistics for the treatment groups are as follows:

	Dosage			
Statistic	0	5	50	500
Mean	32.3092	29.3096	29.8657	29.6461
Std Dev	2.6946	5.0925	3.7624	5.4049
Variance (unbiased)	7.2609	25.9339	14.1557	29.2131
Variance (max. lik.)	6.8978	24.5690	13.3692	27.4947
N	20	19	18	17

Again, the variances appear larger in the treated groups. Table 6.2 displays the unadjusted and adjusted two-tailed p-values for all pairwise comparisons. The resampling-based p-values were calculated using 10,000 data sets $Y_{ij}^* = \epsilon_{ij}^*$, where the ϵ_{ij}^* were

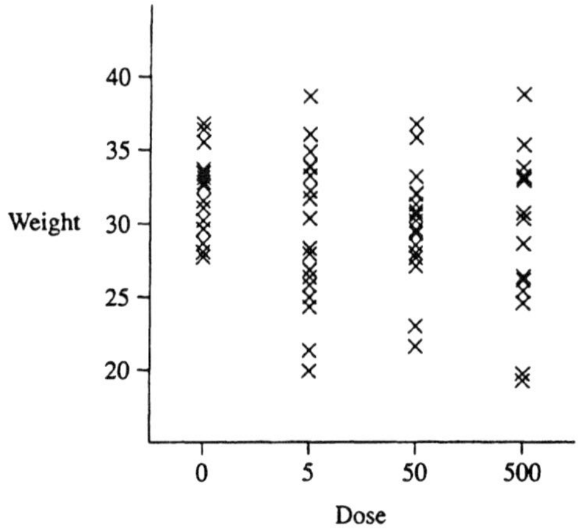

FIGURE 6.1. Average non-birth weights for mice in four different dosage groups.

independent and identically distributed as $\mathcal{N}(0, 1)$. All adjusted p-values are single-step. Note that since the hypotheses are restricted, the restricted step-down adjustments may be used to reduce the amount of adjustment. Nevertheless, the single-step adjustments are valid, though conservative.

Our suspicions are confirmed by the results of the multiple testing analysis: the variances of the dosage groups "5" and "50" *are* different from the variance of the control group. A possible explanation for this phenomenon is that the compound affects some litters but not others. Litters from a non-resilient dam may be affected, lowering overall weights for those litters. For resilient dams, there might be little or no effect of the various doses upon body weights. □

6.2.2 One-Way ANOVA: Location-Scale Model

The assumption of normality is crucial for the validity of tests on variances. Without normality, the p-values are incorrect, not only with finite sample sizes, but also asymptotically. A major distinction between tests for means and tests for variances is that the former are asymptotically valid with nonnormal data, but the latter are not.

Table 6.2 Multiple Testing of Variances in Litter Weight Data (Normality Assumed)

Dose Comparison	F Statistic	df_1	df_2	Raw_p	Adj_p	Sid_p
0 vs. 5	0.2800	19	18	.0084*	.0372*	.0494*
0 vs. 50	0.5129	19	17	.1617	.4941	.6530
0 vs. 500	0.2485	19	16	.0048*	.0219*	.0283*
5 vs. 50	1.8321	18	17	.2185	.6039	.7721
5 vs. 500	0.8877	18	16	.8016	.9949	.9999
50 vs. 500	0.4846	17	16	.1488	.4667	.6198

*p-value < .05.

Resampling can make tests for variances valid, at least asymptotically. Although we will not give a proof of the asymptotic theory, we will offer heuristic justifications for our resampling method. We assume that the distributions differ from normality, but are all related via location and scale transformations. In particular, we assume the one-way ANOVA model (6.1), but with the ϵ_{ij}/σ_i having a common distribution G. This generalizes the previous model, in which G was assumed to be the $\mathcal{N}(0, 1)$ distribution.

Again we wish to create bootstrap data sets Y_{ij}^*, then re-compute all test statistics from the resampled data. Under any partial null hypothesis H_0^K, the subcollection of test statistics corresponding to true nulls has a joint distribution depending only upon G. This suggests that we simulate data having a distribution that approximates G. A natural estimate is the empirical distribution of the *standardized* residuals $\hat{\epsilon}_{ij} = (Y_{ij} - \bar{Y}_i)/s_i$.

Note that in Chapter 3, when discussing the use of bootstrap resampling in location shift models, the data were resampled from the pooled *ordinary* residuals, rather than from the pooled standardized residuals. There is a crucial distinction between the empirical distribution of the pooled standardized residuals and the empirical distribution of the ordinary residuals. When testing variances, small changes in the underlying distribution G can greatly affect the inference—changing G from normal to nonnormal changes the asymptotic null distribution of the p-values. Thus, we wish to resample from a distribution that resembles G as closely as possible. If the original data were heteroscedastic, and if we resampled data from the pooled *ordinary* residuals, the estimated distribution would be a *mixture* distribution, clearly inappropriate. For example, when G is normal, the appropriate inferential method is to resample from the normal distribution. Resampling from standardized residuals should approximate sampling from the normal, resulting in approximately valid p-values. Resampling of ordinary residuals might approximate sampling from a mixture distribution, leading to incorrect p-values.

The resampling of standardized residuals is also motivated by the first "guideline" of Hall and Wilson (1991), given in Section 2.2: "... care should be taken to ensure that even if the data might be drawn from a population that fails to satisfy H_0, resampling should be done in a way that reflects H_0."

Example 6.3 Comparing Variances in Litter Weight Data—Nonnormal Resampling

We continue Example 6.2, but with an alternative resampling plan. The data are both centered and scaled: $\hat{\epsilon}_{1j} = (Y_{1j} - 32.3092)/2.6946$, $\hat{\epsilon}_{2j} = (Y_{2j} - 29.3096)/5.0925$, $\hat{\epsilon}_{3j} = (Y_{3j} - 29.8657)/3.7624$, and $\hat{\epsilon}_{4j} = (Y_{4j} - 29.6461)/5.4049$. These residuals are then pooled, and the Y_{ij}^* are generated as a *with replacement* sample from the pooled data $\hat{\epsilon}_{ij}$. The test statistics and p-values are recomputed from the resampled data, and the proportion of samples for which the minimum resampled p-value falls below a given original p-value is the resampling-based multiplicity-adjusted p-value. We use single-step adjustments for ease of comparison with the normality-assuming method in Example 6.2; however, this application is ideally suited for restricted step-down testing, which gives smaller adjusted p-values.

Table 6.3 Bootstrap Multiple Testing of Variances in Litter Weight Data

			Bootstrap					
	Normal		Pooled Standardized		Pooled Unstandardized		Separately Sampled	
Dose Comparison	Raw_p	Adj_p	Raw_p	Adj_p	Raw_p	Adj_p	Raw_p	Adj_p
0 vs. 5	.0084*	.0372*	.0062*	.0240*	.0123*	.0511	.0018*	.0349*
0 vs. 50	.1617	.4941	.1193	.3860	.1744	.5152	.1550	.3941
0 vs. 500	.0048*	.0219*	.0033*	.0129*	.0063*	.0290*	.0029*	.0228*
5 vs. 50	.2185	.6039	.1693	.5004	.2369	.6195	.2023	.4999
5 vs. 500	.8016	.9949	.7728	.9900	.8093	.9946	.7621	.9923
50 vs. 500	.1488	.4667	.1092	.3604	.1603	.4864	.1602	.3679

*p-value $< .05$.

Table 6.3 displays the results of this analysis under the "Pooled Standardized" heading. It is interesting to note that the p-values are *smaller* for the bootstrap adjustments than for the normality-assuming adjustments, repeated here from Table 6.2. Smaller p-values for the bootstrap method are explained by the tail behavior of the pooled standardized residuals. As indicated in Figure 6.2, this distribution has tails somewhat lighter than the normal distribution. If this tail behavior reflects the true underlying distribution, then the normality-assuming tests for variances are conservative: the actual p-values are smaller than those computed from the normal distribution.

Raw (unadjusted) bootstrap p-values for each test are displayed in Table 6.3 under the "Raw_p" headings. These are computed by comparing the bootstrap p-values with the original p-values for *each test*, rather than by using the minimum p-value. The "Pooled Standardized" bootstrap p-values are smaller than the

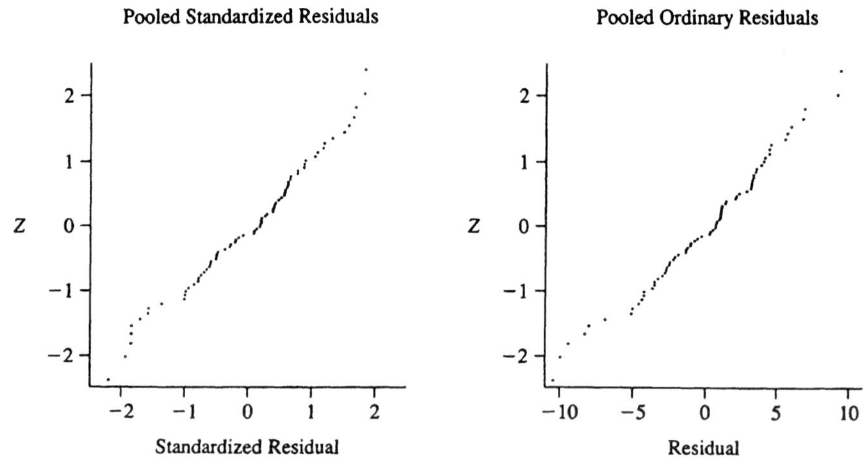

FIGURE 6.2. Normal Q–Q plots of pooled residuals from litter weight data.

corresponding normality-assuming p-values, again because of the empirical distribution of the pooled standardized residuals is slightly light-tailed.

The test results using pooled ordinary residuals appear under the "Pooled Unstandardized" heading in Table 6.3. The p-values are slightly larger, as suggested by the more heavy-tailed appearance of this distribution in Figure 6.2. The adjusted p-value for the "0 vs. 5" variance comparison becomes insignificant under this resampling method.

Finally, the results of resampling the standardized residuals *separately* is given under the "Separately Sampled" heading of Table 6.3. In this resampling scheme, the data Y_{1j}^* are sampled *with replacement* from the standardized residuals $\hat{e}_{1j} = (Y_{1j} - 32.3092)/s_1$, Y_{2j}^* are sampled *with replacement* from residuals $\hat{e}_{2j} = (Y_{2j} - 29.3096)/s_2$, independently of the Y_{1j}^*, and so on. This type of resampling is similar to heteroscedastic resampling, described in Chapter 3, in that unique within-group distributional characteristics are preserved. In this case the standardized residuals ϵ_{ij}/σ_i are allowed to have distinct distributions G_i. These G_i all have common variance (all have unit variance), but possibly different skewness or kurtosis characteristics. The goal is to incorporate the effects of such unique characteristics through resampling the distributions separately.

When using the separately sampled approach, standardization must be done using the *maximum likelihood* estimates, not the *unbiased* estimates. This means that n_i, rather than $n_i - 1$ should be used in the denominators of the estimated variances. The reason for requiring such standardization is that the null hypotheses should all be true in the bootstrap population: resampling under alternative hypotheses causes bootstrap p-values to be biased. In the separate-samples bootstrap probability model:

$$\mathrm{Var}(Y_{ij}^*) = \sum \hat{e}_{ij}^2/n_i - \left(\sum \hat{e}_{ij}/n_i\right)^2$$

When the unbiased estimate of variance $s_i^2 = \sum(Y_{ij} - \bar{Y}_i)^2/(n_i - 1)$ is used, the bootstrap variance of Y_{ij}^* is $(n_i - 1)/n_i$; but when the maximum likelihood estimate of variance $s_i^2 = \sum(Y_{ij} - \bar{Y}_i)^2/n_i$ is used, the bootstrap variance of Y_{ij}^* is exactly 1.0 for all groups. □

For testing variances, both the "Pooled Standardized" and the "Separately Sampled" methods are preferable to the "Pooled Unstandardized" method. Between the "Pooled Standardized" and "Separately Sampled" methods, we suspect that the former works better in two situations, (1) when one or more sample sizes is very small, and (2) when the tail behavior is similar for all underlying distributions. A simple but useful measure of "tail behavior" is the kurtosis statistic.

The data used in the previous example are well-behaved, with no outliers or marked skewnesses. This is expected when data are computed as averages, which is the case here. Nevertheless, there are uncomfortably large changes in the p-values when different resampling schemes are used. The distributions pictured in Figure 6.2 do not differ dramatically from each other, and neither differs dramatically from the normal distribution. Yet there are substantial differences in the "Pooled Standardized" and "Pooled Unstandardized" analyses of Table 6.3. These discrepancies demonstrate how sensitive the inference on variances is to the tail behavior of the underlying distributions. If the data are expected to contain outliers then a robust estimate of variability such as the mean absolute deviation might be used.

6.3 MULTIPLE TESTING FOR CORRELATIONS

It is easy to devise resampling plans for multiple tests of correlations. In multivariate applications, the $(v \times 1)$ data vectors are $\mathbf{Y}$, with correlation matrix

$$\boldsymbol{\rho} = \begin{pmatrix} 1 & \rho_{12} & \rho_{13} & \cdots & \rho_{1v} \\ \rho_{21} & 1 & \rho_{23} & \cdots & \rho_{2v} \\ \rho_{31} & \rho_{32} & 1 & \cdots & \rho_{3v} \\ \vdots & \vdots & \vdots & \ddots & \vdots \\ \rho_{v1} & \rho_{v2} & \rho_{v3} & \cdots & 1 \end{pmatrix}$$

Statistical software packages routinely print p-values for all hypotheses H_{ij}: $\rho_{ij} = 0$. There are $k = v(v-1)/2$ such hypotheses. The multiplicity problem has been recognized, and informal graphical methods have been proposed for its resolution (Hills, 1969, Schweder and Spjøtvoll, 1982). In this section we present formal methods for isolating statistically significant correlations.

The complete null hypothesis states that all variables are uncorrelated. A somewhat stronger statement is that all variables are independent. If the distribution of $\mathbf{Y}$ is multivariate normal, then the two statements are equivalent. Resampling plans are easily constructed under the independence hypothesis, so we will be testing for independence rather than for zero correlation.

To generate a resampled data set under the complete null hypothesis, we sample variables *independently*. Let Y_{ij} denote the jth observation of variable i. As usual, sample $Y_{i1}^*, \ldots, Y_{in}^*$ *with replacement* from the original data $Y_{i1}, \ldots, Y_{in}$. However, to simulate data under the complete null hypothesis, generate all v bootstrap samples independently. This is in sharp contrast with previous analyses, where vectors where resampled to preserve correlations. Here, we are sampling all variables independently to simulate the null hypothesis. By resampling, we incorporate possible nonnormal distributional characteristics.

The Pearson product-moment correlation statistic may be used for measuring association of variables i and j:

$$r_{ij} = \frac{1}{n-1} \sum_{l} \left(\frac{Y_{il} - \bar{Y}_i}{s_i} \right) \left(\frac{Y_{jl} - \bar{Y}_j}{s_j} \right)$$

Here $\bar{Y}_i$ and s_i respectively denote the sample mean and standard deviation of variable i. Assuming normality, the statistic $r_{ij}\sqrt{n-2}/\sqrt{1-r^2}$ has the t distribution with $n-2$ degrees of freedom when $\rho_{ij} = 0$. The two-tailed normal-theory p-value is thus monotonically related to r_{ij}^2: larger r_{ij}^2 implies a smaller p-value.

Replacing Y_{ij} by the bootstrap data Y_{ij}^*, the bootstrap correlations r_{ij}^* are computed. Because of the inverse monotonicity between the squared correlation and the two-tailed p-value, resampling-based adjustments may be performed by comparing the maximum of the bootstrap $(r_{ij}^*)^2$ statistics to the original r_{ij}^2 statistics. This may be done in single-step or step-down fashion.

For rank-based inference, the same general method may be used. First the data **Y** are converted to ranks, separately within each variable, with average ranks assigned in cases of tied observations. The Spearman rank correlations are then computed upon the ranked data. Permutation resampling is appropriate for rank-based inference, so the ranked vectors are permuted (shuffled) independently to create the resampled data. The maximum of the squared resampled rank correlations is then compared with each of the original squared rank correlations when calculating the adjusted p-values.

Sometimes *partial* correlations are used, rather than ordinary correlations. In this case the correlations may be computed using residuals from regression equations. The general ideas of Chapter 4 can be applied in such cases: resample the residuals, and recompute the partial correlations using the resampled residual data.

Example 6.4 Multiple Tests for Correlations Involving SAT Scores

Several years ago a local North Carolina paper created quite a controversy when they gave Scholastic Aptitude Test (SAT) scores for the 50 states and the District of Columbia. North Carolina ranked last. The paper also gave the percent of the students that took the exam and made the point that one of the reasons that the average performance of North Carolina students appeared poor was that many North Carolina students took the exam whereas fewer (presumably) elite students took the exam in other states. Many states in the Midwest use the ACT exam in place of the SAT, so only if students in these states wish to go away to college do they take the SAT.

A plot of Total SAT versus percent students taking the exam, Figure 6.3, confirms a strong relationship between the two variables. Thus it is unfair to evaluate educational effectiveness with the Total SAT score without correction for the percent students taking the exam. The average ACT by state is not publicly available.

We computed the deviation, SATdev, of each state from a linear regression of Total SAT on the square root of the percent students taking the exam. The square root transformation was used to correct for nonlinearity; the estimated function is indicated by the solid line in Figure 6.3. Positive deviations indicate the state is performing better than expected; negative deviations indicate the state is performing worse. We were interested to see if other factors were related to student performance so we collected several additional variables at the local library that are possibly related to student performance including: student/teacher ratio, teacher salary, % Black, and number of crimes per 100,000 population; this data is listed in Table 6.4. We noted two outliers in the data set, the salaries of Alaska teachers are unusually high and the % Black students in Washington, D.C. is high. These observations were removed before correlations and p-values were computed. We computed a correlation matrix for SATdev, student/teacher ratio, teacher salary, % Black, and crime rate. With 5 variables there are 10 pairwise correlations.

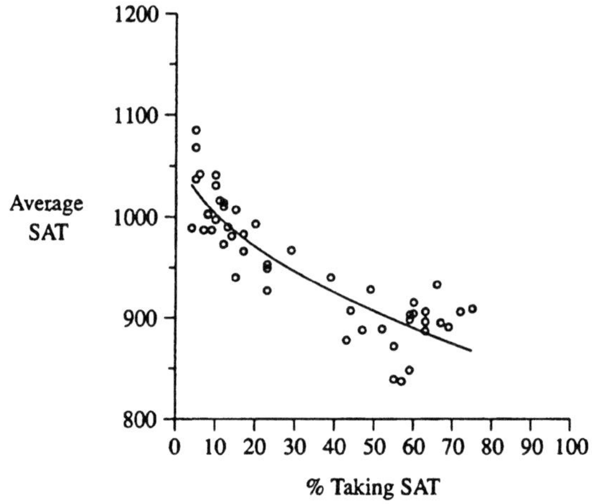

FIGURE 6.3. SAT score vs. percent taking SAT exam for 49 states. Alaska and the District of Columbia are deleted. Linear regression of SAT on $\sqrt{\%}$ is overlaid.

Correlation coefficients, and three types of p-values, ordinary raw p-value, bootstrap raw p-value, and bootstrap adjusted p-value, are given in Figure 6.4. The correlations and p-values are given in the lower triangle, and all pairwise scatterplots are given in the upper triangle of the figure. Data for resampling include the residuals from the regression of SAT on $\sqrt{\%\text{ Taking SAT}}$ and the original four remaining variables. The five variables were resampled independently, and residuals from the SAT data were re-computed to mimic the residual analysis in the original data set (see Chapter 4). Then all 10 pairwise correlations were re-computed, and the free step-down multiplicity-adjusted p-values were computed using 10,000 bootstrap data sets (Algorithm 2.8 of Chapter 2).

There are many correlations of political interest. The correlation between % Black and SATdev is significantly negative, even after multiplicity adjustment. This inverse relationship is well known. Because of this relationship, politicians have urged better education for minorities, and sociologists have charged that the SAT is racially biased.

Another correlation of interest is that between SATdev and student/teacher ratios. We have a negative correlation, as expected, which remains significant even after adjusting for 10 tests of hypotheses. This suggests that smaller class size benefits students. On the other hand, the correlation between teachers' salaries and SATdev is insignificant even *before* multiplicity adjustment, so there is no clear indication that paying teachers more enhances education. □

Example 6.5 Multiple Tests for Fitness Correlations

Over the last 20 years there has been increased interest in human fitness with increasing evidence that fit individuals are less susceptible to heart attacks and have a higher "quality of life" than less fit individuals. Scientists and medical doctors have sought to quantify fitness using expensive laboratory tests and would like inexpensive alternative measures. A. C. Linnerud collected the data given in Table 6.5, which compare

Table 6.4 SAT Data

State	Average SAT	% Taking SAT	S/T Ratio	Teacher Salary	% Black	Crime Rate
Ala	1002	8	19.8	23.6	23.2	559
Alas*	923	42	16.7	42.4	3.3	455
Ariz	952	23	18.4	25.2	2.4	612
Ark	986	7	17.5	19.6	13.6	413
Cali	906	44	23.0	29.8	7.5	918
Colo	966	29	18.2	26.7	3.6	468
Conn	908	75	13.7	27.2	7.1	419
DC*	846	67	14.3	34.0	64.8	1609
Del	903	60	16.0	25.2	16.4	432
Fla	887	47	17.5	22.7	11.8	1024
Ga	847	59	18.9	23.6	24.4	576
Haw	888	52	22.6	26.4	1.8	263
Idah	965	17	20.4	21.4	0.5	214
Ill	982	17	17.4	27.5	14.2	795
Ind	871	55	18.3	24.9	7.5	329
Iowa	1084	5	15.5	22.2	1.6	231
Kan	1040	10	15.4	23.2	5.2	361
Ky	996	10	18.6	21.4	6.9	338
La	986	9	18.5	20.8	27.5	693
Me	897	59	15.5	20.0	0.3	153
Mass	905	72	14.4	27.4	4.2	565
Md	914	60	17.1	27.4	23.9	768
Mich	972	12	20.2	30.7	13.2	780
Minn	1006	15	17.4	27.9	1.3	286
Miss	988	4	19.0	18.9	31.4	269
Mont	992	20	15.6	23.0	0.2	151
Mou	989	13	16.4	22.4	9.7	544
NC	836	57	18.7	22.8	20.2	484
ND	1067	5	15.3	21.3	0.5	57
NH	932	66	15.9	20.7	0.6	149
NJ	894	67	14.7	27.8	12.7	541
NM	1015	11	19.0	22.3	1.6	628
NY	890	69	15.4	31.2	14.5	1007
Neb	1030	10	15.1	21.4	3.0	252
Nev	926	23	20.4	26.2	5.9	695
Ohio	948	23	18.1	25.1	10.0	421
Okla	1001	8	16.9	22.0	6.0	419
Ore	927	49	18.3	26.2	1.4	540
Pa	886	63	16.3	26.4	8.6	368
RI	895	63	15.0	30.1	3.2	360
SC	838	55	17.3	22.1	27.3	665
SD	1041	6	15.6	18.5	0.4	120
Tenn	1009	12	19.9	21.9	14.7	534
Tx	877	43	17.3	25.0	11.2	632
Utah	1036	5	23.4	23.1	0.7	230
Va	902	59	20.5	23.6	17.8	295
Vt	905	63	16.8	21.3	0.3	137
WV	939	15	15.3	21.1	2.9	138
Wash	939	39	20.5	23.6	2.3	439
Wis	1013	12	16.3	26.9	3.9	250
Wyo	980	14	14.0	27.8	0.7	283

*Deleted as outlier.

	SATdev	S/T Ratio	Salary	% Black	Crime Rate
SATdev	r_{ij} Raw p Bootstrap p Adjusted p				
S/T Ratio	−0.3864 .0061 .0049 .0404	r_{ij} Raw p Bootstrap p Adjusted p			
Salary	0.0980 .5030 .5005 .8753	0.0045 .9754 .9759 .9759	r_{ij} Raw p Bootstrap p Adjusted p		
% Black	−0.5089 .0002 .0005 .0019	0.1732 .2341 .2277 .6474	−0.0354 .8090 .8113 .9641	r_{ij} Raw p Bootstrap p Adjusted p	
Crime Rate	−0.3033 .0341 .0327 .1843	0.2290 .1135 .1182 .4485	0.4902 .0003 .0003 .0030	0.4844 .0004 .0000 .0036	r_{ij} Raw p Bootstrap p Adjusted p

FIGURE 6.4. Summary of correlations and multiple tests for SAT data.

fitness as measured on a treadmill, oxygen consumption and maximum pulse rate; and as measured by Dr. K. H. Cooper's 1.5 mile run test, run time (min/1.5 miles) and run pulse. Several additional variables were recorded: rest pulse, age, and weight (kgs). Experts would like to be able to recommend the run test as a replacement for the oxygen consumption determination on the treadmill. Correlations among the other variables are of interest as well, either to suggest alternative measures of fitness, to confirm correlations noted by other investigators, or to suggest future hypotheses for testing.

Correlations among the seven variables are given in Figure 6.5. Also given are the raw and adjusted p-values for pairwise correlations, with the adjusted p-values computed in step-down fashion, with 10,000 resampled data sets. It is clear that run time is highly correlated with oxygen consumption so that the inexpensive 1.5 mile run can serve as a proxy for the expensive treadmill test. It is interesting that a correlation as extreme as −.4329, for the age vs. max pulse, is insignificant with adjusted p-value .2305. This insignificance is partly attributable to small sample size: even when all variables are independent, correlations as extreme as −.4329 are frequent, when 21 correlations are considered from an experiment with a sample size of $n = 31$. □

When testing correlations, the subset pivotality condition does not hold, since the distribution of a correlation coefficient depends on the values of other correlation coefficients. We are sampling under the assumption of complete independence. If

Table 6.5 Fitness Data*

	Tread Mill		1.5 Mile Run		Additional Data		
ID	Oxygen	Max Pulse	Run Time	Run Pulse	Rest Pulse	Body Weight	Age
1	60.05	186	8.63	170	48	81.87	38
2	59.57	172	8.17	166	40	68.15	42
3	54.62	155	8.92	146	48	70.87	50
4	54.30	168	8.65	156	45	85.84	44
5	51.85	170	10.33	166	50	83.12	54
6	50.55	155	9.93	148	49	59.08	57
7	50.54	168	10.13	168	45	73.03	44
8	50.39	168	10.08	168	67	73.37	49
9	49.87	180	9.22	178	55	89.02	38
10	49.16	185	8.95	180	44	81.42	49
11	49.09	170	10.85	162	64	81.19	43
12	48.67	188	9.40	186	56	76.32	49
13	47.92	176	11.50	170	52	61.24	48
14	47.47	172	10.50	170	53	82.78	52
15	47.27	164	10.60	162	47	79.15	47
16	46.77	164	10.25	162	48	91.63	48
17	46.67	168	10.00	162	48	77.91	51
18	46.08	165	11.17	156	62	79.38	54
19	45.79	188	10.47	186	59	73.71	51
20	45.68	180	11.95	176	70	75.98	40
21	45.44	166	9.63	164	48	76.32	52
22	45.31	185	10.07	185	62	75.07	40
23	45.12	172	11.08	172	48	67.25	51
24	44.81	176	11.63	176	58	77.45	47
25	44.75	176	11.12	176	51	66.45	45
26	44.61	182	11.37	178	62	89.47	44
27	40.84	172	10.95	168	57	69.63	51
28	39.44	176	13.08	174	63	81.42	44
29	39.41	176	12.63	174	58	73.37	57
30	39.20	172	12.88	168	44	91.63	54
31	37.39	192	14.03	186	56	87.66	45

*With permission of Dr. A. C. Linnerud.

all variables are independent, all correlation coefficients are asymptotically independent as well (Aitken, 1969, 1971). Thus, step-down resampling-based adjustments under the complete null hypothesis are conservative: we do not take advantage of correlations induced by non-null correlation coefficients.

6.4 INFINITELY MANY TESTS

When all possible linear combinations of parameters are to be examined, then the family of tests has infinitely many elements. This situation is of particular interest when linear combinations *suggested by the data* are examined. Called *post-hoc* tests,

	Treadmill		1.5 Mile Run		Additional Data		
	Oxygen	Max Pulse	Run Time	Run Pulse	Resting Pulse	Body Weight	Age
Oxygen	r_{ij} Raw p Boot p Adj p						
Max Pulse	−0.2368 .1997 .1958 .8616	r_{ij} Raw p Boot p Adj p					
Run Time	−0.8622 .0000 .0000 .0000	0.2261 .2213 .2197 .8616	r_{ij} Raw p Boot p Adj p				
Run Pulse	−0.3980 .0266 .0249 .3579	0.9298 .0000 .0000 .0000	0.3137 .0858 .0834 .6833	r_{ij} Raw p Boot p Adj p			
Resting Pulse	−0.3994 .0260 .0257 .3579	0.3051 .0951 .0930 .6966	0.4504 .0110 .0104 .1804	0.3525 .0518 .0501 .5507	r_{ij} Raw p Boot p Adj p		
Body Weight	0.1435 .4412 .4360 .8889	0.2494 .1761 .1742 .8509	0.1435 .4412 .4360 .8889	0.1815 .3284 .3273 .8889	0.0440 .8143 .8168 .8889	r_{ij} Raw p Boot p Adj p	
Age	−0.3044 .0959 .0960 .6966	−0.4329 .0150 .0144 .2305	0.1888 .3092 .3082 .8889	−0.3379 .0630 .0611 .5979	−0.1641 .3777 .3707 .8889	−0.2335 .2061 .2099 .8616	r_{ij} Raw p Boot p Adj p

FIGURE 6.5. Summary of correlations and multiple tests for treadmill data.

such linear combinations usually are not anticipated, and are therefore not included in an *a priori* enumeration of tests of interest.

To accommodate *all* contrasts, including those suggested by the data, it is safest to define the family to include infinitely many tests. All the methods prescribed so far have assumed a finite collection of tests. The minimum p-value statistics have been calculated from such finite collections. If the family is infinite, it is impossible to compute such a minimum directly. However, there are well-known mathematical theorems that allow such calculations indirectly. A cornerstone of simultaneous inference techniques for univariate and multivariate analysis is a particular matrix maximization result, an extended version of the Cauchy-Schwarz inequality. Commonly used methods relying on this result are "Scheffé's method" for judging all contrasts in an ANOVA, and the "Roy-Bose method" for judging all linear combinations of variables in a multivariate analysis.

Letting A denote a real, symmetric, positive definite ($n \times n$) matrix, and **x** denote an arbitrary ($n \times 1$) real vector, the result is as follows:

$$\max_{\mathbf{a} \neq 0} \frac{(\mathbf{a}'\mathbf{x})^2}{\mathbf{a}'\mathbf{A}\mathbf{a}} = \mathbf{x}'\mathbf{A}^{-1}\mathbf{x} \tag{6.2}$$

From the resampling standpoint, this result is useful as it allows easy calculation of the maximum of infinitely many resampled t-statistics for each resampled data set. The general form of the t-statistic in univariate and multivariate applications is $t = (\mathbf{a}'\mathbf{x})/\sqrt{\mathbf{a}'\mathbf{A}\mathbf{a}}$, where the vector **a** defines the contrast to be tested, $\mathbf{a}'\mathbf{x}$ is the value of the estimated contrast, and $\sqrt{\mathbf{a}'\mathbf{A}\mathbf{a}}$ is the standard error of the estimate. When computing adjusted p-values for two-sided tests, one can use the maximum of the squares of the t-statistics, since they are inversely monotonically related to the two-tailed p-values. Thus, rather than compute the maximum of all squared t-statistics, simply compute the value $(\mathbf{x}^*)'(\mathbf{A}^*)^{-1}(\mathbf{x}^*)$ for each resampled data set and compare its value with the squared t-statistics from the original data.

When the data are properly distributed according to normal or multivariate normal laws, the distribution of the maximum $\mathbf{x}'\mathbf{A}^{-1}\mathbf{x}$ is simply related to the F distribution, and the adjusted p-values can be computed analytically. However, when the data have nonnormal characteristics, the F approximation can be poor. The validity of the F approximation can be judged by resampling, as illustrated by the following example.

Example 6.6 Infinitely Many Contrasts in Repeated Measures—Dog Anesthesia Data Revisited

In Section 4.4.1, we presented an example of a repeated measures analysis for which $n = 19$ dogs were each subjected to four different treatments. In Table 4.6 we tested nine different treatment contrasts using multivariate normal resampling. Now we will consider the problem of the bootstrap resampling of *infinitely many* contrasts, in comparison with bootstrap resampling of only the nine contrasts.

As given in Section 4.4.1, the normal-based pivotal t-statistic for testing a general contrast is

$$T = \frac{\hat{\boldsymbol{\mu}}\mathbf{d}}{(\mathbf{d}'\mathbf{S}\mathbf{d}/n)^{1/2}} \tag{6.3}$$

where $d_1 + d_2 + d_3 + d_4 = 0$. (Recall that $\boldsymbol{\mu}$ is a (1×4) row vector.) The maximization result (6.2) refers to *all* vectors **a**, not to contrast vectors. The standard method for accommodating contrast vectors is as follows: let **C** denote the contrast matrix

$$\mathbf{C} = \begin{pmatrix} 1 & 1 & 1 \\ -1 & 0 & 0 \\ 0 & -1 & 0 \\ 0 & 0 & -1 \end{pmatrix}$$

and redefine the t-statistics as

$$T = \frac{\hat{\mu}\mathbf{Cb}}{(\mathbf{b}'\mathbf{C}'\mathbf{SCb}/n)^{1/2}} \tag{6.4}$$

The linear span of {**Cb**, for all (3 × 1) vectors **b**} is identical to that of {**d**, for all (4 × 1) contrast vectors **d**}, and the maximization lemma may be applied using (6.4). Comparing (6.4) with (6.2), let $\mathbf{a} = \mathbf{b}$, $\mathbf{x} = \mathbf{C}'\hat{\mu}'$, and $\mathbf{A} = \mathbf{C}'\mathbf{SC}/n$. Then

$$\max_{\mathbf{b}\neq 0} \frac{(\hat{\mu}\mathbf{Cb})^2}{(\mathbf{b}'\mathbf{C}'\mathbf{SCb}/n)} = \hat{\mu}\mathbf{C}(\mathbf{C}'\mathbf{SC}/n)^{-1}\mathbf{C}'\hat{\mu}' = Q \quad \text{(say)} \tag{6.5}$$

If the data are from a multivariate normal distribution, the quantity Q is distributed as $\{(n-1)(q-1)/(n-q+1)\}F_{q-1,n-q+1}$ (e.g., Johnson and Wichern, 1988, p. 218). In this case $n = 19$, q = # of variables = 4, and $F_{q-1,n-q+1}$ denotes a random variable having the F distribution with $q-1$ numerator and $n-q+1$ denominator degrees of freedom. Thus, adjusted p-values for infinitely many contrasts may be computed directly, without bootstrap sampling, as

$$\tilde{p}_{\mathbf{d}} = \Pr(\{(n-1)(q-1)/(n-q+1)\}F_{q-1,n-q+1} > t_{\mathbf{d}}^2)$$

where $t_{\mathbf{d}}$ denotes the observed value of the test statistic for contrast **d** of (6.3). These values can be computed exactly, and they are reported in Table 6.6 for nine selected contrasts under the column heading "Multivariate Normal, $k = \infty$." The column "Multivariate Normal, $k = 9$" shows the values reproduced from Table 4.6 for comparison.

The values from Table 4.6 are simulated from the estimated multivariate normal model using 20,000 simulations. As expected, the p-value adjustments are not as large

Table 6.6 Infinitely Many versus Finitely Many Contrast Tests for Dog Anesthesia Data

	Cell Means				Multivariate Normal		Bootstrap	
Effects	μ_1	μ_2	μ_3	μ_4	$k = 9$	$k = \infty$	$k = 9$	$k = \infty$
Pairwise								
1 vs. 2	+1	−1	0	0	.1001	.1476	.1436	.2117
1 vs. 3	+1	0	−1	0	.0000	.0000	.0006	.0116
1 vs. 4	+1	0	0	−1	.0000	.0000	.0000	.0029
2 vs. 3	0	+1	−1	0	.0008	.0024	.0147	.0467
2 vs. 4	0	+1	0	−1	.0000	.0003	.0042	.0283
3 vs. 4	0	0	+1	−1	.2907	.3588	.3208	.3954
ANOVA								
Halothane	−1	−1	+1	+1	.0000	.0000	.0000	.0048
CO_2	+1	−1	+1	−1	.0138	.0286	.0496	.0930
Interaction	+1	−1	−1	+1	.9297	.9459	.9362	.9503

when only nine members of the family are considered, compared to infinitely many. The discrepancy between the $k = 9$ and $k = \infty$ entries indicates the potentially large price one must pay for unlimited data dredging. On the other hand, some contrast tests are significant in the $k = \infty$ column, even though infinitely many tests are considered.

In Chapter 4, Example 4.4, we note that these data are not well behaved: Figure 4.5 in Chapter 4 shows varying degrees of skewness in the measurements. Bootstrap sampling can be used to assess the effect of such nonnormalities upon the multiplicity-adjusted p-values, as discussed in Section 4.2.2. The data are resampled using

$$\mathbf{Y}_i^* = \boldsymbol{\epsilon}_i^*$$

In this case, the residual vectors are computed as $\boldsymbol{\epsilon}_i^* = \mathbf{Y}_i - \hat{\boldsymbol{\mu}}$. With a finite number of contrast vectors $\mathbf{d}_j$, $j = 1, \ldots, k$, the resampling-based multiplicity adjustments require the evaluation of $\max_j(\hat{\boldsymbol{\mu}}^*\mathbf{d}_j)^2/(\mathbf{d}_j'\mathbf{S}^*\mathbf{d}_j/n)$ for each bootstrap data set. These maxima are then compared to each of the observed squared t-statistics in the original analysis. With infinitely many contrasts $\mathbf{d}$, the maximum of the bootstrap squared t-statistics is $Q^* = \hat{\boldsymbol{\mu}}^*\mathbf{C}(\mathbf{C}'\mathbf{S}^*\mathbf{C}/n)^{-1}\mathbf{C}'(\hat{\boldsymbol{\mu}}^*)'$, assuming $\mathbf{C}'\mathbf{S}^*\mathbf{C}$ is invertible. (This matrix should be invertible for most, if not all bootstrap data sets, provided the data are fairly continuous and the sample sizes are not too small. See also the discussion in Section 3.1.4 concerning zero variance estimates in bootstrap data sets.)

The bootstrap multiplicity-adjusted p-values are then computed by comparing the value of Q^* to each original squared t-statistic. For a given contrast, a multiplicity-adjusted p-value is defined as the proportion of bootstrap samples for which the value Q^* exceeds the squared t-statistic for that particular contrast. Table 6.6 displays the bootstrap multiplicity adjustments using $k = 9$ and $k = \infty$ contrasts. The adjustments are larger for $k = \infty$, as they should be. Comparing the "Multivariate Normal" columns with the "Bootstrap" columns, substantial effects of nonnormality are indicated: the bootstrap p-values are much larger. □

An interesting feature of this method is that the adjusted p-values do not depend on the size of the family of tests. Any contrasts can be examined, even contrasts suggested by the data; the multiplicity-adjusted p-values remain conservative since infinitely many tests are allowed. However, when a finite number of tests is considered, the degree of multiplicity adjustment always depends upon the size of the family chosen—larger families mean more adjustment. Infinite-family multiplicity adjustments give the upper bound on the amount of adjustment required for increasingly larger families.

For more general applications, such as tests for multivariate regression of the form $H\colon \mathbf{c}'\boldsymbol{\beta}\mathbf{d} = f$ (Section 4.2.1), the method is similar. The t-statistic for testing this hypothesis is

$$T(\mathbf{c}, \mathbf{d}, f) = \frac{\mathbf{c}'\hat{\boldsymbol{\beta}}\mathbf{d} - f}{(\mathbf{d}'\mathbf{S}\mathbf{d})^{1/2}[\mathbf{c}'(\mathbf{X}'\mathbf{X})^{-1}\mathbf{c}]^{1/2}}$$

and the bootstrap version is

$$T^*(\mathbf{c},\mathbf{d}) = \frac{\mathbf{c}'\hat{\boldsymbol{\beta}}^*\mathbf{d}}{(\mathbf{d}'\mathbf{S}^*\mathbf{d})^{1/2}[\mathbf{c}'(\mathbf{X}'\mathbf{X})^{-1}\mathbf{c}]^{1/2}}$$

The maximum of $\{T^*(\mathbf{c},\mathbf{d})\}^2$ over all possible $(\mathbf{c},\mathbf{d})$ can be computed using the following extended version of inequality (6.2):

$$\max_{\mathbf{a}\neq 0} \frac{\mathbf{a}'\mathbf{B}\mathbf{a}}{\mathbf{a}'\mathbf{A}\mathbf{a}} = \mathrm{ch}_1(\mathbf{A}^{-1}\mathbf{B})$$

where $\mathrm{ch}_1(\mathbf{M})$ denotes the largest characteristic root (eigenvalue) of a matrix $\mathbf{M}$. Applying this result, along with matrix algebra calculations yields

$$\max_{\mathbf{c}\neq 0,\mathbf{d}\neq 0} \{T^*(\mathbf{c},\mathbf{d})\}^2 = \mathrm{ch}_1\{(\hat{\boldsymbol{\beta}}^*)'(\mathbf{X}'\mathbf{X})\hat{\boldsymbol{\beta}}^*(\mathbf{S}^*)^{-1}\}$$

This expression is one form of the "greatest root" so common in MANOVA tests.

When the class of vectors $\mathbf{c}$ and/or $\mathbf{d}$ includes contrasts only, the given maximum must be redefined to incorporate contrast matrices.

Using these expressions for the maximum of the test statistics, resampling-based multiplicity adjustments are obtained by comparing the mathematical maximum (greatest root) with the experimentally observed values of the squared t-statistics. For a specific test, the resampling-based adjusted p-value is then computed as the proportion of bootstrap samples for which the bootstrapped greatest root statistic exceeds the original squared t-statistic for that test.

6.5 POWER

A major criticism frequently leveled at multiple testing methods is their lack of power: it is much more difficult to reject hypotheses, and therefore reach firm conclusions, when multiple testing methods are used. By this point, we have (hopefully) demonstrated the other side of the coin: without adjustment, it is *too easy* to reject null hypotheses, thereby reaching potentially erroneous conclusions. Power is an important issue when comparing alternative methodologies, as well as when determining sample sizes.

In this section we offer some ideas concerning power and multiple testing. It must be emphasized that power is difficult to compute, since the test statistics are based upon simulations. One may always "simulate the simulation." This is a cumbersome and potentially time-consuming process, although a recent article by Oden (1991) shows ways to reduce the level of computational effort.

In univariate testing applications, the power of a test is defined as

$$\text{Power} = \text{Probability of rejecting } H \text{ when } H' \text{ is true}$$

For typical univariate applications, the hypothesis H has the form $H\colon \theta = 0$, and the power may be computed for any given $\theta \neq 0$. Graphing the power for all $\theta \neq 0$ yields the power curve.

In multiple testing situations, power is not so easily defined. Assume there are k null hypotheses H_i. Suppose for some subset $K \subseteq \{1, \ldots, k\}$, all H_i are true when $i \in K$, and all remaining hypotheses are false (false for $i \in K'$). Various definitions of power include

(1) the probability of correctly rejecting *at least one* H_i, for $i \in K'$,
(2) the probability of rejecting *all* H_i, for $i \in K'$,
(3) the probability of rejecting *exactly one* H_i, for $i \in K'$, and
(4) for a *specific* $i \in K'$, the probability of rejecting H_i.

The appropriate definition of power in multiple testing situations depends upon the researcher's goals. Researchers ideally would like a high power using definition (2); however, this is a rather stringent definition, and power is usually very low using this definition. When a particular hypothesis is labelled important prior to data collection, definition (4) is attractive. Definition (1) is most compatible with multiple testing methods that aim to control the FWE at α, since the power function approaches the nominal FWE level α as the parameters approach the complete null configuration. In the case of rodent carcinogenicity testing, where a compound is indicted as a carcinogen if *any* test is positive, definition (1) is appropriate. An example of the use of definition (1) occurs in Figure 5.4, Section 5.4.2, where the power functions of bootstrap and permutation multiple testing are compared.

Definition (3) has little value, but is included to show that many definitions are possible. We also could have considered the probability that *exactly two* hypotheses in K' are rejected, and so on.

6.5.1 Power Comparisons for Alternative Testing Methods

Use of definition (4) highlights an important drawback of multiple testing. Suppose that the hypothesis $H_i: \theta_i \leqslant 0$ is tested against the alternative $H_i': \theta_i > 0$. Let us make the simplifying assumption that the hypotheses are tested using z-statistics Z_i, all independent, with Z_i distributed as $\mathcal{N}(\theta_i, 1)$. The θ_i may denote noncentrality parameters. For example, consider testing whether the mean of a normal population is equal to μ_{0i}, assuming known variance σ_i^2. Then $Z_i = \sqrt{n}(\bar{Y}_i - \mu_{0i})/\sigma_i$, distributed as $\mathcal{N}(\theta_i, 1)$, where $\theta_i = \sqrt{n}(\mu_i - \mu_{0i})/\sigma_i$.

For given i and $\theta_i > 0$, the unadjusted hypothesis test rejects H_i when $Z_i > z_\alpha$, where z_α denotes the $1 - \alpha$ quantile of the standard normal distribution. The power function of the test is therefore $1 - \Phi(z_\alpha - \theta_i)$, where $\Phi(.)$ denotes the cumulative standard normal distribution function. On the other hand, using multiple testing, H_i is rejected only when $Z_i > z_{\alpha'}$, where $\alpha' = 1 - (1 - \alpha)^{1/k}$ (Šidák method used since Z's are assumed independent). Thus, the power of the multiple testing method is $1 - \Phi(z_{\alpha'} - \theta_i)$, which can be much smaller than $1 - \Phi(z_\alpha - \theta_i)$, implying much less likelihood of rejecting H_i when H_i is false. Of course, this loss of power is balanced by the FWE protection. The experimenter must then decide: which is the lesser evil, claiming a result to be true when it is not true, or failing to claim a result that is true? In this book we have argued that incorrectly claiming results to be true is the greater infraction, but we acknowledge that this may not be true in every instance.

We consider now the comparison of unadjusted tests with multiplicity-adjusted tests, as well as with the meta-analytic test that uses $\sum Z_i$ as a test statistic. Using definition (1), power is defined as the probability of rejecting *at least one* false hypothesis. A situation very favorable to meta-analysis is the case where all θ_i's measure similar quantities. Most favorable is the situation $\theta_i \equiv \theta$. When $\sum Z_i > k^{1/2} z_\alpha$, the meta-analytic test rejects the complete null hypothesis, allowing the conclusion that at least one of the constituent hypotheses is false. As a function of θ, the power of the meta-analytic test is then

$$\text{Power}_{\text{meta}} = 1 - \Phi(z_\alpha - k^{1/2}\theta)$$

For the unadjusted tests, we have

$$\begin{aligned} \text{Power}_{\text{unadjusted}} &= \Pr(\text{Reject at least one } H_i) \\ &= 1 - \Pr(Z_i \leqslant z_\alpha \text{ for all } i) \\ &= 1 - \{\Phi(z_\alpha - \theta)\}^k \end{aligned}$$

Similarly,

$$\text{Power}_{\text{adjusted}} = 1 - \{\Phi(z_{\alpha'} - \theta)\}^k$$

Now consider a situation less favorable to the meta-analysis test: $\theta_1 = \theta$, and $\theta_2 = \theta_3 = \cdots = \theta_k = \theta/100$. In this situation, there is one dramatically false hypothesis, H_1, and the deviations from the null are small at all other sites. Here,

$$\begin{aligned} \text{Power}_{\text{meta}} &= 1 - \Phi\{z_\alpha - k^{-1/2}\theta(1 + (k-1)/100)\} \\ \text{Power}_{\text{unadjusted}} &= 1 - \Phi(z_\alpha - \theta)\{\Phi(z_\alpha - \theta/100)\}^{k-1} \end{aligned}$$

and

$$\text{Power}_{\text{adjusted}} = 1 - \Phi(z_{\alpha'} - \theta)\{\Phi(z_{\alpha'} - \theta/100)\}^{k-1}$$

Figure 6.6 displays the power functions for these three methods when $k = 10$ and $\alpha = .05$, using the two different parameter configurations. For both parameter configurations, the unadjusted method has the most power. However, when $\theta = 0$, the power function of the unadjusted method intercepts the y-axis well above .05, indicating a large probability of claiming false significances.

Both the meta-analytic test and the multiplicity-adjusted tests have power functions intercepting the y-axis at .05, indicating that both properly control the FWE rate. In the case where all θ_i are equal, the power function for the meta-analytic test is clearly superior to that of the multiplicity-adjusted tests; however, for the second parameter configuration, the power function for the multiplicity-adjusted tests is clearly superior. (Actually, the meta-analytic test has slightly higher power near $\theta = 0$ in the second parameter configuration, but the difference is practically negligible.) Further comparisons concerning power functions of combined tests are given by Louv and Littell (1986).

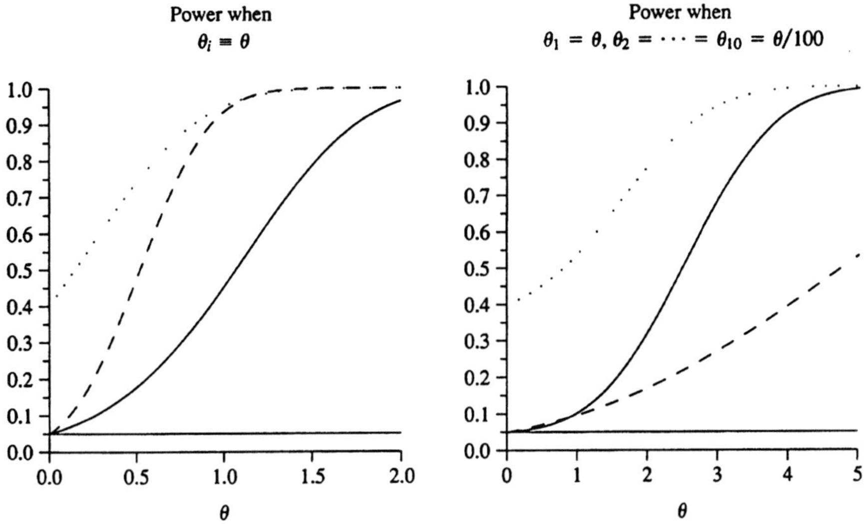

FIGURE 6.6. A comparison of power functions for multiple testing methods: no adjustment (dotted line), meta-analysis (dashed line), and multiplicity-adjusted (solid line).

The methods of this book are geared specifically towards the multiplicity-adjusted tests, which may be less powerful than combined, meta-analytic tests. With meta-analysis, *all* the data are directed toward answering a single question, in effect increasing the sample size. However, the power is not always greater with meta-analysis, as illustrated by the second parameter configuration of Figure 6.6. Additionally, we have provided real examples in this book where multiplicity-adjusted tests have smaller p-values than meta-analytic tests (e.g., Example 5.1).

Use of alternative definitions of power highlights an additional benefit of using multiplicity-adjusted p-values. Consider definition (4), where power is defined as the probability of correctly declaring a *particular* hypothesis to be false. Because meta-analytic tests *may not* declare particular hypotheses false, the power of the meta-analytic test is *identically zero* using definition (4), and the multiplicity-adjusted test is uniformly more powerful.

Again, the specific focus of multiplicity adjustment is the evaluation of *individual hypotheses*, while controlling overall error rates. The power comparisons given in this section highlight the benefits and pitfalls of multiplicity adjustment. On the positive side, it allows site-specific conclusions, and controls overall FWE error rates. On the negative side, multiplicity-adjusted tests lack power compared to meta-analytic tests for the types of parameter configurations suggested by Figure 6.6, and they certainly lack power relative to unadjusted test procedures, also shown in Figure 6.6.

6.5.2 Power and Sample Size

For continuous applications and well-behaved data, the power of resampling-based multiple tests can be approximated by assuming normality and independence of all tests. Incorporation of dependence structures will reduce the p-values, so any sample sizes determined using such approximations may be considered conservative. The degree of conservativeness depends upon the nature of the correlations among variables and tests.

Consider the following hypothetical example: multivariate data will be observed in each of two groups, treatment and control. Upper-tailed two-sample comparisons will be constructed for each variable. The alternative hypotheses are that the treated measurements are greater, on average, than the corresponding control measurements.

If the data are reasonably well-behaved (no outliers), then the results of bootstrap multiple testing and multivariate normal resampling-based multiple testing should agree fairly well, as shown by examples in Chapters 3 and 4. We have also noted that the effect of independence can be slight, particularly in applications where there is only one test per variable (no multiple contrast tests). Thus, we can assume for convenience that the distributions are multivariate normal, homoscedastic, and having all variables independent.

Suppose that an equal number n of treated and control vectors are to be sampled. The test statistics are

$$T_j = \sqrt{\frac{n}{2}}\,\frac{\bar{Y}_{2j} - \bar{Y}_{1j}}{s_j}$$

where s_j is the pooled standard deviation estimate for variable j. This statistic has the central t-distribution with $\nu = 2(n-1)$ degrees of freedom under the null hypothesis H_{0j}: $\mu_{1j} = \mu_{2j}$. Under the alternative hypothesis, the statistic is distributed as noncentral t with $\nu = 2(n-1)$ and noncentrality parameter

$$\delta_j = \sqrt{\frac{n}{2}}\,\frac{\mu_{2j} - \mu_{1j}}{\sigma_j}$$

where σ_j^2 is the variance for variable j, assumed constant for both groups.

Assuming an interest in individual tests, we will adopt definition (4) for power computation. Suppose then that we wish to declare test j significant when in fact $(\mu_{2j} - \mu_{1j})/\sigma_j >$ Constant, where "Constant" denotes a *scientifically significant* increase. That is, if the difference in the means exceeds an important fraction of within-group standard deviation, we wish to declare a significance for variable j. Given all our assumptions, the adjusted p-value for variable j will be significant, i.e. $\tilde{p}_j < .05$, when $T_j > t_{\nu\alpha'}$, where $\alpha' = 1 - (1-\alpha)^{1/k}$, and where $t_{\nu\alpha'}$ is the $1 - \alpha'$ quantile of the central t-distribution with ν degrees of freedom.

The probability that test j will be declared significant can be computed using the noncentral t distribution, available in many statistical software packages. Taking $k = 10$ and $\alpha = .05$, Figure 6.7 graphs the power as a function of n for different constant proportions of the standard deviation.

While a graph such as Figure 6.7 is easily constructed and can be used to guide sample size selection when multiple tests are planned, there are many assumptions implicit in constructing this graph that may not be true when resampling-based methods are to be used. First, the assumption of independence among variables makes adjusted p-values larger, hence the power smaller, than when correlations are

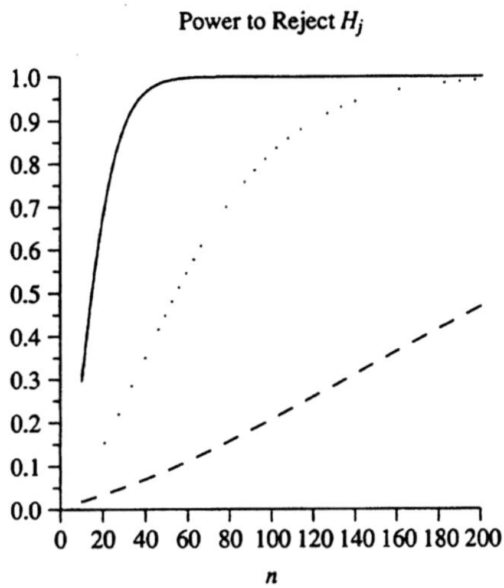

FIGURE 6.7. Power functions for detecting H_j when $\mu_{2j} - \mu_{1j} = .25\sigma_j$ (dashed line), $\mu_{2j} - \mu_{1j} = .50\sigma_j$ (dotted line), and $\mu_{2j} - \mu_{1j} = \sigma_j$ (solid line).

incorporated. It is expected that the power curves will be higher when correlations are incorporated, thereby reducing the needed sample size. This effect will not be very large, however, unless many correlations are extremely large. (As a rough guide, we suggest that "many correlations are extremely large" if most exceed 0.70 in absolute value, or if a principle components analysis of the correlation matrix shows that more than 80% of the variation is explained by a small number of components.) Second, the assumption of normality may not be satisfied. If bootstrap adjustments are performed, possible nonnormalities may affect the adjusted p-values. As indicated in Chapter 2, Section 2.5, the effect of nonnormality on the performance of the multiplicity adjustments depends upon the nature of the skewnesses of the variables. This effect can be quite complex in general applications, and its assessment difficult. Our best suggestion is to simulate data having skewness and other distributional characteristics similar to that expected of the data; in addition to a *scientifically important* alternative hypothesis configuration. Then one may compute resampling-based adjusted p-values, and simulate the entire process to evaluate power. Historical data can be very helpful in creating data with distributional characteristics similar to those of the data to be sampled.

If resampling-based multiple testing is to be used with multivariate binary data, one should ask: what is the *effective* number of tests? As discussed in Chapter 5, when a variable has a very small binomial probability, it will not contribute to the adjusted p-values. Of course, if the treatment affects a particular variable, then the probability of occurrence becomes much higher, and that variable may well contribute to the multiplicity adjustment when data from control and treatment groups are pooled. The analyst needs to determine three things: (1) how many total binary variables are to be considered, say k_1; (2) of these, approximately how many have background rates (in the absence of treatment) small enough not to affect the multiplicity adjustments, say k_2; and (3) of those variables with small background rates, approximately how many are likely to be affected by the treatment, raising the number of sample occurrences enough to make a difference in the multiplicity adjustment, say k_3? It will be expected that $k = k_1 - (k_2 - k_3)$ variables in the study will have observed rates large enough to affect multiplicity adjustments. Power may be approximated using standard normal approximations for k test statistics, as well as the assumption of independence.

Alternatively, one may create pseudo-populations from historical data bases to reflect scientifically important alternative hypothesis configurations and expected discreteness characteristics. Pseudo-data may be simulated from such pseudo-populations, resampling-based p-values computed, and the entire process simulated to evaluate power (see Chapter 5, Section 5.6, on the use of historical data). This is admittedly a cumbersome approach, but it has the advantage of conceptual simplicity. The problem of evaluating power with resampling-based tests from multivariate binary data is quite complex, both from the analytical perspective and from the computational perspective.

6.6 ASYMPTOTIC RESULTS

Frequently, resampling methods are justified using asymptotic arguments, where "asymptotic" refers to behavior of the procedure as the sample size (or sizes) becomes large. Since we are willing to assume that careful analysts will choose simulation sample sizes large enough to guarantee stable conclusions, we refer only to experimental sample sizes, not to simulation sample sizes.

Intuitively, as sample sizes get larger, empirical distribution functions of the original data should better approximate the true underlying distribution functions. Since resampling often involves sampling from such empirical distribution functions, the behavior of resampling-based methods should approximate methods based upon sampling from the true unknown distribution functions, as sample sizes become large. These intuitive concepts have very firm mathematical underpinnings. An early reference is Bickel and Freedman (1981); numerous discussions of resampling-based asymptotics followed in journals such as *The Annals of Statistics*.

In this section we provide some asymptotic results, one in a general setting and two in binary data settings. We begin with a discussion of single-sample bootstrap asymptotics for binary data, demonstrating that bootstrap coverage estimates for the multiple confidence intervals are consistent. In the case of two-sample binary data, we give an example to show that the single-step multiplicity adjustments are at least asymptotically conservative, even under partial null hypotheses.

While the single-step testing methods are usually asymptotically conservative, the step-down methods are often asymptotically *exact*, or more specifically, asymptotically *consistent*. By consistency, we mean the probability of falsely rejecting any hypothesis that is a member of the set K of true null hypotheses approaches α. We discuss situations for which the step-down resampling-based method is asymptotically consistent under any subset hypothesis, when certain mild conditions hold. This property is a major advantage of resampling-based testing. Virtually all competing methods are not asymptotically consistent, rather, they are asymptotically conservative.

6.6.1 Multivariate Binary Asymptotics—Single Sample Confidence Intervals

In the single-sample binary data application of Section 5.5, the bootstrap method was used to estimate the simultaneous confidence level

$$C_\alpha = \Pr(\pi_j \in I_\alpha(Y_j), \text{ for all } j)$$

for Clopper-Pearson intervals $I_\alpha(Y_j)$, when the observed data are $Y_j \sim \mathcal{BIN}(\pi_j, n)$. Use of bootstrap vector-based resampling yields the coverage estimate

$$\hat{C}_\alpha = \Pr(\hat{\pi}_j \in I_\alpha(Y_j^*), \text{ for all } j) \tag{6.6}$$

Because we are considering the behavior of the estimate as the sample size n becomes large, we denote the actual simultaneous coverage rate by C_α^n and the bootstrap estimate by $\hat{C}_\alpha^n$. This estimate assumes there is no Monte Carlo error. The bootstrap estimate obtained in practice is $\widehat{\hat{C}_\alpha^n}$, which differs from $\hat{C}_\alpha^n$ because of binomial simulation error. The difference becomes negligible as the simulation sample size *NBOOT* increases. We will ignore this source of error, which in practice means a very large value of *NBOOT*, as well as a good random number generator.

Assuming the binary vectors $\mathbf{B}_i$ are independent and identically distributed according to a $\mathcal{MVB}_k(\boldsymbol{\pi}, 1, \mathbf{D})$ distribution, and that the marginal probabilities π_j are strictly between zero and one, the following convergence result is established by Westfall (1985):

$$\hat{C}_\alpha^n - C_\alpha^n \longrightarrow_{\text{a.s.}} 0$$

Here, "$\longrightarrow_{\text{a.s.}}$" denotes "almost sure convergence" (e.g., Rao, 1973, p. 110).

Thus, with larger and larger sample sizes, the bootstrap estimate provides increasingly accurate estimates of the true coverage level. This property is not shared by competing simultaneous coverage estimates, such as those using the Roy-Bose-type confidence ellipsoids, or Bonferroni-style inequalities: these methods instead provide asymptotically conservative coverage estimates.

A similar use of the bootstrap to estimate the true coverage levels of confidence intervals is given by Loh (1987). He considers ordinary (e.g., parametric) confidence intervals for parameters such as variances, and uses the bootstrap to estimate the true confidence level when the underlying distributions are nonnormal. Again, the bootstrap estimates are consistent, with the difference between estimated coverage probability and true coverage probability tending to zero.

6.6.2 Multivariate Binary Asymptotics—Two-Sample Multiple Tests

In this section we wish to show that the single-step resampling-based method described in Section 5.3.1 asymptotically controls the FWE, even under subset hypotheses. Following Westfall and Young (1989a), consider the multivariate binomial two-sample case where $\mathbf{B}_{11}, \ldots, \mathbf{B}_{1n_1}$ are i.i.d. $\mathcal{MVB}_k(\boldsymbol{\pi}_1, 1, \mathbf{D}_1)$, and $\mathbf{B}_{21}, \ldots, \mathbf{B}_{2n_2}$ are i.i.d. $\mathcal{MVB}_k(\boldsymbol{\pi}_2, 1, \mathbf{D}_2)$, with the $\mathbf{B}_{1j}$ independent of the $\mathbf{B}_{2j}$. This model may apply, for example, to vectors of binary adverse event indicators for individual patients in control and treated groups.

Consider the null hypotheses H_{0j}: $\pi_{1j} = \pi_{2j}$ versus the single-tailed alternatives H_j^l: $\pi_{1j} < \pi_{2j}$, for $j = 1, \ldots, k$. These may be tested using the two-sample Z-statistic

$$Z_j = \frac{\hat{\pi}_{2j} - \hat{\pi}_{1j}}{\sqrt{\hat{\pi}_j(1 - \hat{\pi}_j)(\frac{1}{n_1} + \frac{1}{n_2})}}$$

which uses the pooled probability estimate $\hat{\pi}_j = (Y_{1j} + Y_{2j})/(n_1 + n_2)$.

If a particular subset K_0 of null hypotheses is true, then there are $m = |K_0|$ true hypotheses. Assume $1 \leqslant m \leqslant k$. Without loss of generality, we may order the parameter vectors $\boldsymbol{\pi}_1$ and $\boldsymbol{\pi}_2$ so that the m variables corresponding to true nulls come first:

$$\boldsymbol{\pi}_1 = (\,\pi_1 \quad \ldots \quad \pi_m \quad \pi_{1(m+1)} \quad \ldots \quad \pi_{1k}\,)$$

and

$$\boldsymbol{\pi}_2 = (\,\pi_1 \quad \ldots \quad \pi_m \quad \pi_{2(m+1)} \quad \ldots \quad \pi_{2k}\,)$$

When the resampled data are generated from the same set of pooled data, the multivariate resampling distributions are *identical* for each group. This suggests that the resampling-based test assumes that the true distributions are *identical* for both groups. In reality, only the assumption of common covariance matrices is required for the subvectors corresponding to the first m components: letting $\mathbf{B}_i = (\mathbf{B}_i^{(1)} : \mathbf{B}_i^{(2)})$ denote the partitioned vector, we assume that

$$\text{Cov}(\mathbf{B}_1^{(1)}) = \text{Cov}(\mathbf{B}_2^{(1)}) = \Sigma_{11} \tag{6.7}$$

This is much less restrictive than the assumption that the multivariate binary distributions of $\mathbf{B}_1^{(1)}$ and $\mathbf{B}_2^{(1)}$ are identical. Since the diagonal elements of $\text{Cov}(\mathbf{B}_1^{(1)})$ and $\text{Cov}(\mathbf{B}_2^{(1)})$ are identical under the subset hypothesis K_0, (6.7) requires only equality of the off-diagonal elements. Thus, under the null hypothesis $H_0^{K_0}$, the covariance matrix of a multivariate binary vector from group i, $i = 1, 2$, is assumed to be

$$\text{Cov}((\mathbf{B}_i^{(1)} : \mathbf{B}_i^{(2)})) = \begin{pmatrix} \Sigma_{11} & \Sigma_{i12} \\ \Sigma_{i21} & \Sigma_{i22} \end{pmatrix}$$

With two-sample data, asymptotic results require that both n_i's tend to infinity. So let n_1 and n_2 tend to infinity such that $n_1/(n_1 + n_2)$ tends to a constant between zero and one. In this asymptotic setting, with the given assumptions concerning the common covariances of the binary vectors, Westfall and Young (1989a) show that the probability of the resampling method reaching a false positive conclusion is bounded asymptotically by α. Specifically, if the single-step resampling method of Section 5.3.1 is used, with infinitely many simulations, and with FWE significant results declared whenever $\tilde{p}_j \leqslant \alpha$, then

$$\Pr(\text{reject at least one of } H_{0j}, j = 1, \ldots, m) \longrightarrow_{\text{a.s.}} d_\alpha$$

where $d_\alpha \leqslant \alpha$. When all hypotheses are true ($m = k$), the method is *consistent*: $d_\alpha = \alpha$.

6.6.3 Consistency of Step-Down Resampling-Based Tests

In the previous section, an application was given for which single-step resampling-based testing with multivariate binary data is at least asymptotically conservative. When *all* null hypotheses are true, the method is *asymptotically consistent* since the actual false rejection probability approaches the nominal FWE level. This improves upon the Bonferroni-style adjustments, for which the true false rejection probability usually converges to a number *strictly smaller* than the nominal FWE level. However, when using step-down resampling-based methods, asymptotic consistency may be proven *even under partial null hypotheses*. That is, assuming *any* non-empty subset K_0 of hypotheses $H_0^{K_0}$ is true, the probability of falsely rejecting any hypothesis H_{0j}, $j \in K_0$, converges to the nominal FWE level α. In what follows we give a specific situation for which this consistency property holds.

Consider the notation of Section 2.8: p-values P_i are used to test hypotheses H_{0i}. Assume the subset pivotality condition also holds, and that the subset hypothesis $H_0^{K_0}$ is true, which implicitly means that all hypotheses H_{0i}, $i \in K_0'$, are false. Assume further that the p-values are continuous random variables, and that adjusted p-values of Section 2.8 may be evaluated without error. These assumptions hold for parametric models such as the homoscedastic normal model.

Consider an asymptotic sequence of p-values indexed by a sample size n for which

(1) $\max_{i \in K_0'} P_i^{(n)} \mid H_0^{K_0'} \longrightarrow_w 0$ as $n \longrightarrow \infty$

Here, "$\longrightarrow_w$" means "convergence in probability" (e.g., Rao, 1973, p. 110). Condition (1) typically is satisfied in most hypothesis testing applications: when the alternative hypothesis is true, the probability of correctly rejecting any H_{0i}, $i \in K_0'$, converges to one as the sample sizes get larger.

Also, for given $\alpha \in (0, 1)$, define $x_K^{(n)\alpha}$ to be the α *quantile* of $\min_{i \in K} P_i^{(n)} \mid H_0^C$, and assume there exists $\epsilon_\alpha > 0$ such that

(2) $\min_{K \subseteq S} x_K^{(n)\alpha} \geqslant \epsilon_\alpha$, for all n

When the marginal null distributions are uniform for all n, this property is trivially satisfied by taking $\epsilon_\alpha = \alpha{=}k$.

Finally, assume that

(3) $\min_{i \in K_0} P_i^{(n)} \mid H_0^{K_0} \longrightarrow_d P^\infty$ as $n \longrightarrow \infty$, where "$\longrightarrow_d$" denotes convergence in distribution (e.g., Rao, 1973, p. 116), and P^∞ is a continuous random variable.

This third condition is commonly satisfied since the joint distribution of test statistics from which the p-values are calculated is commonly asymptotically multivariate normal. This leads to asymptotically uniform marginal distributions of the p-values.

Consider the testing rule "Reject H_{0i} if $\tilde{p}_i \leqslant \alpha$," where the adjusted p-values are either the free step-down adjustments of (2.10), or the restricted step-down adjustments (2.13). Given the preceding assumptions, we have the following consistency property:

$$\Pr(\text{Reject at least one } H_{0i}, \text{ for } i \in K_0 \mid H_0^{K_0}) \longrightarrow \alpha$$

To demonstrate this property, choose any $\epsilon \in (0, \epsilon_\alpha)$, and notice the following nesting of probabilistic events:

$$\begin{aligned} &\{\text{Reject at least one } H_{0i}, \text{ for } i \in K_0\} \\ &\quad \supseteq \{\{\max_{i \in K_0'} P_i^{(n)} < \epsilon\} \cap \{\epsilon \leqslant \min_{i \in K_0} P_i^{(n)} < x_{K_0}^{(n)\alpha}\}\} \end{aligned}$$

In words, this nesting states that the event "Reject at least one H_{0i}, for $i \in K_0$" will occur when all false hypotheses are rejected first in the step-down algorithm, then one of the true hypotheses is rejected. This is not the only way that the event "Reject at least one H_{0i}, for $i \in K_0$" can occur, but it is asymptotically the most plausible scenario.

Conditions (1) and (3) imply

$$\begin{aligned} &\Pr\left(\left\{\max_{i \in K_0'} P_i^{(n)} < \epsilon\right\} \cap \left\{\epsilon \leqslant \min_{i \in K_0} P_i^{(n)} < x_{K_0}^{(n)\alpha}\right\}\right) \\ &\quad \longrightarrow \Pr(\epsilon \leqslant P^\infty < x_{K_0}^\alpha) \\ &\quad = \alpha - \Pr(P^\infty < \epsilon) \end{aligned}$$

where $x_{K_0}^\alpha$ denotes the α quantile of the distribution of P^∞.

Thus, for arbitrary $\epsilon \in (0, \epsilon_\alpha)$, the probability of falsely rejecting at least one H_{0i}, $i \in K_0$, is asymptotically between α and $\alpha - \Pr(P^\infty < \epsilon)$. Taking ϵ arbitrarily small, the probability of falsely rejecting at least one H_{0i}, $i \in K_0$, is seen to converge to α.

6.7 MORE COMPLEX MODELS

The models considered so far have been simple, where resampling plans can be constructed with relative ease. However, more complicated models demand more complex resampling plans. Resampling is *model-driven*: the model determines the resampling plan, not vice versa. Users of resampling should first decide which model best suits the available data, then construct a resampling plan to mimic the properties of this model, using the data to estimate unknown model characteristics.

With simple models, simple resampling plans work very well. For example, the empirical distribution of the pooled multivariate binary vectors is a reasonable estimate of the underlying multivariate distribution under the complete null hypothesis, so resampling of these data provides reasonable multiplicity adjustments. For continuous location-shift data, the model dictates that the data be centered before resampling. And when testing variances using the location-scale model, the model dictates that centering and scaling are appropriate. These resampling plans are simple, but they incorporate important model assumptions.

In more complex models, such as logistic regression and proportional hazards models, simple "sampling with replacement" plans will not do. Rather, parameters of the model should be estimated, then data sets should be *generated from the estimated model*. Simple *with replacement sampling* will not generate data satisfying the assumed form of the model in such situations.

For example, in a logistic regression model, the dependent variable Y is binary, 0 or 1, satisfying

$$\text{logit}\{\Pr(Y = 1 \mid X_1, \ldots, X_v)\} = \beta_0 + \beta_1 X_1 + \cdots + \beta_v X_v$$

Here, $\text{logit}(p) = \ln(\frac{p}{1-p})$. The observed data are $Y_i, X_{i1}, \ldots, X_{iv}$, $i = 1, \ldots, n$. As in ordinary regression analysis (e.g., Chapter 4), inference in logistic regression is *conditional on the observed* X's; thus we wish to create resampled data sets of the form $Y_i^*, X_{i1}, \ldots, X_{iv}$, $i = 1, \ldots, n$. Having obtained estimates $\hat{\beta}_j$, $j = 0, 1, \ldots, v$ (preferably maximum likelihood), "resampled" data sets may be constructed using

$$Y_i^* = \begin{cases} 1, & \text{with probability } \dfrac{\exp(\hat{\beta}_0 + \hat{\beta}_1 X_{i1} + \cdots + \hat{\beta}_v X_{iv})}{1 + \exp(\hat{\beta}_0 + \hat{\beta}_1 X_{i1} + \cdots + \hat{\beta}_v X_{iv})} \\[2ex] 0, & \text{with probability } \dfrac{1}{1 + \exp(\hat{\beta}_0 + \hat{\beta}_1 X_{i1} + \cdots + \hat{\beta}_v X_{iv})} \end{cases}$$

Note that this data-generating process may not be replicated by simple *with replacement* sampling of the raw Y's, or of the residuals, as for ordinary regression analysis.

For multiple testing purposes, the parameters β_j should be computed under the complete null restriction H_0^C; thus they are restricted maximum likelihood estimators. Standard asymptotic p-values may then be used, Wald or likelihood ratio, for the individual tests of hypotheses and multiplicity adjustments. The subset pivotality condition is problematic: the joint distribution of the test statistics corresponding to true null hypotheses will differ under partial and complete null hypotheses. Strictly speaking, resampling-based adjustments test only the complete null hypothesis H_0^C in this case; however, the adjusted p-values still provide an assessment of the chance variability inherent in the raw p-values.

Logistic regression is one example of a complex model requiring more complex resampling plans. Another example is given by Soper and Westfall (1990), who use the general resampling method to analyze multiple tumor data in animal carcinogenicity studies. While such data are simply analyzed using a multivariate binomial probability model, the actual process is much more complex. Differing treatment doses commonly cause different mortality; therefore, the simple analysis of binomial counts may underestimate tumor frequencies in groups with higher mortality. Another problem in this analysis is that tumors classified either as *lethal* or *incidental* should be analyzed differently. Methods for analyzing these data are described in Peto et al. (1980).

Peto et al. consider the univariate testing problem: for a given tumor, they test the null hypothesis that the mortality-adjusted tumor prevalence and tumor lethality functions are identical for all treatment groups. The derived test statistic for tumor l is Z_l, having an approximately standard normal distribution under the null hypothesis. Larger values of Z_l suggest that the compound causes more tumors of type l. There are assumed to be L tumor types, so $l = 1, \ldots, L$. Formulas for calculating Z_l are given by Soper and Westfall (1990), and in Appendix A.III.2 of this book.

Soper and Westfall consider the distribution of the max Z_l under the *complete null hypothesis* that tumor lethality and prevalence functions are identical for all treatment groups. It often happens that higher doses affect survival, but survival is not always of primary interest in tumorigenicity studies. The complete null hypothesis of Soper and Westfall allows the survival functions to vary by treatment group.

The model for such data includes the functions

$$\boldsymbol{\theta} = [F_1, \ldots, F_L; G_0, \ldots, G_J; H_1, \ldots, H_k]$$

Here, F_l is the distribution of the variable "time to death from tumor type l" (assumed constant for all treatment groups). The distributions G_j denote the time until death from non-tumor causes for animals in treatment group j, $j = 0, 1, \ldots, J$. The distributions H_k, $k = 1, \ldots, K$ denote multivariate tumor prevalence functions for the K time strata, allowed to vary by time strata, but assumed constant for all treatment groups under the complete null hypothesis.

Assuming multivariate versions of the *non-informative censoring* and *representativeness* assumptions (the assumptions reduce to the univariate assumptions required for the validity of the Peto test), Soper and Westfall estimate all distribution functions, obtaining $\hat{\boldsymbol{\theta}}$. Data are then generated from these estimated distribution functions, yielding "resampled" data sets. The Peto test statistics are then computed from the resampled data sets, and the p-values are adjusted using the resampled minimum p-value statistic.

One problem with this method is the many parameters to be estimated in $\boldsymbol{\theta}$. When data are sparse, the estimate $\hat{\boldsymbol{\theta}}$ is expected to be inaccurate. This inaccuracy may explain the empirical result noted by Soper and Westfall: the resampling-based p-values tend to be too small. To reduce this problem, they consider an analysis conditional upon observing two or more tumors for each variable.

Further work is needed to develop adequate resampling methods for such complex applications. The model used by Soper and Westfall is one application where the resampling method cannot be reconciled with simple *with replacement* and *without replacement* resampling of data, which is why there is no resampling option in place for use with the "PETO" tests in PROC MULTTEST (see Appendix III). In general, resampling methods should be model-driven. A naive resampling scheme may implicitly estimate an underlying model that differs greatly from the analyst's conception of the model.

CHAPTER 7

Practical Applications

The most important consideration in using any statistical method is whether the method can solve or give insight into important practical problems. Until now we have emphasized theory over applications, showing that the resampling approach works well in the sanitized world of clean data where idealized probability models hold. In this chapter, we discuss through examples the basic issues involved in the application of multiple testing concepts to specific disciplines.

Multiple testing considerations are relevant to real world problems in important disciplines—medicine, business, physical and engineering sciences, and social sciences. In any situation where there is sampling or other natural variation, users and consumers of statistics must be aware that estimates are subject to variation. Further, when the data are examined from multiple angles and the most extreme features are selected, users must also be aware that the reported significances are possibly overstated. Since natural and other variation occurs in every field of study in which statistics is used, this problem is ubiquitous.

Of course, if one is absolutely certain that all null hypotheses are false, then the selection issue is not one of reporting distorted significance levels, but rather a possible problem of reporting distorted *effect sizes*. For example, if we run an experiment where we *know* (God tells us!) that each of five treatments, T1–T5, is 2% better than the control, C. In an empirical determination of efficacy, it is expected that one of the treatments will have a claimed efficacy far in excess of the actual 2% figure, due to sampling variability. The actual estimated efficacy will be related to sample sizes and process variation. Reporting simultaneous interval estimates can alleviate this problem to some extent. However, when many hypotheses are planned to be tested, there often is substantial doubt whether there are any effects at all. If in fact many or most null hypotheses are true, then use of unadjusted p-values will easily lead to one or more false significances.

When there is doubt concerning whether a hypothesis is true or false, then a hypothesis testing procedure is in order. This notion is Bayesian—"doubt" can be expressed as a prior probability that H is true. In multiple testing situations, there is often substantial doubt concerning a *collection* of null hypotheses, and it is reasonable to consider it *a priori* probable that many or all null hypotheses are in fact true. In this case one may *expect, a priori*, that some unadjusted p-values will erroneously indicate statistical significance.

This chapter is a collection of examples. We set out to illustrate three points: First, there are subject areas where multiple questions are asked and it is probable that many or all null hypotheses hold. Second, using the data analysis paradigm, complex data analysis problems can be greatly simplified. And third, data analysis with multiplicity adjustment can be depicted using various tabular and/or graphical summaries to provide a quick and compact assessment of experimental or survey results.

Section 7.1 is a collection of examples where multiple testing is an important issue. These examples are discussed, and common elements are identified. In Section 7.2–7.4, we provide worked examples with continuous, categorical, and ordered categorical data. An application of the methods to a large and complex epidemiological survey is analyzed and discussed in detail in Section 7.5.

7.1 MOTIVATING EXAMPLES

Often, substantial doubt exists concerning the collection of hypotheses tested. In this section we identify some of these situations. We give a wide variety of examples where multiple testing is a problem, and where p-value adjustment is a solution.

Example 7.1 ESP Studies

In experimental parapsychology, a subject may participate in a battery of tests to assess various types of extra-sensory perception (ESP), such as telepathy or precognition. A particular null hypothesis may assert that a subject, or collection of subjects, does not possess a particular type of ESP ability. There is ongoing research in this field, with journals such as *Journal of the Parapsychology* and *Journal of the American Society for Psychical Research* devoted to the topic. As with most scientific journals, these require that claims be substantiated using observed data, which often entails significance testing.

Most will agree that multiplicity is a definite problem in the area of psychical research. In *most* tests of such hypotheses, the null hypothesis is actually true, that is, there is no ESP in the subject or subjects. Thus, there will be many false indicants of ESP—approximately 5% of all such tests will be significant at the 5% level. If ESP journals required statistical significance and did not report non-significant studies, then a reporting bias would result. To reduce reporting bias, *Journal of Parapsychology* has formally adopted the editorial policy of *not* rejecting a manuscript just because it does not find statistical significance. The term "file drawer effect" was coined to describe the fact that non-significant studies remain in the "file drawers" of researchers, while a select group of "significant" studies are published. The more insignificant studies that exist in "file drawers," the more suspect are the few published significant studies. Rosenthal (1979) describes this phenomenon in more detail, and provides an estimate of the number of insignificant studies in "file drawers" that would be needed to overturn the significance of the many published statistically significant results.

Using familywise significance levels is one way to reduce the proportion of false significances. If there are 100 tests performed in a given study, then there is only a 5% chance that *any* test among the 100 will be declared significant after multiplicity adjustment using a familywise significance level of 5%. Without multiplicity adjustment, it is highly probable that at least one out of the 100 tests will be declared significant.

In parapsychology experimentation, such conservativeness seems mandatory. Some may feel that the usual multiplicity adjustments are not conservative enough: if 20 independent experiments are done, with 100 tests performed within each experiment, there remains a $1 - .95^{20} = .64$ chance of falsely declaring at least one test significant, even when the tests are adjusted for multiplicity within each experiment. This probability calculation assumes that all null hypotheses in every experiment are in fact true, which might be reasonable to assume in ESP studies.

William Feller (1969) was skeptical of statistically significant differences in parapsychology and gave the following illustration:

> Another favorite method of producing "significant differences" is to test only selected portions of the data. The procedure is extensively used in many fields, and is illustrated by an example taken from an early volume of the *Journal of Parapsychology.* To test his extrasensory abilities, a "prominent experimental physicist" made a series of 3,500 card guesses in which the probability of a hit was 1/5. The expected number of guesses was therefore 700, whereas the experimenter obtained 711 hits. The difference of 11 was not judged to be significant. However, the series was broken up into three subseries according to the time of day and the feeling of the author. He presented the outcome as follows:

	Number of guesses	Number of correct guesses exceeding chance
Morning, well:	2,100	56
Evening, tired:	1,100	−29
Morning, ill:	300	−16

> The subseries "morning well" has a significant critical ratio of 2.99, and because of this the experimenter would have us believe that he is endowed with extrasensory perception despite his poor overall performance. In this sense, clairvoyance is, of course, absolutely universal. The trouble is that breakfast or lunch, fitness or illness, coffee or beer, or combinations of these may be required to produce the desired effect—and the nature of the favorable conditions is subject to instant change.*

This is not to claim that all ESP research is worthless. Diaconis (1978) discusses various statistical problems in ESP research impartially. There are interesting, apparently repeatable, small effects found in the ESP literature. Utts (1991) provides a provocative and controversial overview of replicated experimental results, and

*W. Feller, "Are Life Scientists Overawed by Statistics?" *Scientific Research* (February 3, 1969), with permission of McGraw-Hill, Inc.

suggests that telepathy, clairvoyance, and/or precognition, collectively called "psi" abilities, are possessed by some individuals. □

Consideration of multiple testing in ESP brings to mind the farcical question often voiced by detractors of multiplicity adjustment: "Why not multiplicity adjust all tests performed in a researcher's lifetime?" This practice ordinarily would be considered completely unacceptable, but may actually have merit with ESP experiments. Suppose, for example, that a particular ESP researcher intends to prove that a particular type of ESP exists, when, in reality, it does not. Then all null hypotheses tested by that experimenter in his or her lifetime will be true, and multiplicity adjustment "over the researcher's lifetime" will provide the correct conclusions with 95% probability.

In most cases, of course, it is not expected that all null hypotheses tested in the lifetime of a researcher will be true. Even in ESP studies, there are apparently repeatable results which suggest that certain null hypotheses are false. We do not suggest that all ESP tests ever performed be thrown into the same family for multiplicity adjustment; however, it is only appropriate that some multiplicity adjustment be performed in such studies. The level of "surprise" indicated by a small p-value in a study should definitely incorporate the fact that many tests were performed. It is also quite reasonable to calculate multiplicity adjustments under the complete null hypothesis, as the resampling-based methods do, as it is likely that all null hypotheses tested might actually be true.

The issue of prior likelihood concerning collections of null hypotheses is important. The ESP area represents an extreme case, for most people doubt whether *any* of the effects tested are real and repeatable. Given the importance of *a priori* reasoning in multiplicity adjustment, we suggest the following guidelines for determining the "family" of tests over which multiplicity adjustment is to be performed:

Guidelines for Selection of a "Family" of Tests

- The questions asked form a natural and coherent unit. For example, they all result from a single experiment, such as indicated in Section 1.6.
- All tests are considered simultaneously. (For example, when the results of a large study are summarized for publication, all tests are considered simultaneously. Usually, only a subset of the collection is selected for display, but the entire collection should constitute the "family" to avoid selection effects.)

Perhaps most important:

- It is considered *a priori* probable that many or all members of the family of null hypotheses are in fact true.

The previous example concerning ESP studies suggests that such *a priori* concerns are of great importance in identifying when to consider multiple testing issues and multiplicity adjustment strategies.

Example 7.2 Animal Carcinogenicity Studies

For the U.S. Food and Drug Administration (FDA) to approve a new drug, it requires evidence of the drug's safety and effectiveness. As one part of the demonstration of drug safety, the FDA usually requires that the company submit results of a study in which animals (usually rats and mice) are subjected to varying doses of the compound for most of their lifetime. At the end of the trial, which usually lasts two years, any surviving animals are killed and the frequencies of many untoward outcomes, such as death, cancer, blindness, etc., are compared for the various treated groups.

In such two-year rodent carcinogenicity studies, cancer is often a primary outcome of interest, lending the name "rodent carcinogenicity study" to such experiments. Many tumors will be present naturally due to the advanced age of the animals and will not be caused by the test compound. The difficulty is to distinguish tumors caused by the drug from those that occur spontaneously. Typically there are 20 to 50 tissues—heart, liver, brain, lymphatic system, etc.—which are examined for any of several possible cancers. In the evaluation of a compound, these experiments are commonly performed using every combination of a two sexes × two species plan, greatly increasing the number of endpoints in the evaluation of the carcinogenic nature of the compound.

It is useful to count the number of statistical tests computed in a two-year rodent study. In two-year rodent tests, a compound is evaluated to see if it induces any of a large number of tumors. Four National Toxicology Program long-term rodent studies serve as examples of the problem's magnitude. Within each study, there are four subexperiments—males and females for both rats and mice. Several statistical tests, e.g. linear trend or control versus each treated group, are computed on each tumor type within all studies. Table 7.1 taken from Young (1989) summarizes the statistical testing on four NTP studies. For example, study 225 analyzes tumor data for rats, mice, males and females. There are 60 tests of hypotheses for male rats; there were 11 statistically significant increases and 11 significant decreases. Summaries for each study are given in the right most column. And a summary for each species/sex is given at the bottom of the table. For example, in the four studies there were a total of 2179 tests (about 545 tests per study!) and 6.06% were positive and 4.22% were negative. Given the number of tests per study, the correlations among the tests, and the sparseness of count data, it is difficult for expert toxicologists to judge the repeatability of observed results.

There is a reason that so many endpoints are considered in these studies: it is a common belief among toxicologists that if a compound is carcinogenic, its effects will be specific to one or a few sites. Therefore, it is considered important to isolate particular site-specific effects of a compound. A single omnibus (meta-analytic) test for "carcinogenicity" would not identify such particular effects, and might not be a powerful test for specific effects. It is usually not sufficient to claim that a new compound causes an increase in tumors at some unspecified site; rather, the report should identify the particular sites where unusual increases are noted. Such experimental considerations dictate separate tests for each site.

In these studies, the multiplicity problem is real and important. It is possible that a particular compound does not cause any tumors, yet when hundreds of separate tests

Table 7.1 Number of Significant $p < .05$ Results in Four NTP Long-Term Rodent Studies*

		Rats		Mice		Summary	
Report	Result	Male	Female	Male	Female	Total	%
225	Positive	11	1	3	1	16	9.8
	Negative	11	8	0	0	19	7.3
	Total Tests	60	45	33	24	162	
250	Positive	17	0	23	12	52	8.5
	Negative	7	0	1	0	8	1.3
	Total	207	162	135	108	612	
281	Positive	0	6	5	3	14	2.3
	Negative	27	8	8	3	46	7.9
	Total Tests	207	162	153	63	585	
309	Positive	28	13	9	0	50	6.1
	Negative	16	0	0	3	19	2.3
	Total Tests	279	180	198	162	819	
Summary	Positive	56	20	40	16	132	6.06
	Negative	61	16	9	6	92	4.22
	Total Tests	753	549	519	357	2179	

*From Young (1989), with permission from Academic Press.

are examined, one for each endpoint within each sex/species combination, there is a large chance of declaring the compound to have a carcinogenic effect. If erroneous, such a determination can be costly for the company planning to market the drug, in terms of money spent on the two year study and other research and development expenses, as well as in terms of potential lost revenue. The societal costs of lives not saved or pain suffered also are potentially large as well.

In order to alleviate this problem of false significances, some type of multiplicity adjustment seems mandatory. Clearly, such adjustments will make demonstrating *safety* of a drug easier for pharmaceutical companies. On the other hand, if multiplicity adjustment is performed blindly or unscrupulously, then unsafe drugs might be mistaken to be safe.

To give due consideration to Type II errors, that is, the problem of declaring compounds safe when they are in fact not safe, any particular site-specific associations that are considered probable *a priori* should not be included in the family over which multiplicity adjustment is to be performed. This will greatly improve the power of the procedure at that particular site. Among all remaining tests, there should be substantial doubt whether an effect is present in *any* location, justifying the practice of performing multiplicity adjustments for that family under the assumption of the complete null hypothesis. Of course, such actions need to be justified *before* collecting or analyzing data. Further, if there are *many* such suspected sites, then we are back where we started: the collection of suspected sites forms a "family" of tests, and multiplicity adjustment remains desirable, perhaps using simultaneous intervals rather than tests. □

Like the case of ESP experiments, there are many questions asked within the study of a compound, and there is some doubt whether *any* of the effects tested are real. We conducted a small survey of statisticians familiar with animal testing, including representatives from both industry and government, and asked the following question:

> Consider a "typical" long-term rodent carcinogenicity experiment. Suppose you are considering only one sex/species combination. Still, there are 30–50 tumor types that must be analyzed. Suppose you first remove from consideration any tumor types that are suspected, *a priori*, to be induced by the compound under study.
>
> The *remaining* tumor types are not suspected, *a priori*, to be associated with the compound, yet it is certainly possible that one of them will be associated with the compound. I ask you to give me a *probability* that *at least one* of these tumor types will *in fact* be associated with the compound. (I am *not* asking for the probability that at least one *p*-value will be found to be less than .05, I am asking whether there is in fact a *real* association for any of these tumor types.)
>
> Please answer "generically." Feel free to give your answer as a range of values, if you like. Also, feel free to comment.
>
> Probability Guess: ______

Twelve statisticians who were judged experts were surveyed, and nine responded. Among the respondents, eight offered a probability "guess," and all guesses were less than 50%. Some guesses were quite sophisticated, involving calculations with historical data bases. The educated guesses were between 10% and 40%. Many respondents were uncomfortable providing such blatantly subjective (Bayesian) responses, and one respondent flatly refused to give any such probability value.

While the survey is small, it suggests doubt whether tested compounds have *any* effect at nonsuspected sites. It is therefore reasonable to assess the probability of extreme occurrences by adjusting *p*-values under the complete null hypothesis of no tumorigenic effect at any site. Such adjustments are likely to be very useful to toxicologists and pathologists in the evaluation of the tumorigenic potential of a compound.

As an historical note, it is interesting that the original designers of the long-term rodent tests partially justified these experiments by arguing that they expected most compounds to be completely devoid of carcinogenic effect! Recently, Ames and Gold (1990) argued that most positive results are due to excessive doses that induce tumors through toxic cell killing, and that at moderate doses they would expect most compounds to be without carcinogenic effect.

Example 7.3 Adverse Events in Clinical Trials

Another example from the pharmaceutical field is the examination and comparison of adverse events experienced by patients in drug and placebo groups. This problem is similar to the problem of animal carcinogenicity testing with regard to multiple testing characteristics. Patients in both groups report various side effects; the list may include more than 20 possible ailments such as headaches, fatigue, and vomiting. Many or most of these side effects are unanticipated. Nevertheless, frequencies of side effects must be compared between the treatment and control groups as part

Table 7.2 An Example of Reported Side Effects

Treatment-Emergent Adverse Experience Incidence in Placebo-Controlled Clinical Trials (Percent of Patients Reporting)*

Adverse Experience	BuSpar $N = 477$	Placebo $N = 464$
Cardiovascular		
Tachycardia/Palpitations	1	1
CNS		
Dizziness	12	3
Drowsiness	10	9
Nervousness	5	1
Insomnia	3	3
Lightheadness	3	—
Decreased Concentration	2	2
Excitement	2	—
Anger/Hostility	2	—
Confusion	2	—
Depression	2	2
EENT		
Blurred Vision	2	—
Gastrointestinal		
Nausea	8	5
Dry Mouth	3	4
Abdominal/Gastric Distress	2	2
Diarrhrea	2	—
Constipation	1	2
Vomiting	1	2
Musculoskeletal		
Musculoskeletal Aches/Pains	1	—
Neurological		
Numbness	2	—
Paresthesia	1	—
Incoordination	1	—
Tremor	1	—
Skin		
Skin Rash	1	—
Miscellaneous		
Headache	6	3
Fatigue	4	4
Weakness	2	—
Sweating/Clamminess	1	—

*Events reported by at least 1% BuSpar patients are included.

—Incidence less than 1%.

Copyright © Physicians' Desk Reference®, (1991) ed., published by Medical Economics Data, Montvale, New Jersey 07645.

of the evaluation of the compound's safety. A listing of side effects of BuSpar® (published in the *Physicians' Desk Reference*®, Barnhart, 1991), given in Table 7.2 is representative of this type of data. A large number of side effects are reported. Considering that the events occurring in less than 1% of the BuSpar® or placebo patients are not included in the list, the actual number of side effects is potentially much larger. With such large numbers of reported side effects, statistically significant results can easily occur by chance alone. □

When one considers that many side effects are unanticipated, the need for multiplicity adjustment becomes clear. By "unanticipated" we mean that it is thought *a priori* probable that the incidence of a side effect is the same for treatment and control groups. In such a case it is reasonable to compute multiplicity adjusted p-values under the complete null hypothesis of no differences in treatment and control frequencies. When particular side effects are suspected, *a priori*, they should be excluded from the "family" over which multiplicity adjustment is to be performed.

One clinical statistician summarized the need for multiplicity adjusted p-values in clinical adverse events data as follows: "If a patient calls in about a particular side effect, upset stomach, then use the raw p-value. If a tort lawyer is looking to put his children through college, then use the adjusted p-value." The point here is that the reporting of statistical results should depend on how individuals plan to use the statistics. When a patient calls in with a specific complaint, then the raw p-value addresses the specific question. On the other hand, if a tort lawyer wishes to snoop through the data to find any potentially adverse events, then he or she may easily find a number of "significant" results that are actually spurious, and instigate a bogus legal proceeding based on these findings.

Example 7.4 HLA–Disease Associations
The human leukocyte antigen (HLA) system consists of several genetic markers on chromosome number 6. Particular markers have been associated with particular diseases. For example, the *B*5 marker has been associated with Hodgkin's disease (Amiel, 1967), and the *B*27 marker with anklylosing spondylitis (Svejgaard et al., 1975).

Demonstrating such associations has been very exciting for geneticists interested in inherited diseases, and many have attempted to discover other associations. As reported by Smouse and Williams (1982)

> literally scores of HLA–disease associations have been published. Each of these studies has attempted to find an HLA association with a particular disease, and a surprising number have succeeded.... Most of the reported associations, however, have not withstood closer examination.*

A major difficulty with such studies once again is the multiplicity problem. Usually, there are 20 or more markers identified, each of which is tested individually for

*From P. E. Smouse and R. C. Williams, "Multivariate Analysis of HLA-Disease Associations," *Biometrics* **38** 757–768, (1982), with permission of the Biometrics Society.

association with the particular disease in question. In many studies, the disease in question may not be inherited, or at least not inherited from the HLA complex, meaning that most null hypotheses tested are in fact true. Thus, when all markers are examined individually, it is probable that *at least one* will be erroneously declared significant, leading to published associations which are unrepeatable.

This problem is now recognized to the extent that multiplicity adjustment is considered mandatory by researchers in this area. In *The Lancet*, D'Amaro et al. (1984) make explicit use of multiplicity-adjusted p-values to demonstrate association of HLA C antigens with acute leukemia. On the same subject, Peter and Hawkins (1981) comment,

> One aspect of this [hypothesis testing] procedure that has generated some confusion is that if 20 antigens are tested in a single study and there is no prior reason to expect any one of them to occur more frequently in patients, then an antigen showing significance at the 1-in-20 level could be the one comparison in 20 that would be expected to show significance by chance alone. To overcome this problem, the probability associated with each comparison is frequently multiplied by the number of antigens tested for. This procedure has led to the appearance of the term, "corrected p-value," to refer to the product of the initial p-value and the number of antigens tested. This in turn has propagated the notion that there is something inherently incorrect about the "uncorrected" p-value. Consequently, researchers, editors and reviewers often demand that all p-values be corrected regardless of context. In fact, adjustment is necessary only if there is no prior hypothesis that a particular antigen should be associated with the disease. If there are prior reasons to suspect that a particular association exists, then no adjustment should be necessary.* □

The preceding quote reinforces two points: (1) when there is substantial prior doubt whether an association exists for *any* test, then multiplicity adjustment should be performed, and (2) particular hypotheses suspected to be false *a priori* should not be included in the family over which multiplicity adjustment is to be performed. The quote also contains an early reference to the idea of an "adjusted p-value" (called "corrected p-value" by Peter and Hawkins). The adjustment/correction these authors propose is the single-step Bonferroni adjustment $\tilde{p}_i = kp_i$. As we have shown, resampling-based adjustments are preferred, because of the discrete nature of such data.

Why is multiplicity adjustment viewed so differently by researchers in different fields of study? In the study of HLA-disease associations multiplicity adjustment is "demanded" by researchers, editors and reviewers; yet in other fields of study, simple unadjusted tests are frequently performed without any consideration for the multiplicity issue. Again, the answer may lie in prior reasoning. Whenever substantial doubt exists concerning the existence of *any* effect among the many effects that are tested, then multiple testing and multiplicity adjustment are important issues. In many HLA-disease association experiments there is such doubt, particularly if the researcher is the first to try to demonstrate a particular association.

*From Peter and Hawkins, "HLA Antigens and Disease," *Diagnostic Medicine*, (Jan/Feb 1981).

Example 7.5 Tests for Inefficiency in Financial and Betting Markets

We find examples for multiple testing adjustment in other areas of empirical study. The theory of "efficient markets" as developed by Fama (1970) states that investors cannot earn *excess* returns on their securities by using historical and/or publicly available information. If investors learn that certain public information leads to better decisions, then that information becomes incorporated into the price. The current price will then reflect accurately the risk and expected return of an investment, and the market price will be an efficient reflection of true worth.

Much empirical work has supported the hypothesis that markets are efficient. But if the theory is tested in many diverse situations, chance may lead to a few "significant" contradictions.

When testing for the validity of this theory, one typically constructs models to predict the return of a particular security. Excess returns are defined as returns over and above the average market return. The predictability of such excess returns can be tested by adding historical data and/or publicly available information into a regression model that includes the market return, then by determining whether the added variables' coefficients are statistically significant.

There is great opportunity for false indicants of market inefficiency to pop up in market efficiency studies. These studies often involve several distinct predictors of excess returns, with all such predictors tested in a variety of situations, including different companies and/or portfolios. If the market really is efficient with respect to the variables in question, and if many such tests of efficiency are performed, then it is likely that at least one test will show inefficiency by chance alone. However, if the collection of tests is specified in advance, and if multiplicity adjustment is performed over this collection, then it is more likely that the statistical analysis will correctly indicate that the market is efficient.

It has been argued that the market forces that make financial markets efficient also make sports betting markets efficient (Pankoff, 1968). That is, a bettor will not be able to use any historical and/or publicly available information to win money on sporting events.

As an example, the "betting line" used in a football game predicts the winning margin of the favored team. In statistical terms, "betting line" is the number of points given to the underdog so that the odds of winning are even. When a bettor places a bet on the favored team, he or she is predicting that this team will win by more points than given by the line; in other words, the favored team will "beat the spread." If the bettor places a bet on the underdog team, he or she is predicting that this team either will win outright or will lose by fewer points than indicated in the betting line.

If the football betting market is efficient, then the betting line should be an unbiased predictor of the actual difference (Favored team's final score) − (Underdog team's final score). Letting D denote this difference and L denote the line, one may attempt to prove market inefficiency by modeling a dependent variable $Y = D - L$ as a function of various candidate predictors in a multiple regression model. The various predictors used must be publicly available information. For example, Dare (1992) considered such variables as fan characteristics and team rankings when constructing regression models.

Table 7.3 p-Values for Regression Analysis of College Football Betting Market[1]

Year	t-Tests				Combined
	β_1	β_2	β_3	β_4	F-Test
1981	.603	.826	.872	.787	.267
1982	.582	.114	.936	.099	.390
1983	.999	.810	.834	.779	.949
1984	.936	.121	.617	.039*	.308
1985	.030*	.734	.749	.704	.040*
1986	.267	.772	.352	.303	.525
1987	.049*	.352	.159	.184	.224
1988	.889	.453	.497	.194	.487
1989	.276	.129	.667	.889	.156
All Years	.393	.180	.711	.638	.179

[1]Reprinted with permission of W. H. Dare.
*p-value $< .05$.

If *any* regression coefficient, including the intercept, deviates substantially from zero, there is evidence of market inefficiency. The issues of multiple testing and multiplicity adjustment are evident: if many candidate regressors are considered, then one or more tests may easily be found significant by chance alone. If such data are subdivided into different years, conferences, professional vs. collegiate, etc., then the multiplicity problem multiplies. One would like to consider as well the overall hypothesis test (F-test) that no parameter differs significantly from zero; therefore, the family under consideration should include this particular test in addition to all tests for individual regression coefficients.

Russo et al. (1989) compound the multiplicity problem of many predictors by introducing a "filtering constant," λ, which simply omits observations. The larger the λ, the more observations are deleted. Since it is not clear which λ should be used, they try several. This greatly increases the number of hypotheses tested, and the probability of finding significant results increases dramatically.

Dare (1992) performed a regression analysis of football betting markets using the model

$$Y = \beta_1 I_H + \beta_2 I_A + \beta_3(I_H L) + \beta_4(I_A L) + \epsilon$$

Here, the response variable Y is the difference in actual spread and the line:

$$Y = (\text{Favored team's final score} - \text{Underdog team's final score}) - \text{Closing line}$$

The variables I_H and I_A are $(1, 0)$ indicator variables to indicate whether the favored team is the home or away team, and L denotes the actual value of the closing line.

Table 7.3 displays some results obtained from this model. While four out of fifty p-values achieve significance at the $\alpha = .05$ level, they are easily explained as false significances. Indeed, the p-value plot of Schweder and Spjøtvoll (1982) presented

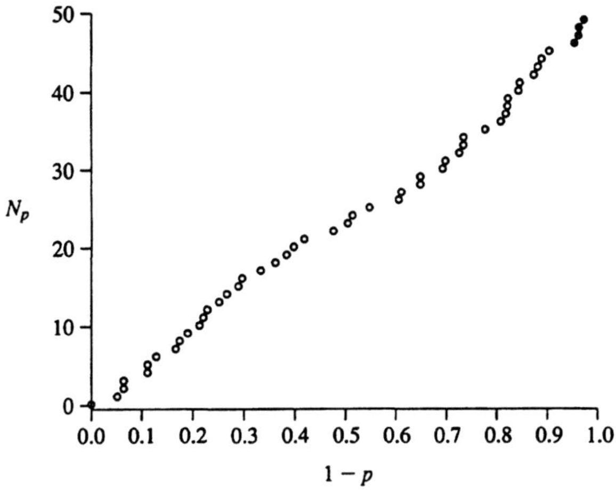

FIGURE 7.1. p-value plot for regression coefficients in a study of the efficiency of college football betting markets. Unadjusted p-values less than .05 indicated by •.

in Figure 7.1 shows very little deviation from a straight line, consistent with the statement that all null hypotheses are true. □

While this example is quite different from the preceding examples in terms of the subject area, it is similar in that there is good reason to believe that all hypotheses tested might actually be true. Once again, the use of multiplicity adjustments under the complete null hypothesis provides a fair evaluation.

Lovell (1983) concurs that data mining is an important concern for applied econometricians, and offers some Bonferroni-type methods for reducing multiplicity-driven conclusions.

Example 7.6 Exposure Effects at Low Toxicity Levels–Leukemia and Power Lines Researchers are continually seeking to demonstrate adverse effects of low levels of certain substances known to be toxic at high levels, e.g. secondary cigarette smoke. Much of their concern is in the best interest of public health, but a certain amount may be misplaced. Often researchers try desperately to prove an association, driven either by the "publish or perish" mentality, or by a "do good" attitude.

One controversial study purporting a link between childhood leukemia and proximity to power lines was reported in the February 8, 1991, *Wall Street Journal*, p. B4, and was also discussed by Morgan (1989). The results of the study were somewhat ambiguous: the apparent correlation between proximity to power lines and leukemia did not hold up when careful measurements of magnetic-field strength were used. Certainly, exposure to high levels of electro-magnetic fields might be dangerous, but the specific physiological effects, if any, at low levels are uncertain. If leukemia is increased at high exposure levels, then one also might expect increased leukemia at low exposure levels. The size of the effect at low exposure levels might be negligible from a practical standpoint, but when many tests and various forms of data manipulation are used, there is a good chance of finding a "statistically significant" or "highly statistically significant" result. Indeed, the *Wall Street Journal* article reports that the study included "a wide range of factors such as household chemicals, cigarette smoke, and parent's occupations." With multiple variables, various analyses can be performed, perhaps using alternative regression formulations, all of which can inflate the chance of finding a "significant" result, even when the true effect size is negligible. □

Applications involving low-level exposure differ somewhat from the previous applications in that there is *a priori* reason to suspect that many null hypotheses tested might be false, in other words, that effects really do exist. The question is whether the effect *sizes* are large enough to be considered important. While this question is best analyzed using multiplicity-adjusted interval-based inference, analysts often perform significance tests. Letting statistics substitute for ethical and economic analysis, analysts often act as if a "highly statistically significant" result reflects an effect size that is large enough to be considered important. When the effect sizes of all multiple tests are actually very small, multiplicity adjustment under the complete null hypothesis offers some protection against the possibility of declaring an effect large enough to be important, when it is not important at all. To improve the hypothesis testing procedure, researchers could specify in advance what a "large" (or scientifically important) effect size is for each hypothesis, then test whether the actual effect sizes exceed the given thresholds. In this instance the *a priori* plausibility of all null hypotheses (lower-tailed, for example) is relatively larger, enhancing the utility of multiplicity adjustments computed under the complete null hypothesis.

Example 7.7 Exposure Effects at Low Toxicity Levels—IQ and Lead

A very controversial example concerning small but "statistically significant" effects involves a current controversy over the effects of low levels of lead on the IQ's of school children (Palca, 1991, see Appendix II of this book). In *The New England Journal of Medicine*, Needleman et al. (1979) claim that "Children with high lead levels scored significantly less well on the Wechsler Intelligent Scale for Children (Revised) than those with low lead levels." This is not an inconsequential assertion; the Environmental Protection Agency is preparing changes in lead exposure standards that will be very expensive. Ernhart, et al. (1981) assert:

> However, incorporating a brief measure of parent IQ into the analysis decreased variance associated with lead and leads to a strong suspicion of the remaining significant results. For this and other admittedly difficult methodologic reasons, conclusions from prior studies are questioned.

Among the methodological reasons was the question of multiplicity:

> ...methodologic difficulties...(d) inadequate identification and handling of other confounding variables that affect development. To this we add an additional issue, multiple comparisons.*

*From *Pediatrics* **67** 911, (1981), with permission from *Pediatrics*.

Table 7.4 Teacher Ratings of Behavioral Effects of Lead

	Ernhart*		Needleman		
	Low Lead	Mod. Lead	Low Lead	High Lead	p-Value
Easily frustrated	23%	19%	11%	25%	.04
Hyperactive	20	9	6	16	.08
Impulsive	37	19	9	25	.01
Distractable	33	31	14	36	.003
Daydreams	13	25	15	34	.01

*From *Pediatrics* **67** 911–919, (1981), with permission from *Pediatrics*.

Ernhart et al. go on to claim that the statistically significant conclusions claimed by Needleman et al. were "probably unwarranted in view of the number of nonsignificant tests."

Unfortunately, a protracted and bitter dispute has developed between Needleman and Ernhart. Needleman et al. are cognizant of the multiplicity issue. As a footnote to one of their tables they state,

> It should be remembered that when a large no. of simultaneous comparisons are made between two groups of subjects, the probability that a "significant" result may be found is larger than the P value for a single test. An approximate & conservative adjustment for this fact may be obtained if the reported P value is multiplied by the no. of simultaneous tests.
>
> In this sense the "full-scale IQ" above constitutes a single test, & the "verbal IQ" and "performance IQ" constitute a pair of simultaneous tests.*

However, they do not state the total number of tests considered. This footnote suggests that each class of responses stands on its own: there may be adjustment within a class, but there is no adjustment over the seven classes of neurological response variables they examine. They also consider 39 potential confounders, and the various methods for adjusting for such confounding variables greatly increases the multiplicity problem.

Ernhart, et al. (1981) essentially repeat the study and find no evidence for decrease in IQ. The IQ finding with a p-value of .03 is not confirmed. Behavioral effects with p-values as small as .003 are also not confirmed; see Table 7.4. □

There is no doubt that high levels of lead are extremely harmful. A particularly troublesome aspect of reports on the effects of toxic substances at low levels is that the biological plausibility has already been established. This means that researchers may be extra-zealous in their attempts to discover associations—they are sure that an effect exists, so they may go to great lengths, including data mining, to prove it. Compounding this problem is the fact that journal editors and referees are more likely to review a finding favorably when there is biological plausibility, and tend to

*Reprinted, by permission of *The New England Journal of Medicine*, **300** 689–695, (1979).

be less concerned about the extent of data manipulation needed to prove the result. So sampling variability along with multiple testing can easily lead to false reports of low-level toxicity associations.

On the other hand, it is plausible that vitamins can lead to *improved* health, so the following example reported by Hamilton (1991) in *Science* is an interesting follow-up to the lead/IQ controversy.

Example 7.8 Brain Food? Vitamins and IQ

The following material is excerpted from *Science*:

> A report recently published in Britain has created new controversy over an old subject: attempts to improve children's IQs through nutritional supplements.
>
> Criminologist Stephen Schoenthaler of California State University in Turlock claims the study demonstrates that children's nonverbal IQs can be raised significantly by giving them double the current Recommended Daily Allowances (RDAs) of vitamins and minerals. The findings have stirred up considerable public attention in England, where they were announced on 28 February [1991] and published in the journal *Personality and Individual Differences*. The journal's editor, psychologist Hans Eysenck, is a coauthor.
>
> The double-blind experiment involved three groups of children and a control group. The subjects were given a dietary supplement of 23 vitamins and minerals over 13 weeks. Schoenthaler claims that vitamin and mineral supplements raised nonverbal IQ scores by an average of 3.7 points. (The gains were greatest for children in the lowest socioeconomic groups.) Oddly, the effect is reported only for supplements equaling 100% of the Recommended Daily Allowance and did not show up for the groups receiving 50% or 200% of RDAs.
>
> Schoenthaler is now planning a follow-up study with a group of 538 people aged 18–25. This time he's going to triple the vitamins and reduce the minerals in the group getting the most supplements. He says he suspects that too many minerals have a toxic effect.
>
> Although Schoenthaler says the findings have been widely hailed, the journal *Nature* sharply criticized the study in its 7 March [1991] issue, saying that of 87 comparisons conducted between subjects and controls, only seven yielded significant results. And the findings have so far been virtually ignored in the United States.* □

Finally, we briefly note two news-making results, both likely false positives, from the physical and astronomical sciences. In the physical sciences, there was a recent sensational claim by Fleischmann and Pons of the University of Utah that "Cold Fusion" could be made to occur in the laboratory. Dettmer (1989) discusses attempts at reproducing the result, finding "no statistically significant evidence of cold fusion." The fact that statistical evidence was considered at all suggests that the measurements are quite variable, and that the original "finding" of Fleischmann and Pons might be explained as a random occurrence.

In the astronomical sciences, Dessler (1991) disputes claims of Frank, Sigwarth and Craven (1986) that comets are bombarding the earth's atmosphere; concluding instead that the satellite images contained *randomly* darkened pixels that were

*From D. Hamilton, "Brain Food," *Science* **251** 1566, (1991).

misinterpreted as real phenomena. Again, the presence of randomness increases the possibility of claiming that a particular scientific theory is fact, even when the phenomenon under study has no real effect.

7.2 CONTINUOUS EXAMPLES

When data are continuous, regression models are commonly used for the analysis and attempted resolution of research questions. We begin this section with a pure continuous application to baseball hitting statistics, Example 7.9; we re-examine the biochemical levels data in Example 7.10, and finish continuous examples with an application to the simultaneous testing of means and variance in an industrial experiment, Example 7.11.

Example 7.9 Baseball Hitting Statistics
Baseball is a national sport where many statistics are kept and evaluated, both by team managers and by fans. Offense statistics are of particular interest. Regular players come to bat several hundred times over the course of a season; they get hits, score runs, get home runs, and drive in runs. Since many offense statistics are tabulated for each player, the data are multivariate.

Each player also plays a defensive position, typically one position, although there are some utility players that can field more than one position. The arrangement and position names of baseball players are given in Figure 7.2. Letting the defensive position be the grouping variable, the multiple offense statistics have the one-way MANOVA data structure.

Using this example we will go through the steps of the data analysis paradigm of Chapter 1, Section 1.6, and generate the necessary SAS code (DATA step and PROC MULTTEST) as we go along. Our purpose is to compare the offense statistics

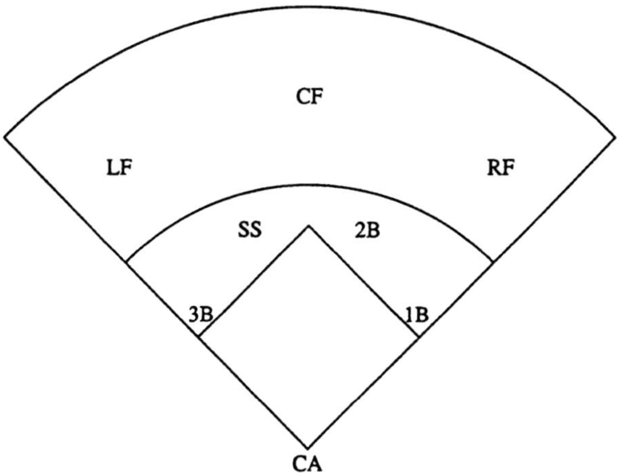

FIGURE 7.2. Players' positions on the baseball field: catcher (CA), first base (1B), second base (2B), third base (3B), shortstop (SS), left field (LF), centerfield (CF), and right field (RF).

for different defensive positions, to understand better the relationships between these variables. In particular, we wish to investigate whether offensive skills tend to be worse for positions requiring more defensive skill.

Our raw data set, $\mathbf{Y}_1$ of Section 1.6, consists of the batting statistics for all the National League hitters with more than 200 "at bats" (AB) in 1985. Six players were eliminated because their defensive position was unspecified, or because they played multiple positions. The raw data contains the name of the player, his team, his defensive position, the number of hits, at bats, runs scored, home runs, and runs batted in. As a first step in the analysis, we constructed the responses of interest, $\mathbf{Y}_2$, by dividing hits, runs, home runs and runs batted in by AB to make the players performance comparable in opportunity. The SAS code for this standardization, $\mathbf{Y}_1$ to $\mathbf{Y}_2$, is

```
data baseball;
   infile 'baseball.dat';
   input ab r h hr rbi name $ team $ pos $ ;
   h_ab = h/ab;
   r_ab=r/ab;
   hr_ab=hr/ab;
   rbi_ab=rbi/ab;
   if pos=' ' then delete;
```

The resulting $\mathbf{Y}_2$ data for a selected group of players is given in Table 7.5.

The second step in the data analysis paradigm is to create **S**, summary statistics. We are interested in the hitting statistics for the various positions, so we will summarize the player performance by position. The mean, standard deviation (Std Dev) and sample size (N) for each position are given in Table 7.6.

Table 7.5 $\mathbf{Y}_2$, Hitting Statistics for Baseball Players

Player	Team	Position	Hits/AB	Runs/AB	HR/AB	RBI/AB
McGee	StL	CF	.353	0.133	0.0165	0.186
Guerrero	LA	LF	.321	0.181	0.0688	0.202
Raines	Mon	LF	.318	0.070	0.0193	0.196
Gwynn	SD	RF	.316	0.075	0.0098	0.145
Hernandez	Mets	1B	.310	0.155	0.0170	0.148
Parker	Cin	RF	.309	0.196	0.0510	0.137
Sandberg	Chi	2B	.308	0.138	0.0431	0.187
Moreland	Chi	RF	.308	0.180	0.0241	0.127
Herr	StL	2B	.302	0.184	0.0135	0.162
Murphy	Atl	CF	.301	0.181	0.0605	0.191
Cruz	Htn	LF	.299	0.140	0.0131	0.123
Scioscia	LA	CA	.298	0.125	0.0165	0.109
Orsulak	Pit	RF	.296	0.054	0.0000	0.134
Oester	Cin	2B	.296	0.066	0.0019	0.114
Marshall	LA	RF	.294	0.183	0.0546	0.140

The third step is to write down the contrasts that address questions of importance. We are interested in the relationship between a player's defensive position and hitting performance. The players through the center of the playing field have more defensive responsibility: they move in both directions and cover more defensive area. Coaches might give defensive ability some preference over hitting ability in these positions. Infielders have more opportunities to catch and throw the ball, and might be selected for these skills rather than hitting. A notable exception is the first baseman: most hitters are right handed and hit the ball to the left side of the playing field, away from the first baseman. Also, he must typically play close to his base to receive throws.

Table 7.6 S, Summary of Hitting Statistics

Variable	Statistic	Position			
H_AB		1B	2B	3B	CA
	Mean	.2658	.2760	.2533	.2434
	Std Dev	.0243	.0254	.0196	.0276
	N	13.	12.	15.	13.
		CF	LF	RF	SS
	Mean	.2668	.2656	.2744	.2566
	Std Dev	.0345	.0290	.0285	.0154
	N	12.	14.	14.	14.
R_AB		1B	2B	3B	CA
	Mean	.1446	.1051	.1217	.1298
	Std Dev	.0403	.0376	.0278	.0218
	N	13.	12.	15.	13.
		CF	LF	RF	SS
	Mean	.1011	.1216	.1242	.0934
	Std Dev	.0430	.0430	.0499	.0284
	N	12.	14.	14.	14.
HR_AB		1B	2B	3B	CA
	Mean	.0340	.0145	.0266	.0275
	Std Dev	.0180	.0138	.0145	.0165
	N	13.	12.	15.	13.
		CF	LF	RF	SS
	Mean	.0206	.0288	.0270	.0127
	Std Dev	.0174	.0177	.0208	.0066
	N	12.	14.	14.	14.
RBI_AB		1B	2B	3B	CA
	Mean	.1288	.1376	.1161	.1005
	Std Dev	.0207	.0313	.0270	.0286
	N	13.	12.	15.	13.
		CF	LF	RF	SS
	Mean	.1438	.1429	.1397	.1122
	Std Dev	.0395	.0366	.0242	.0214
	N	12.	14.	14.	14.

Table 7.7 Treatment Contrasts

	Field Position							
Linear Contrast	1B	2B	3B	CA	CF	LF	RF	SS
Edge vs. Center	1	−1	1	−1	−1	1	1	−1
In vs. Outfield	−3	−3	−3	−3	5	5	5	−3
Right vs. Left	1	0	−1	0	0	−1	1	0
1st vs. the rest	7	−1	−1	−1	−1	−1	−1	−1

Agility may therefore be less of a premium, and a manager may place a strong hitter in this position.

We give linear contrasts to compare offensive statistics for various defensive positions in Table 7.7. While the coefficients in each row sum to zero, we have not attempted to make the contrasts orthogonal to one another. (The PROC MULTTEST code for this example is given in Appendix IV. One must be careful to make the order of the coefficients in the trend statements the same as the order of the groups in the input data, so the tests are correctly computed.)

The variables of interest are

H_AB	hits per at bat,
R_AB	runs per at bat,
HR_AB	home runs per at bat, and
RBI_AB	runs batted in per at bat.

There are $4 \times 4 = 16$ tests of hypothesis. As we are prepared to be incredulous of differences in the direction opposite to the presumed effects, we use upper-tailed testing.

A summary of the output from PROC MULTTEST output is given in Table 7.8. We mark any p-values less than .05 with an asterisk. Four tests have adjusted p-values smaller than .05. Players on the edge of the field hit more home runs and score more runs. The outfield has more runs batted in than the infield. And the first baseman scores more runs that the other players. □

Additional data displays are useful for assessing hitting performance of baseball players. Figure 7.3 gives H_AB, HR_AB, RBI_AB, and R_AB arranged on the playing field diagram. Examination of this figure in combination with the testing results helps determine the size of the effects—it is important to examine the size of the tested effects as well as the p-values. Table 7.9 gives the scaled contrast means with the raw and adjusted p-values. In this case the scaled contrast means are the difference between the average of one group of players and another. For example, the average H_AB for the edge players is $(.2658 + .2533 + .2656 + .2744)/4 = .264775$; for the center players it is $(.2760 + .2434 + .2668 + .2566)/4 = .2607$. The difference of .0041 batting average points, given in Table 7.9, is not statistically significant.

Table 7.9 is a particularly compact presentation of the results. Contrasts among the groups that match the experimental questions form one margin and the important variables form the other margin of this two-way table. Each cell in the table gives the

Table 7.8 Summary of *p*-Values

Contrast and Variable	*p*-Value			
	Raw_p	Adj_p	Sid_p	Bon_p
Edge vs. Center				
H_AB	0.2092	0.8801	0.9766	1.0000
R_AB	0.0027*	0.0375*	0.0418*	0.0426*
HR_AB	0.0007*	0.0117*	0.0114*	0.0115*
RBI_AB	0.0709	0.5429	0.6918	1.0000
In vs. Outfield				
H_AB	0.0291*	0.2943	0.3770	0.4664
R_AB	0.6685	0.9996	1.0000	1.0000
HR_AB	0.2302	0.9012	0.9848	1.0000
RBI_AB	0.0001*	0.0011*	0.0011*	0.0011*
Right vs. Left				
H_AB	0.0632	0.5028	0.6482	1.0000
R_AB	0.1036	0.6677	0.8263	1.0000
HR_AB	0.2602	0.9269	0.9919	1.0000
RBI_AB	0.2733	0.9358	0.9940	1.0000
1st vs. the rest				
H_AB	0.3230	0.9615	0.9981	1.0000
R_AB	0.0033*	0.0457*	0.0516	0.0528
HR_AB	0.0092*	0.1126	0.1378	0.1476
RBI_AB	0.4427	0.9896	0.9999	1.0000

*p-value < .05.

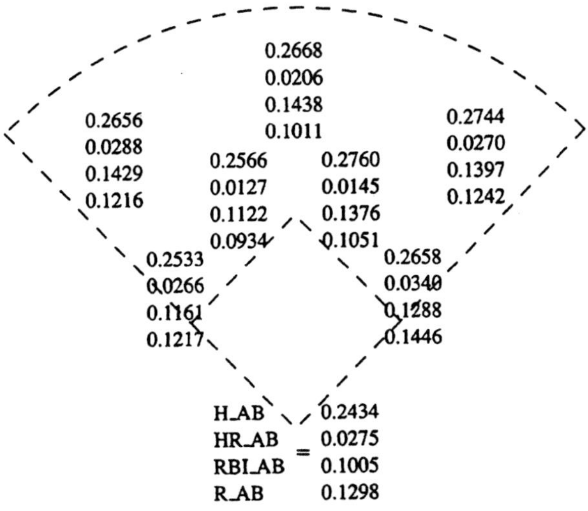

FIGURE 7.3. Offensive statistics for different defensive positions. The statistical key is given for the catcher's position.

Table 7.9 Average Sizes of Estimated Effects, with p-Values

		Variable			
Contrast	Statistic	H_AB	R_AB	HR_AB	RBI_AB
Edge vs. Center	Mean	0.0041	0.0207	0.0103	0.0084
	Raw_p	0.2092	0.0027*	0.0007*	0.0709
	Adj_p	0.8801	0.0375*	0.0117*	0.5429
In vs. Outfield	Mean	0.0099	−0.0033	0.0024	0.0231
	Raw_p	0.0291*	0.6685	0.2302	0.0001*
	Adj_p	0.2943	0.9996	0.9012	0.0011*
Right vs. Left	Mean	0.0107	0.0128	0.0028	0.0048
	Raw_p	0.0632	0.1036	0.2602	0.2733
	Adj_p	0.5028	0.6677	0.9269	0.9358
1B vs. Others	Mean	0.0035	0.0308	0.0115	0.0013
	Raw_p	0.3230	0.0033*	0.0092*	0.4427
	Adj_p	0.9615	0.0457*	0.1126	0.9896

*p-value < .05.

difference tested of an important question by important variable and the unadjusted and adjusted p-values. In complex analyses it is very useful to have all pertinent results on one page so everything can be viewed in context. With multiple tests there is a risk that particularly small p-values are isolated and possibly given more credence than they deserve.

Some comments on the correlations among the variables are in order. The offensive statistics are clearly correlated—better hitters tend to excel in many areas. The correlations are

	H_AB	R_AB	HR_AB	RBI_AB
H_AB	1.00	0.29	0.16	0.57
R_AB	0.29	1.00	0.78	0.24
HR_AB	0.16	0.78	1.00	0.35
RBI_AB	0.57	0.24	0.35	1.00

The bootstrap resampling method incorporates these correlations. Comparing the "Adj_p" and "Sid_p" entries, we see the effect of incorporating these correlations is slight, particularly for smaller p-values. However, incorporating correlation allows an additional significance at the FWE = .05 level: the comparison of runs per at bat for first basemen versus all others shows significance with the bootstrap adjustment ($\tilde{p}$ = .0457). The Šidák adjusted p-value is insignificant $\tilde{p}$ = .0516, but with a difference of .0059, there is very little distinction between these two multiplicity adjustments.

Example 7.10 Biochemical Levels Revisited

In Section 4.2 we analyzed the continuous data on biochemical levels in urine specimens of male subjects. In this section we consider this data set in more detail, and discuss some of the more practical issues surrounding the data and its analysis. We also show how to devise and test a collection of contrasts that better reflects the execution of the experiment than the collection of pairwise contrasts.

Table 7.10 Summary of Personal Data on 17 Subjects*

Group	ID	Age	Height (in)	Weight (lbs)	Wt/Ht
W4	1	29	71.5	270	3.776
	2	29	72.0	255	3.542
	3	20	76.0	262	3.447
W3	4	28	70.0	230	3.286
	5	25	68.0	215	3.162
	6	31	73.0	230	3.151
	7	26	69.0	205	2.970
W2	8	22	76.5	162	2.118
	9	27	71.5	141	1.972
	10	25	72.0	140	1.944
	11	28	71.5	135	1.888
	12	25	72.0	135	1.875
W1	13	21	72.0	130	1.806
	14	25	68.0	122	1.794
	15	25	71.5	128	1.790
	16	26	70.0	115	1.643
	17	29	64.0	98	1.531

* From Smith, Gnanadesikan, and Hughes, "Multivariate Analysis of Variance (MANOVA)," *Biometrics* **18** 22–41, (1962), with permission of the Biometric Society.

There are 17 subjects in the study, with each subject evaluated 1–3 (possibly 4) times with a total of 45 observations. The analyses performed by Smith et al. (1962) and Morrison (1990, pp. 234–236) fail to account for this fact, using $n = 45$ as the number of independent replicates. Their tests might be too liberal because they fail to account for dependence among repeated observations. We will have to consider the same basic model, assuming all 45 observation vectors are independent, and our analysis will suffer the same defect since observations are not identified with particular individuals. The point is that the resampling method, like many nonparametric and semiparametric methods, cannot rescue the analysis from basic flaws in the assumed form of the model.

The body weights and heights are given for the 17 subjects in Table 7.10. Clearly, W4 and W3 combined are the heavy weights, while W2 and W1 combined are the light weights. Most interesting is the large gap in body weight between the W3 and W2 groups! Although not stated in the Smith et al. and Morrison analyses, the original study probably used many people, then discarded the "normals," 50%–67% from the middle of the weight distribution, leaving a heavy group and a light group. The analysis of Smith et al. further subdivided these groups into the (W4, W3) and (W2, W1) classes shown in Table 7.10, presumably to demonstrate how to deal with multiple groups.

Based on this description of the data, the following set of contrasts seems more appropriate to the structure of the data:

	W1	W2	W3	W4
C_1 (Heavy vs. Light)	1	1	−1	−1
C_2 (Weight Linear)	4	3	−3	−4
C_3 (Very Heavy vs. Very Light)	1	0	0	−1

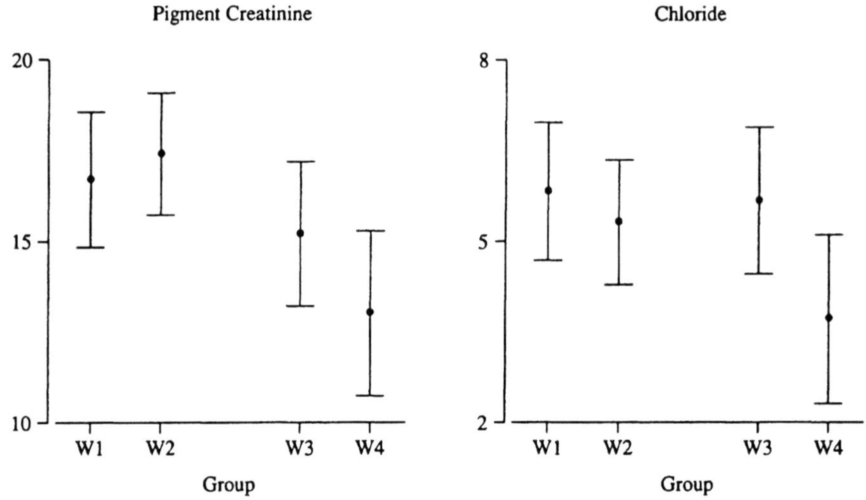

FIGURE 7.4. Adjusted means for biochemical levels, ±2SE.

These contrasts are more selective and subject-matter driven than the collection of all pairwise contrasts. The multiplicity is reduced, and the chosen contrasts directly address the interesting research questions.

Figure 7.4 displays the covariate-adjusted means (covariate is *Urine Volume*) for each treatment group, for the variables *Pigment Creatinine* and *Chloride*. The results of the various hypothesis tests are displayed in Table 7.11. For this analysis, the multivariate normal resampling method of Chapter 4 was used, single-step, with 20,000 simulations. Not only are the indicated contrasts more interesting from a subject matter perspective, but also they are more significant than the results from all pairwise contrasts. There is clearly a substantial amount of correlation among these tests due to the very similar nature of the contrasts. The effect of incorporating these correlations may be seen by comparing the "Adj_p" and "Sid_p" entries of Table 7.11: the "Adj_p" p-values incorporate the correlations via resampling, whereas the "Sid_p" entries are the independence-assuming Šidák adjustments. □

Table 7.11 Summary of Testing for Biochemical Levels

Variable	Contrast	t-Value	Raw_p	Adj_p	Sid_p
Pigment creatinine	Heavy vs. Light	3.14	.0032*	.0123*	.0191*
	Weight Linear	3.16	.0030*	.0112*	.0179*
	Very Heavy vs. Very Light	2.59	.0133*	.0464*	.0774
Chloride	Heavy vs. Light	1.50	.1416	.3879	.5999
	Weight Linear	1.76	.0868	.2603	.4200
	Very Heavy vs. Very Light	2.41	.0206*	.0676	.1173

*p-value < .05.

Table 7.12 Pipe Data*, Original Order

		Inspector				
Rep	Loc	1	2	3	4	5
1	1	3.85	−1.05	−0.55	−0.45	−0.65
	2	−1.95	−0.65	−2.25	−0.35	0.25
	3	−1.10	−0.10	0.10	0.40	−0.30
	4	−4.95	−15.55	−20.25	0.95	3.15
2	1	−2.25	−0.55	−0.95	0.05	9.65
	2	−2.05	−4.55	−0.25	−0.15	−0.85
	3	−1.00	0.00	−2.50	−1.00	−0.30
	4	−23.15	2.55	−19.95	−0.75	5.85
3	1	−1.75	0.05	0.55	−0.45	0.65
	2	−0.45	−0.35	−0.15	0.15	−0.25
	3	0.70	−0.60	−1.90	1.00	1.10
	4	3.45	−24.15	−0.35	3.15	−13.75

*From J. A. John and N. R. Draper (1980), *Applied Statistics* **29** 190–197, (1980), with permission of the Royal Statistical Society.

Example 7.11 Testing Means and Variances in a Factorial Experiment
Recently, there has been considerable interest in the analysis of variances in industrial experiments. Usually, experimental *means* are analyzed using ANOVA models—ANOVA, or "analysis of variance," is actually a misnomer since the method is used to test *mean* values. However, experimental variances might be equally important: much of the recent effort in quality improvement is directed toward the reduction of process variability. In the quality improvement arena, low variation is an essential component of a high quality product.

John and Draper (1980) analyze the data given in Table 7.12. They describe the data as follows:

> ...part of the results of an experiment carefully planned to compare the performance of expert inspectors in assessing the thickness of certain types of piping. Four different locations were marked on a pipe and an ultrasonic reader was used to give estimates of the wall thickness at each location. These values were considered the "true" measurements. Five equally qualified inspectors were included in the test. They each independently tested all locations on the pipe and [the results given] measure the difference (multiplied by 100) of the inspector reading from the "true" reading. A negative value means that the inspector value was below the "true" value. There were three replications of the entire test.*

Note that the experiment is a (4 locations)×(5 inspectors) design with three replications in each cell.

We will use this data set for testing for location and inspector effects for both mean response and variance of the response—a total of four questions. To this end,

*From J. A. John and N. R. Draper, *Applied Statistics* **29** 190–197, (1980), with permission from the Royal Statistical Society.

Table 7.13 Replicate Values for 5 Inspectors Measuring at 4 Locations

Loc	Inspect	Rep1	Rep2	Rep3	Mean	LogVar
1	1	3.85	−2.25	−1.75	−0.050	2.440
1	2	−1.05	−0.55	0.05	−0.517	−1.193
1	3	−0.55	−0.95	0.55	−0.317	−0.505
1	4	−0.45	0.05	−0.45	−0.283	−2.485
1	5	−0.65	9.65	0.65	3.217	3.449
2	1	−1.95	−2.05	−0.45	−1.483	−0.219
2	2	−0.65	−4.55	−0.35	−1.850	1.703
2	3	−2.25	−0.25	−0.15	−0.883	0.339
2	4	−0.35	−0.15	0.15	−0.117	−2.759
2	5	0.25	−0.85	−0.25	−0.283	−1.193
3	1	−1.10	−1.00	0.70	−0.467	0.023
3	2	−0.10	0.00	−0.60	−0.233	−2.270
3	3	0.10	−2.50	−1.90	−1.433	0.617
3	4	0.40	−1.00	1.00	0.133	0.052
3	5	−0.30	−0.30	1.10	0.167	−0.426
4	1	−4.95	−23.15	3.45	−8.217	5.220
4	2	−15.55	2.55	−24.15	−12.383	5.224
4	3	−20.25	−19.95	−0.35	−13.517	4.868
4	4	0.95	−0.75	3.15	1.117	1.341
4	5	3.15	5.85	−13.75	−1.583	4.726

we compute the means and variances over the three replications for each cell, and transform the variances using a log transformation. Thus our basic data is a 4×5 table of cell means (MEAN), and log variances (LOGVAR), and is given in Table 7.13. Careful inspection indicates that location 4 is rather unusual. Figure 7.5 gives the cell means and log variances using two-way profile plots that are easier to examine. Inspection of this figure also indicates that location 4 is unusual.

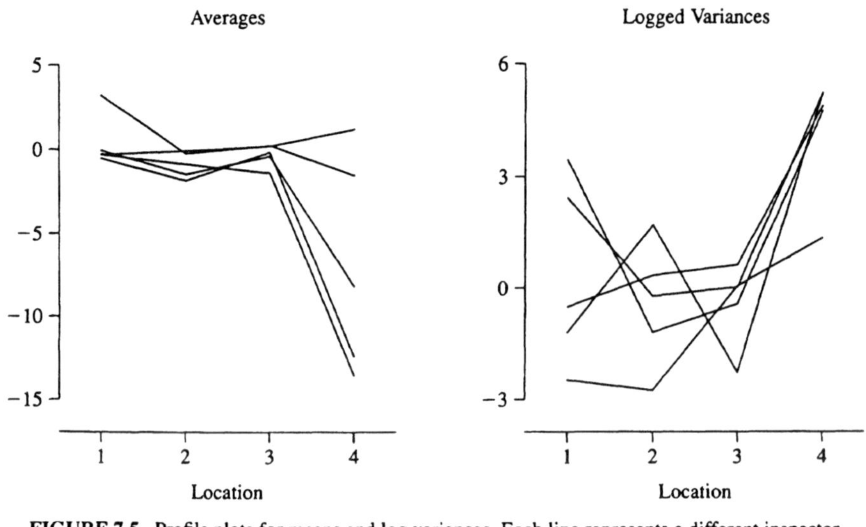

FIGURE 7.5. Profile plots for means and log variances. Each line represents a different inspector.

Table 7.14 Analysis of Variance for Pipe Data

Variable	Source	DF	Sum of Squares	Mean Square	F Value	p-Value $> F$
Mean	Loc	3	168.991	56.330	6.15	0.0089
	Inspectors	4	72.196	18.049	1.97	0.1630
	Error	12	109.807	9.150		
	Total	19	350.996			
LogVar	Loc	3	75.746	25.248	9.56	0.0017
	Inspectors	4	20.494	5.123	1.94	0.1683
	Error	12	31.687	2.640		
	Total	19	127.928			

Computation of an analysis of variance on MEAN and LOGVAR, Table 7.14, gives raw p-values for location and inspector effects and confirms our impression that the locations differ from one another both in their means and in their variances. The original observations are deviations from "true" values for each location on the pipe, and so are somewhat centered. There are various ways that adjusted p-values might be simulated; here the p-values were simulated by bootstrap resampling from the 20 residual vectors of a two-way MANOVA model without interaction. This general method is discussed in Chapter 4. For each resampled data set, the p-values from all four F tests are recomputed, and the adjusted p-values are computed by comparing the original p-values with the minimum for each resampled data set. As seen in Table 7.15, the adjusted p-values also confirm that locations differ in both MEAN and LOGVAR. □

A simple rearrangement of Table 7.12 by location, given in Table 7.16, is instructive. It is very clear that most of the large deviations from the "true" values are concentrated at location 4. One or more large values occur for all but one of the inspectors. A normal probability plot of residuals is given in Figure 7.6. While the shape of the plot suggests that the data are heavy-tailed (outlier-prone), it also indicates that most of the extreme observations are coming from location 4. When using locations 1–3 only, the heavy-tailed appearance remains, suggesting nonnormality as well as heteroscedasticity in the data.

Without input from a subject matter expert, we cannot be definitive about the cause of the extreme values concentrated at location 4. Several comments are in order. Since 4 of 5 inspectors reported extreme values for location 4, it seems most

Table 7.15 p-Values for MEAN and LOGVAR for Location and Inspector

		Variable	
Factor	p-Value	MEAN	LOGVAR
Location	Raw_p	0.0089	0.0017
	Adj_p	0.0357	0.0051
Inspector	Raw_p	0.1630	0.1683
	Adj_p	0.4710	0.4844

Table 7.16 **Pipe Data, Ordered by Location**

		Inspector				
Rep	Loc	1	2	3	4	5
1	1	3.85	−1.05	−0.55	−0.45	−0.65
2	1	−2.25	−0.55	−0.95	0.05	9.65
3	1	−1.75	0.05	0.55	−0.45	0.65
1	2	−1.95	−0.65	−2.25	−0.35	0.25
2	2	−2.05	−4.55	−0.25	−0.15	−0.85
3	2	−0.45	−0.35	−0.15	0.15	−0.25
1	3	−1.10	−0.10	0.10	0.40	−0.30
2	3	−1.00	0.00	−2.50	−1.00	−0.30
3	3	0.70	−0.60	−1.90	1.00	1.10
1	4	−4.95	−15.55	−20.25	0.95	3.15
2	4	−23.15	2.55	−19.95	−0.75	5.85
3	4	3.45	−24.15	−0.35	3.15	−13.75

likely that there is something odd about this location rather than some problem with the inspectors. The presence of several other extreme values at other locations may also be the result of unusual location effects. In any case, the concentration of extreme values at one location points to problems at that location that should be examined. A possible explanation is that the sonic equipment may give false readings if there is an imperfection in the pipe, and that location 4 has an imperfection that may give a short reading rather than a long reading. Inspector 4 had no large negative observations, which may have been chance. Or this inspector may have used a different technique: inspector 4 may have moved the sonic probe until consistent values were obtained, i.e., inspector 4 may have practiced outlier detection and removal during the measuring process!

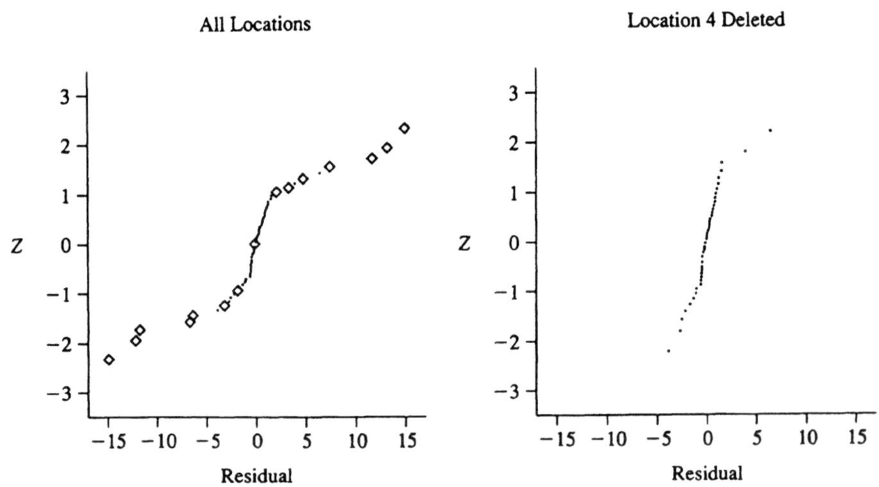

FIGURE 7.6. Normal Q–Q plots of residuals from pipe data. Diamonds (◇'s) are location 4 residuals.

John and Draper (1980) analyze this data with the goal of selecting a transformation to normalize the residuals. They assume that the distributions are equally heavy-tailed in all cells of the ANOVA. We offer an additional explanation: the variances themselves differ by cells. Even if data are normally distributed within cells, the pooled residuals may exhibit heavy tails under heteroscedasticity. The construction of a transformation to normalize the data does not account for the fact that the extreme observations are largely concentrated at one location, which we detect by comparing variances.

Multiple testing of variances for situations including unequal sample sizes are given in Chapter 6. In this example, we exploit the balanced structure of the data to derive an alternative method for comparing variances. Denote the estimated variance for cell (i,j) by s_{ij}^2. We have implicitly assumed that $s_{ij}^2 \sim \sigma_{ij}^2 V_{ij}$, where the V_{ij} are independent and identically distributed random variables. This assumption is motivated as follows: If the original data y_{ijk} are normally distributed, then the V_{ij} are distributed as $\chi^2_{(n-1)}/(n-1)$. For nonnormal data, the distribution of V_{ij} is unknown, but we assume that it is the same distribution for every cell.

Thus, $\ln s_{ij}^2 \sim \ln \sigma_{ij}^2 + \ln V_{ij}$, where the $\ln V_{ij}$ have identical distributions for every cell (i,j). We have assumed that the log transformed variances may differ by cell, but that there is no interaction between location and inspector. These assumptions justify the location-shift model used in our analysis.

The balanced design allows this particular analysis. Had the sample sizes been unequal, the distribution of s_{ij}^2 would be $\sigma_{ij}^2 V_{ij}$, but where the V_{ij} are not identically distributed. With normally distributed y_{ijk}, the V_{ij} are distributed as $\chi^2_{n_{ij}-1}/(n_{ij}-1)$, and the location-shift model for $\ln s_{ij}^2$ would be invalid.

7.3 BINOMIAL AND CATEGORICAL EXAMPLES

Chapter 5 discussed the analysis of binomial data. In this section we provide further examples of the use of binomial models. We start with an application to survival analysis, in which the binary variables have a unique correlation structure: each variable is a "1" until death, then all variables are 0's. This structure gives very high correlations. Resampling is a convenient method for incorporating such correlation structures into hypothesis testing.

Categorical data applications abound in books such as *Discrete Multivariate Analysis: Theory and Practice*, by Bishop, Fienberg, and Holland (1975); *The Analysis of Cross-Classified Categorical Data*, by Fienberg (1980); and *Applied Statistics: Principles and Examples*, by Cox and Snell (1981). A common problem with cross-classified categorical data is multiple classification variables, so the possible number of subdivisions and comparison of proportions can be very large. We give two "textbook" examples of categorical data, and analyze them using binomial methods. Using the fixed-margins binomial model, we can formulate and test hypotheses easily. Log-linear model analysis for categorical data is an attractive alternative, but research questions are often addressed more directly using simple contrasts that compare binomial proportions.

In this section we adopt the multiple testing strategy of Chapter 1, Section 1.6, in which many subdivisions are made and various contrasts of interest are defined. The collection of tests is then adjusted for multiplicity using resampling methods, allowing for the discrete nature of the tests: variables with low frequency outcomes do not contribute as much to the multiplicity adjustment factor (e.g., k for the Bonferroni adjustment), as do variables with frequency outcomes in moderate ranges. In addition to correlations among variables, the resampling method incorporates correlations among tests.

Example 7.12 Multiple Survival Analysis

Rodents are often used to assess the safety of new drugs. A control dose and progressively higher doses are administered to animals over a large portion of their lifetime, usually two years. One of the principal measures of interest is the survival of the animals; another is the number of tumors that an animal develops. We analyze a study for which the time of death was recorded to the nearest week. There were four treatment groups, coded as 0, 1, 2, 3. The study was terminated after 105 weeks, and surviving animals were killed and examined by pathologists for tumors. Animals that died or were killed in a moribund condition were also examined for tumors. The study duration was divided into 27 four-week periods to assess survival. The variables of interest were as follows:

Classification Variables	Treatment	0, 1, 2, 3
	Sex	Female, Male
Response Variables	Survival	0 or 1, over 27 Periods
	Number of Tumors	0, 1, 2, etc.

The classification variables form a 2 × 4 factorial experiment. There are 28 response variables, 27 binomial variables measuring survival and one variable measuring the tumor load (or count), which we treat as continuous. Toxicologists typically analyze the sexes as two separate studies. We will follow that convention here even though the complete evaluation of a compound includes both rats and mice, females and males, greatly increasing the number of comparisons. The compound might be indicted if in any of the three treated doses there is a "statistically significant" deviation from the control for either survival or tumors.

Note that the termination of the study censors the survival distribution, but this poses no computational problems since we will be analyzing each of the time periods. Also, if an animal is removed from study at some intermediate time, for example by accidental death, there is also no computational problem, since both the numerator and denominator in the survival ratio can be decremented. For each animal, the survival variables will be a string of 1's followed by a string of 0's after the death of the animal, or a string of 1's followed by a string of missing value indicators. Typical patterns for the survival vector are as given in the following sample of the data records for such a study:

Variable	*Columns*
Obs	1–2
Animal ID	4–8
Sex	10
Dose	12
Weeks	14–16
S1–S27	18–44
Tumor Count	46

1	14961	F	0	97	111111111111111111111111100	3
2	14962	F	0	105	111111111111111111111111111	1
3	14963	F	0	105	111111111111111111111111111	2
4	14964	F	0	105	111111111111111111111111111	1
5	14965	F	0	97	111111111111111111111111100	2
6	14966	F	0	105	111111111111111111111111111	2
7	14967	F	0	105	111111111111111111111111111	3
8	14968	F	0	55	111111111111110000000000000	1
9	14969	F	0	105	111111111111111111111111100	1
10	14970	F	0	73	111111111111111111100000000	1

For example, the first animal is a female, ID 14961, in the control group (Dose = 0). There were three tumors found in this animal upon necropsy. The animal died after 97 weeks, so was alive at the beginning of the 25th four-week period (S25 = 1), but dead at the beginning of the 26th period (S26 = 0). The data suggest extremely high correlations among the binary survival indicators, particularly for adjacent time periods. These correlations are incorporated by resampling the *entire vectors*. Additionally, these resampled data sets contain only plausible animal records: as entire vectors are resampled, no animal may have a "0" followed by a "1" in any bootstrap data set.

The treatment groups are a control group and groups treated with progressively higher doses of drug. Contrasts of interest are as follows:

Name	Contrast				Description
0 vs. 1	−1	1	0	0	Control vs. Low Dose
0 vs. 2	−1	0	1	0	Control vs. Mid Dose
0 vs. 3	−1	0	0	1	Control vs. High Dose
Trend	−3	−1	1	3	Linear Trend

These various treatment contrasts are correlated since the first three contrasts all include the control group. The fourth contrast heavily emphasizes the control and high dose group and is expected to be highly correlated with the third contrast. Because of the extremely high correlations among binary variables, the correlations induced by treatment contrasts, and the problems involving missing values, the problem of obtaining reasonable multiplicity adjustments for the entire collection of tests seems extremely difficult. However, resampling provides a simple method for handling these problems, while addressing all the unusual characteristics of the data. In particular, the entire set of vectors, including missing values, is resampled. This method of resampling mimics the type of randomness that occurs in repeated experiments where differing numbers of animals will be removed from the study. A very important (and often unjustifiable) assumption is that the reason for removal, or the *censoring mechanism*, should be unrelated to the treatment variable. A test of this assumption is easily incorporated into the testing process; simply create a binomial variable indicating if an animal is removed from the study.

Contributing to the correlation is the addition of the "tumor load" variable, which is related to survival. If multiplicity adjustment is to be performed over all variables, then resampling must preserve such dependence structures as well as to reflect the null hypothesis. Recall that centering is important for continuous variables, but binary variables are simply pooled in bootstrap resampling. One way to accommodate both types of data is to center only the continuous variables within each treatment group, then resample the pooled data. The resampled binary variables remain binary, with the relevant discreteness (or sparseness) characteristics preserved. Thus, the complete null hypothesis of no treatment effects is maintained across all binary and continuous variables, and dependence structures among binary and continuous variables are incorporated. This is the default mode of analysis of PROC MULTTEST (see Appendix III).

Let's count the multiplicities. There are 28 response variables; the 27 survival variables are expected to be highly correlated. There are four treatment contrasts; two of them are highly correlated and three are somewhat correlated. We are formally testing $28 \times 4 = 112$ hypotheses, but there are high correlations both among the variables and among the tests. In this analysis we do not formally address the multiplicity generated by the two rodent species or the two sexes within each species. Nor do we consider the other several hundred variables that are typically measured in these studies.

Let us examine the p-values for comparing survival in this experiment.

See Table 7.17 for the survival p-values at 3-month intervals. The raw p-values were computed using the Cochran-Armitage trend test, discussed in Chapter 5, for the binary survival indicators. The ANOVA-style contrast test, discussed in Chapter 3, was used for the continuous tumor load variable. A given adjusted p-value is the proportion of data sets for which the smallest resampled p-value was smaller than the observed p-value. The number of bootstrap samples was 1,000, and the calculations were performed using PROC MULTTEST.

What can we conclude from an examination of the p-values? While many unadjusted p-values are significant at the 5% level, there is only one adjusted p-value that is smaller than .05. From the adjusted p-values there is evidence that survival decreases in males at the 12th month of the study. There are incidences when the unadjusted p-values are less than .05, but these are to be expected with many statistical comparisons. There is some evidence that the tumor count decreases linearly with dose for females (unadjusted p-value of .002) but this decrease is not significant when multiplicity adjustment is applied, $p = .070$. □

Table 7.17 Multiple Comparison p-Values for Rat Survival at Selected Months

Females

	Contrast							
	0 vs. 1		0 vs. 2		0 vs. 3		Trend	
Month	Raw	Adj	Raw	Adj	Raw	Adj	Raw	Adj
3	.184	.990	.184	.990	.184	.990	.113	.933
6	.641	1.000	.044*	.607	.214	.996	.090	.860
9	.196	.993	.003*	.070	.010*	.198	.002*	.070
12	.395	1.000	.013*	.270	.031*	.496	.009*	.188
15	.516	1.000	.058	.684	.008*	.173	.003*	.086
18	.395	1.000	.209	.993	.023*	.409	.021*	.351
21	.089	.857	.322	1.000	.016*	.305	.037*	.555
24	.010*	.201	.165	.980	.016*	.305	.053	.663
27	.081	.818	.316	1.000	.118	.938	.206	.993

Males

	0 vs. 1		0 vs. 2		0 vs. 3		Trend	
Month	Raw	Adj	Raw	Adj	Raw	Adj	Raw	Adj
3	.786	1.000	.107	.917	.786	1.000	.489	1.000
6	.069	.782	.224	.998	.069	.782	.119	.937
9	.075	.803	.296	.999	.012*	.260	.029*	.486
12	.097	.870	.172	.980	.002*	.046**	.004*	.084
15	.237	.998	.342	1.000	.101	.879	.132	.960
18	.044*	.644	.009*	.200	.069	.784	.052	.694
21	.093	.862	.007*	.154	.064	.759	.034*	.548
24	.539	1.000	.262	.999	.433	1.000	.359	1.000
27	.439	1.000	.439	1.000	.687	1.000	.698	1.000

*Raw p-value $< .05$.
**Adjusted p-value $< .05$.

As indicated in Table 7.17, the degree of adjustment is large, though perhaps not so surprising considering the large number ($k = 112$) of statistical tests. Raw p-values as small as .002 are adjusted to just above .05 but less than .10, but a Bonferroni adjustment would adjust a p-value of .002 to .224. The large discrepancy between the Bonferroni adjustment and the resampling-based adjustment indicates the need for incorporating distributional characteristics into the multiplicity adjustment.

Another way to think of the adjustment is that in just over 5 percent of the experiments involving benign compounds, a p-value as small as .002 would occur. Raw p-values as small as .01 would occur in about 20 percent of experiments of this type, as can be seen by comparing the "Raw" and "Adj" columns for "0 vs. 1" in month 24 for Females. Raw p-values of .05 would occur in well over half of experiments of this type.

Visual examination of the raw and adjusted p-values gives a good indication of the randomness that is inherent in this experiment. There is a tendency on the part of some toxicologists to take raw p-values as tantamount to rigid proof. When tables of this sort are examined, the effects of multiplicity and randomness are apparent.

An important point to be made here is that the large number of time-specific survival endpoints (3 months, 6 months, etc.) has an important effect upon multiplicity. If the experimenters do not wish to claim time-specific significances, then the survival tests for each indicated contrast are better conducted using "logrank tests" (Peto et al., 1980). Using these methods, the multiplicity problem is far less, as the 27 time-specific tests are combined into one. Furthermore, because the logrank test combines data from all time points, the test will tend to be more powerful when consistent survival patterns are apparent for many individual time points. For example, the long string of significant "Raw" p-values in Table 7.17 for the "0 vs. 3" contrasts in Females would contribute to an overall test.

Even when combined survival tests are used, multiplicity problems remain if multiple contrasts are used for comparison of the treatment groups. Sex/species subdivisions result in further multiplicity, and the multiplicity problem can be greatly compounded when many tumor types are considered individually, with "cause of death" information (see Peto et al., 1980). Such multiple tests can be adjusted for multiplicity using resampling methods as discussed by Soper and Westfall (1990) and in Chapter 6, Section 6.7.

Our analysis has the benefit that it allows us to draw conclusions concerning *time-specific* survival comparisons. When only omnibus survival tests are considered, it is poor practice to scan the table of time-specific unadjusted p-values, claiming significance whenever $p < .05$. When an omnibus survival test is selected, it can be sensitive to a difference in the survivals at some particular area of time, so some investigators might use two tests, one sensitive to early differences and one sensitive to late differences. Other investigators test at several time points. These practices lead to multiplicity; our technique addresses the time-sensitive nature of the experimenters' questions while maintaining the correct error rate.

Example 7.13 Market Survey Data for Detergent Preference

This experiment compares two detergents, a new detergent "Z" and a standard product, "M." Table 7.18 is originally from Ries and Smith (1963), but the data has been analyzed repeatedly by many other authors including Cox and Snell (1981), Bishop, Fienberg and Holland (1975), and Fienberg (1980). In Table 7.18, X is the incidence of detergent users preferring product Z to M, n is the sample size in each cell, and % is the percent that prefer Z. Several factors were examined to see if they influenced the preference for the new product:

Water Softness: Soft, Medium, or Hard

Water Temperature: Low or High

Previous Use of the Standard Product M: Yes or No

When arranged in this fashion, the factors form a $3 \times 2 \times 2 = 12$ factorial experiment. The individual observations are univariate binary indicators of preference, with observations coming from 12 distinct "treatment" groups. Note that we are analyzing preference probabilities that are *conditional upon* group membership.

Table 7.18 Detergent Preference Data*

		M Previous User			
		No		Yes	
		Temperature		Temperature	
Water		Low	High	Low	High
Hard	X	68	42	37	24
	n	110	72	89	67
	%	61.8	58.3	41.6	35.8
Medium	X	66	33	47	23
	n	116	56	102	70
	%	56.9	58.9	46.1	32.9
Soft	X	63	29	57	19
	n	116	56	106	48
	%	54.3	51.8	53.8	39.6

*Adapted from Table L.1, D. R. Cox and E. J. Snell, *Applied Statistics* p. 107, (1981), with permission of Chapman & Hall, London.

Here we have three factors that form 12 combinations. For each combination the data may be considered binomial. In all the cells the percent favoring brand M fall in the range 20–80%, so the sampling distributions of all test statistics should be approximately normal. This implies that all tests will be weighted equally in the multiplicity adjustments—if some cell had an extremely small frequency, then resampling would give it less weight. Since there are 11 main effects and interactions, there is an increased risk of a false positive result, thus we compute both raw and adjusted p-values.

Contrasts that address the basic questions are given in Table 7.19. The raw and adjusted p-values are given in Table 7.20: raw p-values are all two-tailed, com-

Table 7.19 Contrasts for Detergent Preference Analysis

Water	Hard				Medium				Soft			
Previous User	No		Yes		No		Yes		No		Yes	
Temperature	Low	High	Low	High	Low	High	Low	High	Low	High	Low	High
Contrast												
Linear Water	−1	−1	−1	−1	0	0	0	0	1	1	1	1
Quadratic Water	−1	−1	−1	−1	2	2	2	2	−1	−1	−1	−1
Previous M User	−1	−1	1	1	−1	−1	1	1	−1	−1	1	1
Temperature	−1	1	−1	1	−1	1	−1	1	−1	1	−1	1
Prev × Linear	1	1	−1	−1	0	0	0	0	−1	−1	1	1
Prev × Quad	1	1	−1	−1	−2	−2	2	2	1	1	−1	−1
Temp × Linear	1	−1	1	−1	0	0	0	0	−1	1	−1	1
Temp × Quad	1	−1	1	−1	−2	2	−2	2	1	−1	1	−1
Prev × Temp	−1	1	1	−1	−1	1	1	−1	−1	1	1	−1
Prev × Temp × Linear	−1	1	1	−1	0	0	0	0	1	−1	−1	1
Prev × Temp × Quad	−1	1	1	−1	2	−2	−2	2	−1	1	1	−1

Table 7.20 Analysis of Detergent Preference

Factor	Raw_p	Adj_p
Softness Linear (SL)	.9042	1.0000
Softness Quadratic (SQ)	.7738	1.0000
Previous User (M)	.0000*	.0000*
Temperature (T)	.0584	.4705
(M × SL)	.0639	.5025
(M × SQ)	.5051	.9995
(T × SL)	.6573	1.0000
(T × SQ)	.9106	1.0000
(M × T)	.1407	.7969
(M × T × SL)	.5673	.9999
(M × T × SQ)	.5409	.9998

*p-value < .05.

puted using the Freeman-Tukey arcsine transform method, with standard normal approximation, as described in Section 5.2.3. Figure 7.7 displays the raw p-values graphically, and shows one particular effect with a small p-value, or large $1 - p$ in the graph, while all other effects are insignificant. Single-step adjusted p-values were computed with 10,000 bootstrap samples using PROC MULTTEST. There is a large effect of previous use (factor M): if the person had previously used brand M, then he or she was less likely to prefer the new brand Z. This conclusion remained significant after adjustment for multiple comparisons. Using the raw p-values, Temperature appears marginally important, $p = .0584$, with low temperatures favoring the new product. And there was a marginally significant interaction, $p = .0639$; the water softness linear effect of product preference changed depending upon previous use of brand M. Both the Temperature and Interaction were not significant when adjusted

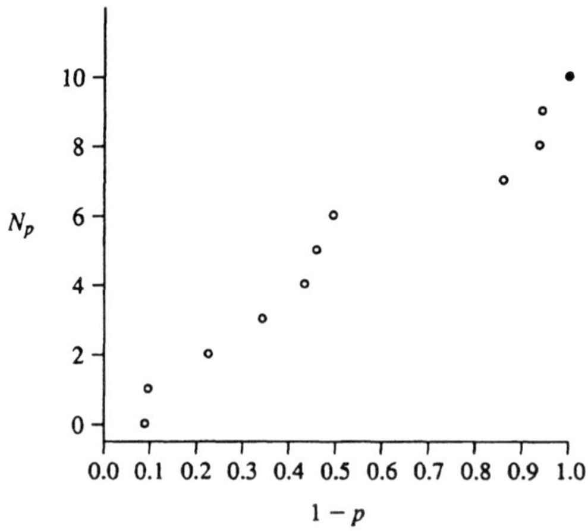

FIGURE 7.7. p-value plot for multiple contrast tests using detergent preference data. Unadjusted significant p-value denoted by •.

for multiple comparisons. So it might be unwise to target previous users of M in an advertising campaign which urges the use of the new brand with low temperature. These users may or may not be more likely to switch to the new brand under those conditions. □

Without multiplicity adjusted p-values it is unlikely that subject matter experts would know how to correct for multiple comparisons appropriately. On the other hand, if the experts have strong *a priori* belief that certain effects such as "Temperature" or "Previous Use × Softness Interaction" are important, then such tests need not be included within the family of tests over which multiplicity adjustment is to be performed. In such cases the focus should be on estimating the size of these effects rather than testing for their existence.

Example 7.14 Housing Satisfaction

The data given in Table 7.21 were analyzed by Madsen (1976) and give the "degree of satisfaction" with housing conditions and the "influence on management" by tenants in a housing project in Copenhagen. In addition, the 1681 people are classified by the degree of contact with other residents and by the four types of housing in the project: Apartments, Tower Block, or Houses (either atrium or terraced).

With observational, cross-classification studies, it is not always clear which are response and which are explanatory variables. Type of housing might influence satisfaction, or there might be a self-selection of more satisfied people into a particular style of housing. Or assuming housing is fixed, it might be considered a blocking factor in the study of the effects of Contact among tenants and Influence on management on degree of Satisfaction. Satisfaction and Influence are ordered categorical variables so score statistics, e.g., 0, 1, 2, could be analyzed.

We are interested in binomial responses in this section, therefore we will treat this problem first as a categorical data analysis problem. We take *Satisfaction* (three levels), and *Influence on Management* (three levels), as six binomial response variables.

Table 7.21 Cross Classification of Housing Data*

		Influence								
		Low			Medium			High		
	Satisfaction	Lo	Med	Hi	Lo	Med	Hi	Lo	Med	Hi
Housing	*Contact*									
Tower Block	High	14	19	37	17	23	40	3	5	23
	Low	21	21	28	34	22	36	10	11	36
Apartments	High	78	46	43	48	45	86	15	25	62
	Low	61	23	17	43	35	40	26	18	54
Atrium House	High	20	23	20	10	22	24	7	10	21
	Low	13	9	10	8	8	12	6	7	9
Terraced House	High	57	23	13	31	21	13	5	6	13
	Low	18	6	7	15	13	13	7	5	11

*Adapted from Table W.1, D. R. Cox and E. J. Snell, *Applied Statistics*, p. 155, (1981), with permission of Chapman & Hall, London.

In addition, we take all possible cross-classifications of *Satisfaction* and *Influence*, $3 \times 3 = 9$, as response variables. So we have a total of $3 + 3 + 9 = 15$ correlated binomial response variables. Each of these three variables is multinomial, implying a special correlation structure for the 15 binary response variables. The correlations will be estimated implicitly using bootstrap sampling, and incorporated into the analysis.

The variables *Type of Housing*, with four levels, and *Degree of Contact* with other residents, with two levels, are the explanatory variables. Combining the two explanatory variables, we create a single classification variable having $4 \times 2 = 8$ levels.

We now define the linear contrasts that address our experimental questions. Table 7.22 gives the contrasts of interest. We compare low and high contact among tenants. We examine three comparisons among the four types of housing: (1) apartment versus house, (2) tower apartment versus non-tower apartment, and (3) atrium house versus terraced house. Finally, we examine the interactions of Contact and the three housing comparisons. There are a total of 7 contrasts over the 8 groups. Abbreviations are used so the classifications and results can be displayed in a compact form.

The specifications for PROC MULTTEST are given in Table 7.23. Freeman-Tukey tests of linear contrasts are computed and bootstrap adjusted p-values are computed from 10,000 bootstrap samples. Since we have no reason to exclude effects in one direction, all tests are two-tailed.

There are 15 response variables and 7 contrasts, a total of 105 computed raw p-values; these are given in Table 7.24 with the adjusted p-values. Observe that the large number of tests lead to large adjustments in the p-values. For example, the raw

Table 7.22 Contrast Descriptions, Treatment Codes, and Linear Contrasts

Contrast	
Name	Description
contact lo vs. hi	Low versus high contact with other tenants
apart vs. house	Tenant lives in apartment versus a house
apt:tower vs. non	Tenant lives in tower versus non-tower apartment
house:atr vs. ter	Tenant lives in an atrium versus terraced house
cont × (apt/house)	Interaction of contact and apartment versus house
cont × (tower/non)	Interaction of contact and tower vs non-tower apt
cont × (atr/ter)	Interaction of contact and atr vs. ter house

	Class*							
Contrasts	h0c0	h0c1	h1c0	h1c1	h2c0	h2c1	h3c0	h3c1
contact lo vs. hi	−1	1	−1	1	−1	1	−1	1
apart vs. house	−1	−1	−1	−1	1	1	1	1
apt:tower vs. non	−1	−1	1	1	0	0	0	0
house:atr vs. ter	0	0	0	0	−1	−1	1	1
cont × (apt/house)	1	−1	1	−1	−1	1	−1	1
cont × (tower/non)	1	−1	−1	1	0	0	0	0
cont × (atr/ter)	0	0	0	0	1	−1	−1	1

*Housing code: h0 = apartment; h1 = apartment, tower; h2 = house, atrium; h3 = house, terraced; Contact code: c0 = low; c1 = high.

Table 7.23 PROC MULTTEST Specifications for Binomial Testing

Option	Specification
Test for DISCRETE variables:	Freeman-Tukey trend test
Test tails:	Two-tailed
Block adjustment?	No
p-value adjustments:	Bootstrap
Number of resamples:	10000
Seed:	85393

Table 7.24 *p*-Values for Binomial Variables

	contact: lo vs. hi		apt: apart vs. house		house: tower vs. non		atr vs. ter	
Variable	Raw_p	Adj_p	Raw_p	Adj_p	Raw_p	Adj_p	Raw_p	Adj_p
INF0	0.0037	0.2929	0.0330	0.9179	0.7955	1.0000	0.6390	1.0000
INF1	0.9194	1.0000	0.1840	1.0000	0.1388	0.9999	0.3227	1.0000
INF2	0.0008	0.0826	0.3797	1.0000	0.0483	0.9709	0.0641	0.9898
SAT0	0.0309	0.9041	0.0060	0.4136	0.0001	0.0033*	0.0001	0.0051*
SAT1	0.2801	1.0000	0.0937	0.9991	0.8713	1.0000	0.1741	1.0000
SAT2	0.4599	1.0000	0.0001	0.0040*	0.0002	0.0172*	0.0036	0.2826
INF0SAT0	0.7320	1.0000	0.0022	0.1894	0.0001	0.0012*	0.0041	0.3162
INF0SAT1	0.0466	0.9672	0.2982	1.0000	0.4357	1.0000	0.2298	1.0000
INF0SAT2	0.1079	0.9997	0.3588	1.0000	0.0001	0.0007*	0.0609	0.9879
INF1SAT0	0.0949	0.9991	0.9387	1.0000	0.8889	1.0000	0.0060	0.4104
INF1SAT1	0.6405	1.0000	0.4251	1.0000	0.6476	1.0000	0.8156	1.0000
INF1SAT2	0.6025	1.0000	0.0166	0.7306	0.1411	0.9999	0.1210	0.9998
INF2SAT0	0.0012	0.1211	0.2765	1.0000	0.0520	0.9773	0.6219	1.0000
INF2SAT1	0.1636	1.0000	0.3361	1.0000	0.2035	1.0000	0.1459	1.0000
INF2SAT2	0.2034	1.0000	0.0282	0.8845	0.7219	1.0000	0.3086	1.0000

	cont × (apt/house)		cont × (tower/non)		cont × (atr/ter)	
Variable	Raw_p	Adj_p	Raw_p	Adj_p	Raw_p	Adj_p
INF0	0.5291	1.0000	0.8295	1.0000	0.0580	0.9852
INF1	0.3194	1.0000	0.9255	1.0000	0.3263	1.0000
INF2	0.8205	1.0000	0.8034	1.0000	0.2289	1.0000
SAT0	0.0590	0.9858	0.6586	1.0000	0.0366	0.9338
SAT1	0.6772	1.0000	0.9154	1.0000	0.6938	1.0000
SAT2	0.0179	0.7567	0.7769	1.0000	0.0797	0.9969
INF0SAT0	0.1798	1.0000	0.8993	1.0000	0.0436	0.9592
INF0SAT1	0.4388	1.0000	0.5535	1.0000	0.5711	1.0000
INF0SAT2	0.0959	0.9992	0.7084	1.0000	0.8883	1.0000
INF1SAT0	0.4451	1.0000	0.4405	1.0000	0.3926	1.0000
INF1SAT1	0.9895	1.0000	0.3446	1.0000	0.3016	1.0000
INF1SAT2	0.0179	0.7573	0.7752	1.0000	0.2128	1.0000
INF2SAT0	0.8974	1.0000	0.6927	1.0000	0.6265	1.0000
INF2SAT1	0.7372	1.0000	0.3927	1.0000	0.9356	1.0000
INF2SAT2	0.6485	1.0000	0.9041	1.0000	0.2311	1.0000

*Adjusted p-value $< .05$.

Table 7.25 Incidence Table and (%) for *Influence* and *Satisfaction*

		Tower Satisfaction			Non-Tower Satisfaction		
		0	1	2	0	1	2
	0	35	40	65	139	69	60
		(8.7)	(10.0)	(16.3)	(18.1)	(9.1)	(7.8)
Influence	1	51	45	76	91	80	126
		(12.8)	(11.3)	(19.0)	(11.9)	(10.5)	(16.5)
	2	13	16	59	41	43	116
		(3.3)	(4.0)	(14.8)	(5.4)	(5.6)	(15.2)
Total		99	101	200	271	192	302
(%)		(24.7)	(25.3)	(50.0)	(35.4)	(25.1)	(39.5)

p-value of 0.0037 (which tests for a difference in the incidence of low influence on management between low and high contact) is adjusted to .2929. Raw p-values as small as 0.0008 are adjusted to more than 0.05.

Using adjusted p-values, we conclude that there are no significant interactions between *Contact* and the housing contrasts. The proportion of tenants who are highly satisfied is higher in house dwellers than in apartments. Tower apartment dwellers had fewer unsatisfied and more very satisfied tenants than non-tower apartment tenants. Two of the interaction cells differed significantly between tower and non-tower apartment dwellers: there were more unsatisfied and fewer satisfied low influence tenants. Table 7.25 gives incidences and percentages for these significant effects. Note that confirming patterns of significance occur, as should be expected, given the dependencies among the 15 binomial variables. For example, the increase in highly satisfied tenants in the tower apartments is accompanied by a decrease in the very unsatisfied tenants.

This analysis may lack power because the ordered nature of the categories is ignored. Analysis of this data set using score statistics, $0, 1, 2$, for *Influence* and *Satisfaction* incorporates the ordering of the categories. Table 7.26 gives the PROC MULTTEST specifications for testing the score statistics. Table 7.27 gives the average scores for Influence and Satisfaction for the eight treatment groups, 4 housing classifications by 2 levels of contact. Table 7.28 gives the raw and adjusted p-values for the 7 contrasts of interest. The degree of Influence on management is higher

Table 7.26 PROC MULTTEST Specifications for Testing Scores

Option	Specification
Test for CONTIN variables:	T test of mean
Test tails:	Two-tailed
Block adjustment?	No
p-value adjustments:	Bootstrap
Center CONTIN variables?	Yes
Number of resamples:	10000
Seed:	85393

Table 7.27 Summary Statistics for *Influence* and *Satisfaction* Scores

Variable	Statistic	Class			
		h0c0	*h0c1*	*h1c0*	*h1c1*
INFSCORE	Mean	0.9406	0.7845	0.9905	0.8549
	Std Dev	0.7609	0.7172	0.7935	0.7620
	N	219	181	317	448
		h2c0	*h2c1*	*h3c0*	*h3c1*
	Mean	0.8780	0.8408	0.9158	0.6209
	Std Dev	0.8072	0.7886	0.7532	0.7084
	N	82	157	95	182
		h0c0	*h0c1*	*h1c0*	*h1c1*
SATSCORE	Mean	1.1598	1.3646	0.9401	1.1116
	Std Dev	0.8551	0.7815	0.8712	0.8545
	N	219	181	317	448
		h2c0	*h2c1*	*h3c0*	*h3c1*
	Mean	1.0488	1.1783	0.9053	0.7033
	Std Dev	0.8448	0.7886	0.8639	0.8005
	N	82	157	95	182

among tenants that have low contact with other tenants. Apartment dwellers are more satisfied than housing dwellers. Tower apartment tenants are more satisfied than non-tower apartment tenants. And atrium house tenants are more satisfied than terraced house tenants. As expected with fewer comparisons, the adjustment of p-values is smaller with the use of score statistics. □

Cox and Snell (1981, pp. 155–161) analyze and discuss this example. They emphasize the use of statistical testing to guide the construction of understandable tables. They treat *Satisfaction*, *Influence*, and *Contact* as response variables; their interpretation differs somewhat from that given here, in that various interactions were declared significant, a conclusion we did not find using adjusted p-values.

Table 7.28 p-Values for Influence and Satisfaction Scores

	INFSCORE		SATSCORE	
Contrast	Raw_p	Adj_p	Raw_p	Adj_p
contact lo vs. hi	0.0003*	0.0044*	0.1054	0.7614
apart vs. house	0.0647	0.5814	0.0001*	0.0016*
apt: tower vs. non	0.2047	0.9488	0.0001*	0.0001*
house: atr vs. ter	0.1988	0.9429	0.0001*	0.0016*
cont × (apt/house)	0.8124	1.0000	0.0169*	0.2005
cont × (tower/non)	0.8290	1.0000	0.7498	1.0000
cont × (atr/ter)	0.0692	0.6052	0.0337*	0.3594

*p-value $< .05$.

Table 7.29 Summary of Total Adverse Events Data

	Statistic	Treatment Group A_1	A_2	B_1	B_2
	Mean	0.561	0.744	0.900	0.800
With S_1	Std Dev	0.896	0.938	1.128	1.018
	N	41	39	40	40
	Mean	0.463	0.641	0.600	0.475
Without S_1	Std Dev	0.840	0.843	0.841	0.679
	N	41	39	40	40

Example 7.15 Adverse Events Data Revisited

In Example 5.1 of Chapter 5, adverse events data were analyzed using multiplicity adjustment methods for pure multivariate binary data. A general criticism of this analysis (and of multiplicity adjustment in general) is that when individual variables are tested, and then all tests are multiplicity-adjusted, one loses power to detect a slight but persistent overall change in the variables. For side effects data, a possible remedy is to construct a semi-continuous measure, such as total reported adverse events. This may be computed by *summing* the binary adverse event indicators for each patient.

Table 7.29 summarizes these combined measures. Since there was prior reason to suspect treatment-related adverse events of type "S_1," two totals are computed, one including the binary S_1 indicator, and one excluding it. As would be expected given the highly significant binary S_1 comparison, the mean values for total adverse events that *include* the S_1 data are relatively higher for the B groups than the A groups. When the binary variable S_1 is removed, this difference disappears.

To test statistical significance of the total side effects variable, simple ANOVA-type contrast statistics, as discussed in Chapter 3, may be used to compute unadjusted p-values. In this example, there is strong dependence since the total side effects measure is completely determined by the binary indicators.

Two separate multiple testing analyses are considered, one with the suspected site S_1 and one without it. In the former case, all the original $28 \times 3 = 84$ binary tests of Table 5.2 are considered, and the three continuous variable tests are added for a total of 87 tests. In the latter case, there are 81 binary tests plus three continuous tests, for a total of 84 tests. Note also that in the latter case the continuous variable is different, as it does not include S_1.

The p-value results are displayed in Table 7.30. These analyses were performed using PROC MULTTEST, with 10,000 bootstrap simulations and step-down multiplicity-adjusted p-values. In this example, even the *unadjusted* p-values for the total side effects variable are insignificant. This shows that meta-analytic tests are not necessarily more powerful than site-specific tests with multiplicity adjustment. The suspected site S_1 remains significant, even when the "family" is expanded to allow for a total side effects variable. □

Table 7.30 Selected *p*-Values for Adverse Events Data

	A vs. *B*		Dose Effect, *A*		Dose Effect, *B*	
Variable	Raw_p	Adj_p	Raw_p	Adj_p	Raw_p	Adj_p
Including S_1: 87 tests						
S_1	.001*	.019*	.939	1.000	.813	1.000
S_2	.813	1.000	.798	1.000	1.000	1.000
⋮	⋮	⋮	⋮	⋮	⋮	⋮
Total	.212	1.000	.415	1.000	.655	1.000
Removing S_1: 84 tests						
S_2	.813	1.000	.798	1.000	1.000	1.000
⋮	⋮	⋮	⋮	⋮	⋮	⋮
Total	.908	1.000	.325	1.000	.488	1.000

**p*-value $< .05$.

7.4 ORDERED CATEGORICAL EXAMPLES

With ordered categorical data, distinct categories have a natural trend, such as a *size* category (Small–Medium–Large), or *dose* (Control–Low Dose–Medium Dose–High Dose). Unordered categories are categories such as *race* (Hispanic–Anglo–Black–Other), or *religion* (Protestant–Catholic–Jewish–Other), for which there is no natural ordering. When data are in the form of ordered categories, it is helpful to use the ordered information to reduce the number of comparisons, rather than to treat each category as a distinct entity. Use of the ordering may reduce the multiplicity problem, as well as provide more powerful tests.

Special difficulties arise when handling ordered categorical data. In this section we demonstrate two resampling methods for the analysis. In the first example, we use permutation resampling, which preserves the original data values. In the second example, we create score statistics out of the ordered information (recall the $\mathbf{Y}_2$ matrix of Section 1.6), and then analyze the data as if they were continuous.

Example 7.16 Industrial Screening Experiment, Ordered Categorical Data
An example in which a large number of null hypotheses are considered, many of which are possibly true, is industrial screening experimentation. In such experiments it is common to run a highly saturated experimental design, in which the number of factors examined is almost as large as the number of experimental runs. This type of design has proven interpretable and cost effective for two reasons: (1) often much is known about the experimental variation, so there is an external frame of reference to evaluate effects, and (2) usually most of the examined factors turn out to be unimportant, so in practice variation can be estimated by using a model that excludes the unimportant variables. The emphasis in these designs is not testing of hypotheses; it is finding promising leads for further work. Nonetheless, one must have an understanding of the reproducibility of any identified findings. Because so

many factors are examined in screening designs, there can be a large risk that some results appear real by chance.

In industrial experiments, it is not unusual that 80% of such factors are thought, *a priori*, not to have any meaningful effect on the response; unfortunately, it is seldom clear which particular factors will be without effect, making multiple factor experiments necessary. Given that a large number of factors are considered, and given an *a priori* assumption that many factors are not important (e.g., based on extensive general experience), unimportant factors easily may appear significant. Worse yet, factors with small effects might be determined to have significant effects in the *opposite direction* from their actual effects, all due to multiple testing.

Hamada and Wu (1990) and several discussants go over the difficulties in the analysis of ordered categorical data in screening industrial experiments. There are three fundamental issues. First, the number of treatments or treatment combinations is large relative to the number of experimental units. Second, ordered categorical data are inherently multivariate. Finally, the analysis should account for the ordered nature of the responses.

Hamada and Wu (1990) report the workability (easy, normal, or difficult) of an arc-welded section between two steel plates, originally given by Taguchi and Wu (1980). The data is reproduced in Table 7.31. In this experiment there is interest in nine single factors, A–I, each observed at one of two levels, and four two-factor interactions, AG, AH, AC, and GH, totaling 13 tests of hypotheses. These factors are examined in a fractional factorial experiment having a total of 16 observations.

As discussed by Hamada and Wu, one method for analyzing such ordered categorical data is to assign numerical scores to the ordered categories (see also Kimeldorf,

Table 7.31 Arc-Welding Experimental Design and Response Data

Experimental Design Factors													Response	
A	B	C	D	E	F	G	H	I	AC	AG	AH	GH	Workability	Score
−1	−1	−1	−1	−1	−1	−1	−1	−1	1	1	1	1	Normal	1
−1	−1	1	1	1	1	−1	−1	1	−1	1	1	1	Normal	1
−1	1	1	−1	−1	−1	−1	1	−1	−1	1	−1	−1	Normal	1
−1	1	−1	1	1	1	−1	1	1	1	1	−1	−1	Difficult	2 or 5
−1	1	1	−1	−1	1	1	−1	1	−1	−1	1	−1	Difficult	2 or 5
−1	1	−1	1	1	−1	1	−1	−1	1	−1	1	−1	Easy	0
−1	−1	−1	−1	−1	1	1	1	1	1	−1	−1	1	Difficult	2 or 5
−1	−1	1	1	1	−1	1	1	−1	−1	−1	−1	1	Easy	0
1	1	1	−1	1	−1	−1	−1	1	1	−1	−1	1	Normal	1
1	1	−1	1	−1	1	−1	−1	−1	−1	−1	−1	1	Normal	1
1	−1	−1	−1	1	−1	−1	1	1	−1	−1	1	−1	Normal	1
1	−1	1	1	−1	1	−1	1	−1	1	−1	1	−1	Normal	1
1	−1	−1	−1	1	1	1	−1	−1	−1	1	−1	−1	Difficult	2 or 5
1	−1	1	1	−1	−1	1	−1	1	1	1	−1	−1	Easy	0
1	1	1	−1	1	1	1	1	−1	1	1	1	1	Normal	1
1	1	−1	1	−1	−1	1	1	1	−1	1	1	1	Normal	1

From M. Hamada and C. F. J. Wu, *Technometrics* **32** 119–130, (1990), with permission of the American Statistical Association.

Sampson, and Whitaker, 1992). This reduces a multivariate problem to a univariate one, and thus decreases the number of tests of hypothesis. But they advise that several different scoring systems be tried to examine the sensitivity of the analysis to the scoring system. With this data set we start with 13 factors and three categories for a total of 39 tests. Using a single scoring system would reduce the number of tests to 13. Hamada and Wu use two scorings: $0, 1, 2$ and $0, 1, 5$, for the categories easy, normal, and difficult. Use of a single scoring system reduces the number of tests from 39 to 13, but use of two scoring systems doubles the number of tests to 26.

Examination of the p-values, unadjusted and adjusted, for these 13 factors is instructive, see Table 7.32. In the original experiment, factors F and D appear somewhat important based on their unadjusted p-values of .0887 and .1548. Given these numbers, an experimenter might feel reasonably comfortable proceeding to a confirmatory experiment. If all the "unimportant" factors are eliminated from the analysis so that more degrees of freedom are available to estimate experimental variability, then the unadjusted p-values for factors F and D become .0016 and .0137, which might further convince experimenters that these factors are "real." The adjusted p-values provide a reality check: there is a .373 probability of seeing a p-value as small as .0887 in an experiment of this type.

In this example, the adjusted p-values were computed by permutation resampling of the observed vector of 16 scorings. For all resampled data sets, the design factors were left fixed. This process is somewhat different from the method of resampling the residuals, which is discussed in Chapters 3 and 4. The purpose of resampling the observed data in this case is to preserve the discrete nature of the scores. Note that subset pivotality does not hold. Therefore, all results pertain to a particular complete null hypothesis—the hypothesis that all permutations of the assigned scores are equally likely.

The multiplicity problem may be particularly pernicious in highly saturated designs—a problem that is highlighted by the permutation process of estimating

Table 7.32 Multiple Testing Analysis of Arc-Welding Data

	0, 1, 2 Scoring			0, 1, 5 Scoring		
Factor	$\|T\|$	Raw p	Adjusted p	$\|T\|$	Raw p	Adjusted p
A	0.4472	.6985	1.0000	1.0932	.3884	.8721
B	0.4472	.6985	1.0000	0.1562	.8902	1.0000
C	1.3416	.3118	.8700	1.4056	.2951	.8587
D	2.2361	.1548	.6425	1.7179	.2280	.7846
E	0.4472	.6985	1.0000	0.1562	.8902	1.0000
F	3.1305	.0887	.3735	2.9673	.0973	.3696
G	0.4472	.6985	1.0000	0.7809	.5166	.9606
H	0.4472	.6985	1.0000	0.1562	.8902	1.0000
I	1.3416	.3118	.8700	1.4056	.2951	.8587
AC	0.4472	.6985	1.0000	0.1562	.8902	1.0000
AG	0.4472	.6985	1.0000	0.1562	.8902	1.0000
AH	0.4472	.6985	1.0000	1.0932	.3884	.8721
GH	0.4472	.6985	1.0000	1.0932	.3884	.8721

adjusted p-values. Permuting the observed scores rearranges them relative to each of the design factors. If a preponderance of 0 and 1 scores are put into the low level of a factor (and the 2 scores are put into a high level of the factor), there will be a difference between the means, and the estimate of the variance will be decreased. Both occurrences will accentuate the appearance of a real treatment effect. With 13 factors where this chance grouping might happen, and with relatively few values of a response possible $(0, 1, 2)$, p-values as small as 0.0887 will occur frequently.

If the data set is analyzed as pure categorical data, the situation is even worse. Suppose each of the categories *easy*, *normal*, and *difficult* is coded as a binary 3-tuple, with $1, 0, 0$ denoting *easy*, $0, 1, 0$ denoting *normal*, and $0, 0, 1$ denoting *difficult*. Here there are $k = 3 \times 13 = 39$ tests; there is a complete linear dependence since the response can occur in only one of the three categories, easy, normal, difficult, for any of the 16 observations. Because there are no replications for any design points, we assume there is a potentially different multinomial vector $\pi_i = (\pi_{i1} \quad \pi_{i2} \quad \pi_{i3})$ for each observation i. The contrasts described by the 1's and -1's of Table 7.31 may then be used in the Cochran-Armitage test of Chapter 5. In this example, there are 16 groups—one for each experimental observation.

Table 7.33 gives the most unusual result for the original experiment for each of the categories, normal, difficult, and easy. For example, consider the normal category and factor G: $7/8$ observations were normal for the low level of factor G and $2/8$ were normal for the high level of this factor. The unadjusted p-value for $7/8$ vs. $2/8$ is .0117. Resampling by permuting the observed vectors 10,000 times gives us an adjusted p-value of .5183, computed using PROC MULTTEST. The permutation test clearly demonstrates the problems with the analysis of multiple categories for this experiment: results that appear significant are easily attributed to chance. □

Example 7.17 Efficacy Analysis of a Clinical Trial

A clinical trial for a potential respiratory drug was conducted at two centers, see Koch et al. (1990). Patients were examined before treatment, visit V0, and at four

Table 7.33 Summary of Multinomial Data and the Most Extreme Binary Tests

	Multinomial Classification for Factor F			
	Easy	Normal	Difficult	
F(−1)	3	5	0	8
F(+1)	0	4	4	8
	3	9	4	16

p-Value Summary for Binary Tests

Factor	Category	Raw_p	Adj_p	Low(−1) vs. High(+1)
D	Easy	.0547	1.0000	0/8 vs. 3/8
F	Difficult	.0209	.9203	0/8 vs. 4/8
	Easy	.0547	1.0000	3/8 vs. 0/8
G	Normal	.0117	.5183	7/8 vs. 2/8
	Easy	.0547	1.0000	0/8 vs. 3/8

subsequent visits, V1–V4, and their respiratory health was evaluated categorically as "very poor" = 0, "poor" = 1, "fair" = 2, "good" = 3, or "excellent" = 4.

There were two treatments, active drug (A) or placebo (P). Two characterization variables were also recorded, sex and age. So the computer record for each patient consists of the following variables: center, ID, sex, age, Trt, and 25 ordered categorical variables, (5 visits) × (5 categories), given in Table 7.34.

We are interested in whether drug treated individuals fare better than placebo treated individuals, and need to pose this question using specific tests. Do we want to test for A versus P without regard to the various characterization variables (sex, age, pretreatment status)? Do we test the difference between A and P for each sex separately? If we dichotomize age (old vs. young), we could analyze data separately within the two age groups. Do we treat center as a block (or strata) variable, or do we test separately within each center? How do we treat the ordered categorical data? We could use scores, e.g. $0, 1, 2, 3, 4$ and/or $0, 0, 1, 2, 4$. The first scoring system encodes a linear improvement with increasing category; the second encodes complete ill-health for the first two categories and a geometric improvement over the 3rd to 5th categories. Or we could dichotomize the response into a binomial variable, e.g. 0 if very poor, poor, or fair, and 1 if good or excellent: there are four ways to collapse the five categories into a dichotomous response.

We can make our choices up front and test, or we could try various choices and repeatedly test. With multiplicity adjusted p-values there is a middle ground; we can propose and use several reasonable data analysis strategies, and by using multiplicity adjusted p-values, we can be sure that we are not deceiving ourselves or others. Raw p-values give a probabilistic frame of reference as if each comparison was *the a priori* planned comparison, adjusted p-values give a probabilistic frame of reference over all the tests of hypotheses. We expect to pay a price, but no more than is necessary. If it is really not clear how the experiment should be analyzed, we can use several reasonable methods.

First, how do we deal with all the potential response variables? The five binomial variables for the four post treatment periods could be analyzed separately, thus testing twenty responses. But trying to make sense of 20 correlated binomial variables is not very practical. Instead, we choose one scoring system, $0, 1, 2, 3, 4$; and we choose one dichotomous system, with "very poor," "poor," and "fair" scored as 0, and with "good" and "excellent" scored as 1. These two response variables are highly correlated.

And how do we treat the repeat visits? The pretreatment evaluation should be used. Some measure of overall performance is essential. And what about trends in the responses after treatment begins? Two time contrasts cover the essentials and are given in Table 7.35. The first contrast measures the overall performance as a deviation from the pretreatment evaluation. The second contrast is a check on whether the patient's status gets better (or worse) over time during the treatment period.

We now have to deal with Treatment, Sex, and Age. We choose to dichotomize age as "young," (age less than 32 years), and "old," (age greater than or equal to 32 years). So there are $2 \times 2 \times 2 = 8$ possible classifications. We choose to ignore "Sex" in the analysis as we think the compound will be effective in both sexes. Others might chose to include the binary Sex variable in the analysis. The factors Trt and Age form

Table 7.34 Raw Data*

Center	ID	Sex	Age	Trt	V0	V1	V2	V3	V4	KochID
1	1	M	46	P	2	2	2	2	2	54
1	2	M	28	P	2	0	0	0	0	47
1	3	M	23	A	3	3	4	4	3	9
1	4	M	44	P	3	4	3	4	2	53
1	5	F	13	P	4	4	4	4	4	28
1	6	M	34	A	1	1	2	1	1	23
1	7	M	43	P	2	3	2	4	4	51
1	8	M	28	A	1	2	2	1	2	15
1	9	M	31	A	3	3	4	4	4	20
1	10	M	37	P	3	2	3	3	2	50
1	11	M	30	A	3	4	4	4	3	18
1	12	M	14	A	2	3	3	3	2	4
1	13	M	23	P	3	3	1	1	1	38
1	14	M	30	P	1	0	0	0	0	48
1	15	M	20	P	4	4	4	4	4	37
1	16	M	22	A	1	2	2	2	3	7
1	17	M	25	P	1	1	2	2	2	41
1	18	F	47	A	2	2	3	4	4	2
1	19	F	31	P	2	1	0	2	2	29
1	20	M	20	A	3	3	2	3	1	6
1	21	M	26	A	2	4	1	4	2	14
1	22	M	46	A	4	3	4	3	4	24
1	23	M	32	A	3	4	4	3	3	22
1	24	M	48	A	2	3	2	0	2	25
1	25	F	35	P	1	0	0	0	0	30
1	26	M	26	A	2	2	2	2	2	13
1	27	M	23	P	4	4	2	4	4	39
1	28	F	36	P	2	3	3	2	2	31
1	29	M	19	P	2	3	3	0	0	36
1	30	M	28	A	0	0	1	2	1	16
1	31	M	37	P	1	0	0	0	0	49
1	32	M	23	A	2	3	4	4	4	10
1	33	M	30	A	3	3	4	4	2	17
1	34	M	15	P	2	2	3	3	2	35
1	35	M	26	A	1	2	2	3	2	12
1	36	F	45	P	2	2	2	2	1	32
1	37	M	31	A	0	2	3	2	1	21
1	38	M	50	A	2	2	2	2	2	26
1	39	M	23	P	2	1	1	1	1	46
1	40	M	26	P	1	2	1	2	2	43
1	41	M	14	P	2	2	1	2	3	34
1	42	M	31	A	1	2	3	1	1	19
1	43	M	13	P	3	4	4	4	4	33
1	44	M	27	P	1	2	2	1	2	44
1	45	M	26	P	2	4	2	4	3	42
1	46	M	49	P	2	2	2	2	2	55
1	47	M	63	P	2	2	2	2	2	56
1	48	M	57	A	3	3	4	3	4	27
1	49	M	27	P	3	3	4	3	3	45
1	50	M	22	A	2	1	3	4	4	8
1	51	M	15	A	0	2	3	3	3	5
1	52	M	43	P	1	1	1	3	2	52
1	53	F	32	A	1	2	2	4	2	1
1	54	M	11	A	4	4	4	4	2	3
1	55	M	24	P	3	4	4	4	3	40
1	56	M	25	A	2	3	3	2	3	11

*From Koch, Amara, Stokes, Uryniak, *Statistical Methodology in the Pharmaceutical Sciences*, Chapter 13, (1990), courtesy of Marcel Dekker, Inc., New York.

Table 7.34 *(Continued)* **Raw Data**

Center	ID	Sex	Age	Trt	V0	V1	V2	V3	V4	KochID
2	1	F	39	P	1	2	1	1	2	88
2	2	M	25	A	2	2	4	4	4	70
2	3	M	58	A	4	4	4	4	4	82
2	4	F	51	P	3	4	2	4	4	94
2	5	F	32	P	3	2	2	3	4	85
2	6	M	45	P	3	4	2	1	2	108
2	7	F	44	P	3	4	4	4	4	90
2	8	F	48	P	2	2	1	0	0	92
2	9	M	26	A	2	3	4	4	4	71
2	10	M	14	A	1	4	4	4	4	62
2	11	F	48	P	2	2	2	2	2	93
2	12	M	13	A	4	4	4	4	4	61
2	13	M	20	P	1	4	4	4	4	102
2	14	M	37	A	4	3	2	2	4	76
2	15	M	25	A	3	4	4	3	3	69
2	16	M	20	A	2	1	1	0	0	65
2	17	F	58	P	1	4	2	2	0	95
2	18	M	38	P	4	3	0	0	0	105
2	19	M	55	A	4	4	4	4	4	80
2	20	M	24	A	4	4	4	4	4	67
2	21	F	36	P	3	3	2	1	3	86
2	22	M	36	P	2	4	3	3	4	104
2	23	F	60	A	4	4	3	3	4	59
2	24	M	15	P	3	2	2	3	3	98
2	25	M	25	A	3	4	3	3	1	68
2	26	M	35	A	4	4	4	4	4	75
2	27	M	19	A	3	3	2	3	3	63
2	28	F	31	P	3	4	4	4	4	84
2	29	M	21	A	3	3	4	4	4	66
2	30	F	37	A	1	3	4	4	4	57
2	31	M	52	P	2	3	4	3	4	110
2	32	M	55	A	2	2	3	3	1	81
2	33	M	19	P	4	2	2	3	3	100
2	34	M	20	P	3	2	4	4	4	101
2	35	M	42	P	3	2	2	2	2	106
2	36	M	41	A	3	4	4	3	4	77
2	37	M	52	A	1	2	1	2	2	79
2	38	F	47	P	2	3	3	2	3	91
2	39	M	11	P	3	4	4	4	4	96
2	40	M	14	P	2	1	2	3	2	97
2	41	M	15	P	4	3	3	3	4	99
2	42	M	66	P	3	3	3	4	4	111
2	43	M	34	A	2	4	4	2	4	74
2	44	M	43	P	2	1	0	0	0	107
2	45	M	33	P	3	3	3	2	3	103
2	46	M	48	P	4	4	0	0	0	109
2	47	M	20	A	2	4	4	4	3	64
2	48	F	39	P	3	2	3	0	0	89
2	49	M	28	A	2	3	2	2	1	72
2	50	F	38	P	1	2	0	0	0	87
2	51	M	43	A	3	3	4	4	2	78
2	52	F	39	A	2	3	4	4	4	58
2	53	M	68	A	2	3	3	3	4	83
2	54	F	63	A	4	4	4	4	4	60
2	55	M	31	A	4	4	4	4	4	73

Table 7.35 Time Contrasts over Visits

	Visit				
Time Contrast	V0	V1	V2	V3	V4
PrtTrt vs. Trt	−4	+1	+1	+1	+1
Trt Linear	0	−3	−1	+1	+3

a 2×2 factorial and we will test for Active versus Placebo in the entire data set and within each age classification, Young and Old. Thus we have three "treatment" contrasts that are given in Table 7.36. A more "statistical" approach is to test the main effects (Trt and Age) and their interaction. The contrasts we have selected offer the opportunity to find effects within Young or Old (but with reduced power).

We choose to block on center.

So we will have two response variables, the wellness score, $0, 1, 2, 3, 4$, and a binomial variable, very poor/poor/fair versus good/excellent (0 or 1). We have two time contrasts, PreTrt versus Treated and Treatment Time Trend. Using the data analysis paradigm of Section 1.6, we have four $\mathbf{Y}_2$ variables:

SPT: Score response, pretreatment versus treated
STL: Score response, treatment linear contrast
BPT: Binomial response, pretreatment versus treated
BTL: Binomial response, treatment linear contrast.

It is instructive to digress from the main study and examine how the time contrasts work with the binomial response variables. If the condition of an individual is good or excellent for the pretreatment visit, then the $-4, 1, 1, 1, 1$ contrast will be zero if all visits V1–V4 are good or excellent, negative if any are very poor, poor or fair. The negative number gives the proportion of the visits where the health deteriorates. If the pretreatment condition of an individual is very poor, poor or fair, then the $-4, 1, 1, 1, 1$ contrast gives the proportion of the time that an individual becomes "healthy." So the "cut point" for the binomial variable is determining the division between unhealthy and healthy.

Note that a person who starts with a rating of "4" and deteriorates to a rating of "3" has a zero for the $-4, 1, 1, 1, 1$ contrast. Likewise, an individual who starts with a "0" rating and improves to a "1" or "2" is evaluated to a zero. Examination of the

Table 7.36 Trt Group Testing Coefficients

	Trt Group			
Trt Group Contrast	A_0	A_1	P_0	P_1
Act vs. Plac	−1	−1	1	1
Act vs. Plac, Young only	−1	0	1	0
Act vs. Plac, Old only	0	−1	0	1

Table 7.37 Raw Y_1 Data, with TRT_AGE Classification Variable, and Y_2 Variables: SPT, STL, BPT, and BTL

Center	ID	Sex	Age	Trt	TRT_AGE	V0	V1	V2	V3	V4	SPT	STL	BPT	BTL
1	1	M	46	P	P1	2	2	2	2	2	0.00	0.0	0.00	0.0
1	2	M	28	P	P0	2	0	0	0	0	−2.00	0.0	0.00	0.0
1	3	M	23	A	A0	3	3	4	4	3	0.50	0.0	0.00	0.0
1	4	M	44	P	P1	3	4	3	4	2	0.25	−0.5	−0.25	−0.3
1	5	F	13	P	P0	4	4	4	4	4	0.00	0.0	0.00	0.0
1	6	M	34	A	A1	1	1	2	1	1	0.25	−0.1	0.00	0.0
1	7	M	43	P	P1	2	3	2	4	4	1.25	0.5	0.75	0.1
1	8	M	28	A	A0	1	2	2	1	2	0.75	−0.1	0.00	0.0
1	9	M	31	A	A0	3	3	4	4	4	0.75	0.3	0.00	0.0
1	10	M	37	P	P1	3	2	3	3	2	−0.50	0.0	−0.50	0.0
1	11	M	30	A	A0	3	4	4	4	3	0.75	−0.3	0.00	0.0
1	12	M	14	A	A0	2	3	3	3	2	0.75	−0.3	0.75	−0.3
1	13	M	23	P	P0	3	3	1	1	1	−1.50	−0.6	−0.75	−0.3
1	14	M	30	P	P0	1	0	0	0	0	−1.00	0.0	0.00	0.0
1	15	M	20	P	P0	4	4	4	4	4	0.00	0.0	0.00	0.0
1	16	M	22	A	A0	1	2	2	2	3	1.25	0.3	0.25	0.3
1	17	M	25	P	P0	1	1	2	2	2	0.75	0.3	0.00	0.0
1	18	F	47	A	A1	2	2	3	4	4	1.25	0.7	0.75	0.3
1	19	F	31	P	P0	2	1	0	2	2	−0.75	0.5	0.00	0.0
1	20	M	20	A	A0	3	3	2	3	1	−0.75	−0.5	−0.50	−0.2

response variables BPT and BTL, for various individual in Table 7.37, makes these points clear.

The SAS code for the construction of the Y_2 variables and execution of PROC MULTTEST is given in Appendix IV. There is repeated use of the SAS indicator function, e.g., (age $\geqslant$ 32) becomes 1 if the age is greater or equal to 32 and 0 if the age is less than 32. The contrasts are scaled to make the results more meaningful. For example, dividing the $-4, 1, 1, 1, 1$ contrast variable by four gives the proportion of treatment visits where health has improved (or if a person started as healthy, it gives the proportion of treatment visits where health has deteriorated).

We have three treatment group contrasts: Active versus Placebo for the entire data set and for each age group. And we have four Y_2 variables, so our testing matrix T is 3×4: there are 12 contrasts of interest.

In setting up the analysis, selections of Y_2 variables and contrasts of interest should be specified before the trial is conducted. Obviously, we are doing it *post hoc*. Skeptics might therefore accuse us of data snooping. Clinical trials are often largely confirmatory; even so the analysis should be specified before the trial begins (or the data are scrutinized). At the price of somewhat less power for particular comparisons, multiple comparisons can be put into the analysis as we have done here.

Table 7.38 gives the summary statistics for the response variables, the matrix S of the data analysis paradigm. A_0 is the Active drug, age = young; A_1 is active, age = old; P_0 is placebo, age = young; and P_1 is placebo, age = old. PROC MULTTEST requires that a single classification variable be used; so if the treatment groups form

Table 7.38 Summary of Time Contrasts Averaged for Trt Groups

Time Contrast	Block	Statistic	Trt Groups A_0	A_1	P_0	P_1
Score PreTrt vs. Trt	1	Mean	0.8289	0.4063	−0.0694	0.0000
		Std Dev	0.8082	0.6936	0.9188	0.6892
		N	19	8	18	11
	2	Mean	0.5577	0.5357	0.3438	−0.3625
		Std Dev	1.1733	1.0184	1.3819	1.3560
		N	13	14	8	20
Binomial PreTrt vs. Trt	1	Mean	0.2368	0.1563	0.0556	0.0682
		Std Dev	0.4123	0.2652	0.3157	0.3371
		N	19	8	18	11
	2	Mean	0.2692	0.2500	0.0000	−0.1250
		Std Dev	0.4837	0.5000	0.4818	0.5707
		N	13	14	8	20
Score TrtLinear	1	Mean	−0.0368	0.0375	−0.0500	−0.0182
		Std Dev	0.4017	0.3815	0.3777	0.3157
		N	19	8	18	11
	2	Mean	−0.1000	0.0286	0.2625	−0.2650
		Std Dev	0.4082	0.2128	0.2326	0.5274
		N	13	14	8	20
Binomial TrtLinear	1	Mean	−0.0316	0.0125	−0.0111	−0.0455
		Std Dev	0.1887	0.1642	0.1451	0.1572
		N	19	8	18	11
	2	Mean	−0.0154	−0.0286	0.1500	−0.0500
		Std Dev	0.1519	0.0825	0.1852	0.1638
		N	13	14	8	20

a factorial arrangement, then a single defining variable must be formed. The time contrasts are listed along the left margin of the table. The block mean, standard deviation, and number of observations are given in the body of the table. Under the null hypothesis of no treatment/age/drug effects, the expected value of each cell mean is zero. And we note, for example, that the means for the placebo columns, P_0 and P_1 for all response variables are close to this expectation. On the other hand, the cells means associated with the PreTrt versus Treated time contrast for both the score response and binomial response are all positive. The average increase in wellness score is in the range .40 to .80; the average improvement in the binomial response is in the range .15 to .27. There is little indication that the treatment response either increases or decreases during the treatment period for either the active drug treated individuals or the placebo treated individuals. We now turn to the results of hypothesis testing to see if these impressions are supported.

Table 7.39 gives the p-values for the 12 tests. The increase in wellness score is statistically significant for the PreTrt vs. Treated period using the bootstrap adjusted p-value. Note that neither the Šidák nor Bonferroni p-values are significant at the .05 level. The p-value of .0119 for Binomial, PreTrt vs. Treated period is adjusted to .0980.

Table 7.39 *p*-Values for Time by Treatment Group Contrasts

		p-Value			
Treatment Contrast	Time Contrast	Raw_p	Adj_p	Sid_p	Bon_p
Act vs. Plac, Combined	Score, PreTrt vs. Trt	0.0049*	0.0477*	0.0576	0.0592
	Binomial, PreTrt vs. Trt	0.0119*	0.0980	0.1338	0.1428
	Score, Trt Linear	0.9998	1.0000	1.0000	1.0000
	Binomial, Trt Linear	0.4024	0.9652	0.9979	1.0000
Act vs. Plac, Young only	Score, PreTrt vs. Trt	0.0588	0.3707	0.5166	0.7054
	Binomial, PreTrt vs. Trt	0.0710	0.4223	0.5867	0.8518
	Score, Trt Linear	0.1100	0.5624	0.7530	1.0000
	Binomial, Trt Linear	0.0363*	0.2577	0.3586	0.4359
Act vs. Plac, Old only	Score, PreTrt vs. Trt	0.0340*	0.2436	0.3397	0.4080
	Binomial, PreTrt vs. Trt	0.0751	0.4403	0.6082	0.9013
	Score, Trt Linear	0.1253	0.6088	0.7995	1.0000
	Binomial, Trt Linear	0.3871	0.9594	0.9972	1.0000

*p-value $< .05$.

The PreTrt vs. Treated period unadjusted p-values for Young and Old individuals are relatively small, but upon adjustment are not significant. These tests are not in conflict with the combined tests, as sample sizes are smaller when considering Young and Old separately. It is clear from examination of this p-value table, and from the construction of the variables, that testing of the score variable should be more powerful than testing of the binomial variable. It is also clear, even with the bootstrap method of adjusting for multiplicity, that the adjustment is large: adjustment for the 12 tests caused the smallest raw p-value of .0049 to increase to .0477. Experimenters should draw two conclusions: small raw p-values can occur with fair frequency in large experiments and experimenters should strive to reduce the number of tests of hypothesis.

It is interesting to check on the possible effect of a different dichotimization of the score variable into a binomial variable. The classification $(0, 0, 1, 1, 1)$ was also analyzed. The raw p-value of .0119 for the PreTrt vs. Treated period increased to .1023. As one might suspect, the choice of how to dichotomize a score variable can rather dramatically affect the resulting p-value. So p-value shopping is possible by means of test and variable selection. □

7.5 A COMPLEX EPIDEMIOLOGY SURVEY

In complex epidemiologic studies, hundreds of questions or relationships typically are examined. It is likely that spurious relationships will appear in such studies if there is no adjustment for multiplicity; therefore, there is a need to adjust the statistical procedures to reflect the number of questions asked. In this section we compute and compare unadjusted and multiplicity adjusted p-values in a complex epidemiologic study.

When hundreds of tests are performed, the minimum p-value is expected to be quite small. Heretofore, statistical theory has not been able to say *how small* the minimum p-value is expected to be, owing to difficulties caused by dependencies among the variables of interest and among the particular comparisons performed.

7.5.1 Mental Health Data Description

We examine data from a large survey conducted to assess the mental health of urban and rural individuals living in central North Carolina by Blazer et al. (1985), called the "Duke Epidemiological Catchment Area Project." Our purpose here is not to quarrel with the particular findings of this study or the methodologies used, but to point out how our methodology can help overcome some of the inherent difficulties in epidemiological and survey studies.

The researchers surveyed five contiguous counties in the Piedmont of North Carolina whose residents were distributed nearly equally between urban and rural counties. Psychiatric interviews and questionnaires were given to a randomly selected set of about 3900 people, one per household. We consider eight diagnoses of the original study: agoraphobia, alcohol abuse (and/or dependence), antisocial personality disorder, cognitive deficit, dysthymia (chronic mild depression), major depressive episode, obsessive-compulsive disorder, and schizophrenia (and/or schizophreniform disorder). Drug abuse was diagnosed in only four people and is not considered further in this analysis. There were seven demographic variables collected on each person: age, sex, race (white and non-white), marital status (married with spouse, separated/divorced, widowed, nonmarried), education (non-high school, high school), mobility (moved in last year, did not move), and location (rural, urban).

To illustrate multiplicity adjusted p-values and address questions of the original paper that describes this data set, we consider only three dichotomous demographic variables, rural/urban, male/female, and high school diploma/no high school diploma

Table 7.40 Coefficients Defining Selected Contrasts

	Urban				Rural			
	Male		Female		Male		Female	
	HS^+	HS^-	HS^+	HS^-	HS^+	HS^-	HS^+	HS^-
Urban vs. Rural	−1	−1	−1	−1	1	1	1	1
Male vs. Female	−1	−1	1	1	−1	−1	1	1
HS^- vs. HS^+	−1	1	−1	1	−1	1	−1	1
Urban vs. Rural, HS^-	−1	0	−1	0	1	0	1	0
Urban vs. Rural, HS^+	0	−1	0	−1	0	1	0	1
Urban vs. Rural, Male	−1	−1	0	0	1	1	0	0
Urban vs. Rural, Female	0	0	−1	−1	0	0	1	1
HS^- vs. HS^+, Male	−1	1	0	0	−1	1	0	0
HS^- vs. HS^+, Female	0	0	−1	1	0	0	−1	1

HS^+ completed high school education.
HS^- did not complete high school education.

(abbreviated HS^+ and HS^-). The multiple comparisons we consider are differences in incidences of the eight psychiatric disorders for the following subdivisions: (1) urban vs. rural overall, (2) male vs. female overall, (3) high school graduates vs. non-high school graduates overall, (4) urban vs. rural for non-high school graduates, (5) urban vs. rural for high school graduates, (6) urban vs. rural for males, (7) urban vs. rural for females, (8) high school graduates vs. non-high school graduates for males, and (9) high school graduates vs. non-high school graduates for females. The contrasts that correspond to these questions are given in Table 7.40.

Since there are a total of eight psychiatric disorders and nine major or minor subdivisions of the population, there are $8 \times 9 = 72$ questions asked. Because some of the subdivisions are constructed by splitting other subdivisions, and because human psychological characteristics might be correlated, it is expected that the 72 questions are not independent of one another.

The questions asked and the results obtained can be given succinctly in three tables. Table 7.41 gives the count, number of observations, and percent incidence for each of the eight psychiatric disorders for the nine subdivisions of the three demographic variables. Certain tendencies are apparent in this table. For example, within a sex/location pair, the incidence of agoraphobia is always less among graduates than

Table 7.41 Incidences of Psychiatric Disorders*

Group	Statistic	Agora-phobia	Alcohol/ Substance Abuse	Anti-social	Cognitive Deficit	Dys-thymia	Major Depression	Obsessive/ Compulsive	Schizo-phrenia
Urban	Count	18	19	6	3	2	6	6	3
Male	n	489	489	489	489	489	489	489	489
HS^+	%	3.68	3.89	1.23	0.61	0.41	1.23	1.23	0.61
Urban	Count	19	13	3	15	0	4	7	5
Male	n	281	281	281	281	281	281	281	281
HS^-	%	6.76	4.63	1.07	5.34	0.00	1.42	2.49	1.78
Urban	Count	70	4	1	0	17	26	9	6
Female	n	671	670	671	671	671	670	671	671
HS^+	%	10.43	0.60	0.15	0.00	2.53	3.88	1.34	0.89
Urban	Count	58	4	1	16	11	7	5	7
Female	n	415	414	415	415	414	414	415	415
HS^-	%	13.98	0.97	0.24	3.86	2.66	1.69	1.20	1.69
Rural	Count	6	12	2	0	2	2	4	3
Male	n	298	298	298	298	298	298	298	298
HS^+	%	2.01	4.03	0.67	0.00	0.67	0.67	1.34	1.01
Rural	Count	27	32	3	28	7	2	10	3
Male	n	418	416	418	418	418	418	418	418
HS^-	%	6.46	7.69	0.72	6.70	1.67	0.48	2.39	0.72
Rural	Count	30	1	0	1	8	7	9	2
Female	n	539	539	539	539	539	539	539	539
HS^+	%	5.57	0.19	0.00	0.19	1.48	1.30	1.67	0.37
Rural	Count	90	5	2	19	18	10	30	3
Female	n	676	673	676	677	676	676	676	674
HS^-	%	13.31	0.74	0.30	2.81	2.66	1.48	4.44	0.45

*Data set provided by Dr. Linda K. George of Duke University Medical Center.

nongraduates. High school graduates also have lower incidences of cognitive deficits (as might be expected).

Table 7.40 gives a statistical shorthand for the specific tests (questions) that are computed. The eight ($2 \times 2 \times 2 = 8$) groupings of individuals are given across the top of the table, and the nine questions are given along the left-hand side. The numbers ($-1, 0, 1$) in the body of the table show how the eight groups are combined to address the specific questions. For example in the first row, the first four groups are all urban individuals and are grouped together with a "-1" to be compared with the last four groups (all rural residents) that are grouped together with a "$+1$." A "0" indicates that the particular group is not pertinent to the question asked.

7.5.2 Results

Table 7.42 is a summary of all p-values (raw and adjusted) for the $8 \times 9 = 72$ questions asked. The variables agoraphobia, alcohol abuse, etc., are across the top of the table, and the questions asked are along the left-hand margin. In the body of the table are the p-values. For convenience, unadjusted p-values meeting the "statistical significance" criterion ($p < .05$) are marked with an asterisk and adjusted p-values meeting this criterion are marked with a double asterisk. When one accounts for the number of questions asked, the number of reportable "significant" effects decreases from 25 to 9. Thus, the multiplicity adjustments enforce a stricter standard upon

Table 7.42 Summary of p-Values for Survey Group Contrasts

Contrast	p-value	Agora-phobia	Alcohol/ Substance Abuse	Anti-social	Cognitive Deficit	Dys-thymia	Major Depression	Obsessive/ Compulsive	Schizo-phrenia
Urban vs.	Raw_P	.0229*	.7289	.3170	.7069	.2029	.0149*	.0914	.0820
Rural	Adj_P	.7250	1.0000	1.0000	1.0000	1.0000	.6020	.9970	.9920
Male vs.	Raw_P	.0000*	.0000*	.0012*	.0115*	.0000*	.0139*	.7701	.4509
Female	Adj_P	.0000**	.0000**	.0820	.5200	.0010**	.5840	1.0000	1.0000
HS^- vs.	Raw_P	.0000*	.0204*	.4806	.0000*	.4885	.3677	.0176*	.2446
HS^+	Adj_P	.0000**	.6910	1.0000	.0000**	1.0000	1.0000	.6450	1.0000
Urban vs.	Raw_P	.7278	.3285	.7036	.9546	.0374*	.2297	.0585	.0246*
Rural, HS^-	Adj_P	1.0000	1.0000	1.0000	1.0000	.8730	1.0000	.9670	.7480
Urban vs.	Raw_P	.0032*	.5944	.2912	.6256	.7245	.0229*	.6500	.8804
Rural, HS^+	Adj_P	.1870	1.0000	1.0000	1.0000	1.0000	.7250	1.0000	1.0000
Urban vs.	Raw_P	.0184*	.3040	.5753	.9491	.3767	.0304*	.0081*	.0280*
Rural, Male	Adj_P	.6570	1.0000	1.0000	1.0000	1.0000	.8250	.4100	.8000
Urban vs.	Raw_P	.3034	.2062	.4047	.5947	.0195*	.1654	.9618	.6377
Rural, Female	Adj_P	1.0000	1.0000	1.0000	1.0000	.6810	1.0000	1.0000	1.0000
HS^- vs.	Raw_P	.0000*	.1542	.2251	.0000*	.2865	.1678	.0709	.3442
HS^+, Male	Adj_P	.0000**	1.0000	1.0000	.0000**	1.0000	1.0000	.9840	1.0000
HS^- vs.	Raw_P	.0006*	.0673	.9476	.0000*	.9717	.9590	.1107	.4631
HS^+, Female	Adj_P	.0420**	.9800	1.0000	.0000**	1.0000	1.0000	.9970	1.0000

*Raw p-value $< .05$.
**Adjusted p-value $< .05$.

the criterion for judging "significance." When the raw p-value is significant and the adjusted p-value is insignificant, it is possible that the effect tested is not real, and that the apparent "significance" is a statistical artifact. On the other hand, tests having significant adjusted p-values probably represent real effects.

It is instructive to re-arrange the incidence data into a small table for some of the tests that are marked as significant. The point here is to move from "statistical significance" to consideration of effect size and consistency. Table 7.43 shows the relationships of agoraphobia incidence among the classification variables. The following conclusions are noted: males are less agoraphobic than females ($p < 0.0001$); high school graduates are less agoraphobic than nongraduates ($p < 0.00001$), both for males and females. However, the statistical significance of "urban vs. rural" comparison ($p = 0.02289$ using the raw p-value) could well be the result of chance when the number of tests is taken into account (adjusted $p = 0.725$). Similar conclusions hold when testing urban vs. rural among high school graduates and when comparing urban vs. rural among males: the comparisons are significant by the raw p-value standard and are insignificant by the adjusted p-value standard. For agoraphobia there is a possible interaction. The incidence of agoraphobia for urban, educated females seems too high (or, alternatively, the incidence is too low for urban educated males). Subject matter expertise is essential for the interpretation of such interactions. And interest in this particular interaction is data set generated; it is not among the original 72 questions of our analysis.

7.5.3 Discussion and Conclusions

Table 7.42 shows large differences between raw and multiplicity adjusted p-values. Raw p-values as small as .0012 (normally considered "highly statistically significant") are adjusted up to .08 (usually considered "not statistically significant"). The degree of adjustment would be even greater if additional, alternative and more complex data analyses were performed. In addition to the testing of multiple responses for various subdivisions of the data, another source of multiplicity is the use of different statistical methodologies to address the same questions. For example, chi-square, logistic regression, and other methods could be used to assess statistical significance. While there is some confirmatory benefit to using alternative methodologies for re-analysis, it is not proper to report only the analysis giving the smallest p-value.

The type of question asked has some bearing on whether the result should be flagged as "significant." For example, any result that has clear and immediate ramifications concerning the survival of individuals should be examined closely, even if

Table 7.43 Agoraphobia Incidence (%) Grouped by Sex, Education, and Location

	Male		Female	
	HS^-	HS^+	HS^-	HS^+
Rural	6.46	2.01	13.31	5.57
Urban	6.76	3.68	13.98	10.43

the *unadjusted* p-value is insignificant. However, science and the general public are not well served by reporting results that do not have serious implications, particularly when such results can easily be spurious (e.g., when the adjusted p-value is insignificant). In other words, one should not "cry wolf" when the existence of the wolf is in doubt, particularly when the wolf has no teeth!

An *ab initio* solution to the multiplicity problem is to state explicitly the particular questions to be asked, and to separate preplanned and serious (e.g., life-threatening) questions from those devised through data exploration. Exploratory tests can still be analyzed, but adjusted p-values are preferred for assessing their "significance."

APPENDIX I

Munchausen's Statistical Grid

The following letter appeared in *The Lancet*, December 22/29, 1984, and is reproduced here with permission of Graham Martin.

OCCASIONAL GRID

Munchhausen's Statistical Grid, Which Makes All Trials Significant

THE best-known story of the travels of the famous Baron von Munchausen[1] is his description of the thaw at the beginning of the Russian spring. Stopping to sleep while crossing the trackless snows of the steppe, he hitched his horse to a small post sticking out of the snow. The next morning he found himself in the centre of a village and his horse hanging by its bridle from the cross on the top of the church steeple. To get the horse down he had to shoot the bridle through with his pistol.

The Baron's medical writings were extensive but not much appreciated till this century. In 1951 Asher[2] published his description of the tall-tale-telling, peripatetic, poly-hospital, surgery-seeking-patient—the syndrome now universally known by the Baron's name.

Recently the Baron's statistical writings have come to light, and with them a statistical device, Munchaüsen's statistical grid. The great virtue of this device is that it can resurrect a significant result from any foundering therapeutic trial. First the patients in the trial are allocated to five classes (eg, age groups). Then the five classes are divided by sex, to make ten subgroups. The experimenter declares a probability of $p = 0.05$,—ie, one in twenty—to be significant, and now there is a fifty-fifty chance that one of the subgroups will produce the desired result. Should the experimenter still have bad luck, subdividing the classes along some other line—eg, Scandinavian and non-Scandinavian patients—provides a second chance of winning; it should give twenty subclasses and considerably improve the chances of a significant result.

The Baron's system has substantial commercial applications. Once a drug has been found to have a probably significant effect in one of the cells, say in Scandinavian women aged 51–60, it may be marketed among the Scandinavians in Minnesota. Since it works for those aged 51–60, it probably works for others, and to confine it to Scandinavian women would be both sexist and racist. Thus, Munchausen's grid can rescue any trial from otherwise certain obscurity.

The Baron foresaw the criticisms that future generations might level at his device; according to him the arguments would run like this (see Carnap[3]). Suppose you have an urn containing red and blue balls, evenly mixed but in unknown proportions—possibly even no red balls at all. If in nineteen random tries you pulled out nineteen blue balls, you might suspect that there were few or no red balls in the urn, and that your chance of getting a red ball on the twentieth try was rather small. The Baron refuted this argument; everyone knew that if your wife had nineteen sons in succession the chance of having a daughter the next time was still fifty-fifty. On this basis he asserted the principle that all events were statistically totally independent of all others, and that therefore all the cells in his grid must be treated as though they were totally independent events like tossing a coin or the determination of the sex of a baby. Thus, there was no similarity between the events in any of the cells in this grid. For instance, in drug trials there was no similarity between Scandinavian men aged 51–60 and those aged 41–50. Therefore it was quite unreasonable to expect some evidence that a drug which worked for one group worked for other similar groups. It was mere cynicism to say that if a drug was ineffective in nine of ten cells in his grid, this said something about its potency.

Curiously, despite his profound intellect, the Baron did not carry this principle of the total independence of all events into his personal life. He shot an acquaintance when the coin they were playing with turned up heads nine times in a row, and discarded a mistress when the butler said nineteen times out of twenty that she was "not at home".

Munchausen's grid has another more subtle application when its use is concealed. If you wish to study, for instance, the relation between backache and changes on the lumbar spine X-ray, you examine a large but unspecified number of features (say forty) of the X-ray, and relate them to the presence or absence of pain. Statistical tests are applied, using a probability level of one in twenty, and you find that there is more back pain with square osteophytes than with round ones (a discrimination that only you and your colleagues can make). The main message of your paper then becomes that square osteophytes cause back pain. You ignore the second statistically significant result among the forty or so tested, that there is a relation between the amount of bowel gas on the film and back pain, because you are not a gastroenterologist. This method has been described as "casting the net widely" a technique known for centuries to improve the chances of catching a fish. Moreover, submitting a larger number of factors to statistical examination not only improves your chances of a positive result but also enhances your reputation for diligence.

GRAHAM MARTIN
Neurosurgical Unit
Wellington Hospital,
Wellington, New Zealand

---

1. Raspe RE, et al. (1785) The singular travels, campaigns and adventures of the Baron Munchausen. London: Cressent Press, 1948.
2. Asher R. Munchausen's syndrome. *Lancet* 1951; i: 339–41.
3. Carnap R. Mathematics, an introduction to its spirit and use. San Francisco: Freeman WH, 1979: 153.

APPENDIX II

Get-the-Lead-Out Guru Challenged

The following article appeared in *Science*, 253, News & Comment section, 23 August, 1991, pp. 842–844. December 22/29, 1984.

As AN ENVIRONMENTAL BOGEYMAN, LEAD'S hard to beat. It ranks right up there with asbestos, dioxin, and nuclear waste. Vice President Dan Quayle has even suggested that lead in the drinking water at the vice presidential mansion might have caused the Bushes' bouts with Graves' disease.

But, irrational fears aside, there's no question that high lead levels can cause brain damage—it's only at low levels of exposure that there is still a debate about what amount of lead in the blood can cause detectable behavioral and medical problems. And that debate has been tainted by a festering, 10-year-old dispute over the credibility of data published by Herbert Needleman of the University of Pittsburg, a world-renowned researcher on lead toxicity and leading adviser to the government on lead issues. Now, in the wake of a government lawsuit against the owners of a lead smelter in which Needleman was to have testified—but never did because the case was settled out of court—his critics have filed a complaint with federal investigators alleging that Needleman engaged in scientific misconduct a decade ago. They accuse the government of helping cover up the flaws in his research in order to deflect criticism of its policy decisions.

To Needleman, the charges are nothing more than old mud slung with new vigor—thoroughly debunked criticisms kept alive by a lead industry desperate to discredit his research.

Regardless of who is right, the Needleman saga shows how hard it is to put to rest charges from persistent critics, or, conversely, to prove misconduct against an acknowledge leader in a scientific field. But it also raises additional questions, widely applicable to other scientific disputes, about who should have access to data collected with federal support. And, of course, refocuses attention on a matter that is especially meaningful to a lot of parents: Just how strong is the link between low level lead exposure and intelligence deficits?

The story begins with a paper by Needleman and his colleagues in the 29 March 1979 issue of *The New England Journal of Medicine* showing that schoolchildren with what all would agree were "high," but not actually toxic, lead levels did significantly poorer in the classroom and had measurably lower IQs than those with "low" lead levels. In order to get a clearer picture of exposure, the researchers had looked at lead concentrations in the children's baby teeth, as well as the more labile measure of lead in the blood. Suzanne Binder, chief of the Lead Poisoning Prevention Branch at the Centers for Disease Control (CDC) in Atlanta, says that most people's first reaction to Needleman's study was "so what?" since the drop in IQ was only 3 or 4 points. But Binder says policy makers came to realize that even a small drop would be important if it was affecting millions of children.

Two years after the Journal article appeared, Claire Ernhart, a psychologist now at Case Western Reserve University, and her colleagues fired the first shot across Needleman's bow. Writing in the journal *Pediatrics*, they suggested that there were serious methodological flaws in the Needleman paper. Ernhart argued that Needleman had not done an adequate job of controlling for confounding variables—other factors such as poor schools or parental neglect that might explain the difference in IQ scores and had performed so many comparisons that he was bound to come up with a few that were statistically significant merely by chance. Ernhart's own work suggested that most lead effects were too small to be detected by a crude measure like IQ, except at some of the highest levels of exposure, just below what would be considered toxic.

The appearance of the *Pediatrics* article touched off what has been a decade-long personal feud between Ernhart and Needleman. They have squared off at numerous scientific meetings with a vigor that has left observers shaking their heads. "Personal hostility is putting it mildly," says Binder.

But the Needleman/Ernhart squabble might have remained nothing more than a classic confrontation between scientists with starkly opposing views had it not entered, in 1983, into a new and grander forum. The year before, the Environmental Protection Agency (EPA) had begun a major review of national air-quality standards for lead and wanted to review all recent data on the health effects of lead exposures. In an effort "to resolve major points of controversy concerning some of the most important and controversial" studies, Lester Grant, director of the EPA's environmental criteria and assessment office, convened a special panel to look into both Needleman's and Ernhart's work (*Science*, 25 November 1983, p. 906).

The panel traveled to Needleman's lab, examined some of his data, and decided there were several problems with the study. Specifically, the panel members concluded that Needleman had used inappropriate measures to categorize lead exposure and had not provided sufficient justification for excluding subjects from the study. Moreover, they expressed concern about missing data, and some of the statistical analyses Needleman had employed, all of which led them to conclude that the study results "neither support nor refute the hypothesis that low or moderate levels of [lead] exposure lead to cognitive or behavioral impairments in children." The panel reached the same conclusion about two of Ernhart's papers which they criticized for methodological flaws.

While Ernhart took the panel's rebuke quietly, Needleman fought back. He insisted that the panel's conclusions were flawed, and he wrote a spirited, point-by-point refutation of the criticisms levied at his work. He blasted Grant for printing the report before sending it to him for review accusing him of violating an agreement he said he and Grant had made. Needleman also performed some new analyses of his original data, and by the time the panel's report was presented to the EPA advisory panel that would decide on the new lead standards, both Grant and the advisory panel had made a 180-degree turn. Now they were convinced that Needleman's original conclusions were accurate. Indeed, those conclusions subsequently became part the scientific basis for the revised air lead standards EPA promulgated in 1986.

But Ernhart was not daunted by this setback; she continued to criticize Needleman's work. And her willingness to argue that the link between low-level lead exposure and behavioral problems was being overstated won favor with the lead industry. As early 1982, she had agreed to testify in favor of the industry's position before an EPA panel contemplating phasing out all leaded gasoline. Just last year she wrote to Senator Harry Reid (D-NV) telling him that basing legislative action on Needleman's findings would be a "egregious error... Serious problems in the Needleman work have long been noted by scientists working in this field." And she appeared from time to time as an expert witness in cases involving lead contamination and cleanup, which brought her feud with Needleman into a new arena: the courtroom.

Their latest face-off which has escalated beyond the hazards of lead to the high-stakes "game" of scientific fraud and misconduct charges began in 1990 with a Superfund case brought by the government against Sharon Steel, UV Industries, and Atlantic Richfield Company. Over a period of several decades, each company had had a financial interest in a defunct lead smelter in Midvale, Utah. The government case sought money for the cleanup of some 250 acres of tailings from a milling facility that prepared the lead ore for the smelter. The government intended to show that the tailings posed a health risk to children living in the area and hired none other than Herbert Needleman as an expert witness to testify to the dangers the tailings posed.

For their part, the corporations' lawyers turned to Ernhart as an expert witness. In addition, the defense team brought in University Virginia psychologist Sandra Wood Scarr, whose work focuses on factors affecting children's educational development She had also served as a member of the EPA. panel that had examined Needleman and Ernhart's research back in 1983. Although Scarr had been among the most critical of Needleman's work then, she says she paid no further attention to it after the panel had wrapped up its business. Now, she and Ernhart felt that they could damage the government's case by demonstrating what they had long believed: that Needleman's 1979 paper which they say has been "highly influential in the establishment of regulatory policies" was seriously flawed.

They asked to see Needleman's raw data for the 1979 study. He agreed to release some of the unpublished material, but not the tapes containing his raw data. Needleman argued, in an affidavit dated 27 July 1990 that, in part because he was in the throes of moving his lab, "it would be a substantial hardship for me to find the proper data tape for this 11-year-old study." He added that since the study had been peer reviewed and the data examined by the EPA, there had already been adequate opportunity to establish the legitimacy of his results.

Needleman did say in his affidavit, however, that he would be willing to let "any scientist who wishes to examine the complete printouts of the raw data from the study come to my laboratory in Pittsburgh for as long as he or she wants." So on 20 September last year, Scarr and Ernhart, along with defense lawyers in the lead smelter case, traveled to Pittsburgh to take Needleman up on his offer. When they arrived, they were directed by Justice Department attorney W. Benjamin Fisherow, who was acting for the government, to a bare room where they were given six volumes of computer printouts containing Needleman's initial analyses of his data. Scarr and Ernhart began plowing through the analyses, although they were hampered by the fact that the data were coded, and they were given an incomplete key. Needleman himself would not talk to them.

For his part, Needleman steadfastly insists that he will happily share his data with anyone who has a legitimate interest and will answer any questions he is asked. But, he says, "I'm just not going to make it easy for people who are going to harass me," a category to which he assigns Scarr and Ernhart.

Since Scarr and Ernhart weren't able to get through all the computer printouts in one day, they returned to the lab the next morning. But this time, Fisherow asked them to sign a document saying they would treat all the data they were being shown in absolute confidence and would discuss it only in oral testimony before the court. While such agreements are not uncommon for litigation involving private corporations, Scarr and Ernhart were appalled at what they saw as an attempt to gag them, and they refused to sign. After a few hours, lawyers for both sides decided that the visit would have to end, so Scarr and Ernhart gathered their notes and left.

Scarr says that even with only one day to study the analyses, she felt she had a clear idea of what had happened back in 1979. "It was really the data analysis strategy that I looked at, and that to me [was] outrageous." According to Scarr, the printouts show that Needleman's first set of analyses failed to show a relationship between lead level and subsequent intelligence tests. "Not one single variable came out as statistically different between the top 10% [of lead-exposed children] and the bottom 10% of the sample," she says. It was only by rerunning the analysis, eliminating important variables that might also cause changes in IQ scores, that "he got the results he wanted."

Scarr's harsh view of Needleman might never have become public, however, had it not been for a curious legal twist—one with potentially major ramifications for the availability of government research data. Before Scarr and Ernhart had a chance to present their conclusions about Needleman and his data in the Utah court case, the litigants settled the case. The defendants agreed pay the government $63 million—the cost cleaning up the lead tailings. Losing the chance to put the charges on record in open court, Scarr and Ernhart wrote a report that they planned to send to the National Institutes of Health (NIH) Office of Scientific Integrity (OSI), since Needleman's study had been funded with NIH money. But 4 days before the settlement agreement was announced, the government lawyers took a remarkable step: They asked the court to force Scarr and Ernhart to return their notes on the Needleman data and refrain from speaking about what they had found—essentially the same rules Scarr and Ernhart refused to agree to back in Pittsburgh when they were poring over Needleman's printouts.

Scarr and Ernhart immediately concluded and believe today that the government was trying to protect Needleman because his research forms the backbone of government lead policy.

Government lawyer Fisherow will not say explicitly why the government sought to gag Scarr and Ernhart, but Needleman's affidavit gives a rationale: "Releasing these raw data to the defendants here will mean that the industry will have the capacity, if it so chooses, to manipulate this data as it sees fit. While any credible researcher should be willing to have the accuracy of his published results debated, the standards of conduct in the scientific community do not extend to making raw data available to advocates of opposing views who then are presented with the opportunity to misuse them."

Scarr and Ernhart weren't buying that argument. They hired David F. Geneson of the Washington, D.C. legal firm Hunton & Williams to fight the gag order. Geneson contended to the court that the government's request was an abridgement of Scarr and Ernhart's First Amendment's rights, and that there was no good cause to suppress data that had been gathered with public money.

This argument certainly rings true with lawyers who specialize in misconduct issues. "It's hard to imagine a legitimate basis for the federal government asking for data to be buried," says one such attorney, Barbara Mishkin of the Washington, D.C., firm of Hogan and Hartson.

Needleman, however, has been trying to make just such an argument by saying that the lead industry has tried to twist his data to make it appear to prove things it doesn't actually prove. Mishkin isn't impressed. "The cigarette industry is always putting out its own dubious analysis of data. Nobody pays any attention to it."

On 26 April this year, federal district court judge Bruce S. Jenkins agreed with Mishkin's point, writing that "there is something inherently distasteful and unseemly in secreting either the fruits or seeds of scientific endeavors." And that freed Scarr and Ernhart to tell their doubts about Needleman and his data to anyone they chose. Their first step was to write a report based on their day-and-a-half visit to Needleman's lab, and they have sent a copy to the OSI. OSI officials are considering whether Needleman's actions fall under their jurisdiction, or whether Scarr and Ernhart's analyses fall under the rubric of legitimate scientific difference of opinion. Jane Duffield, a spokeswoman for the University of Pittsburgh, where Needleman did his work, says OSI has not contacted the university concerning any investigation and adds that "we're not investigating Dr. Needleman and we stand behind his work."

From Needleman's point of view, this latest round of charges is nothing more than harassment from the lead industry. "I just do not want to . . . spend the rest of my life responding to trivia," he says. "My reputation is secure at least among people whom I count as important. I'm a forward looking person, and I have much more important questions to answer." And indeed, Needleman has supporters inside and outside the EPA who have seen his data and find nothing to be suspicious of. Frequent co-author David Bellinger, a psychologist and epidemiologist at Harvard University, says there is no substance to Scarr's charge that Needleman eliminated variables from his analyses until he got the results he wanted. "I've worked with the data set, and nothing has ever come to light to make me concerned about that issue." Adds former EPA panel member Larry Kupper, a biostatistician at the University of North Carolina, Chapel Hill, "I never thought there was any misconduct."

Meanwhile, Ernhart vociferously defends herself against Needleman's charges that she is merely serving the lead industry in its attack on him. Yes, she accepts research support from the International Lead Zinc Research organization, she admits, but says that hasn't affected her objectivity. Her objections to Needleman's conclusions began, she argues, long before she received any lead industry money. Scarr, too, bristles at the suggestion that her objectivity is tainted. "I have no ties to the lead industry," she says. "I don't care what happens to them." What bothers Scarr is that policy decisions are being made based on what she is convinced is flawed work that no one wants to take the time to examine closely. "There's just something wrong about the procedure here, and the role that science is playing in this."

Not so, says EPA's Grant. "The particular studies that are at issue there, and the publications that they are fighting about, are more or less passé," he maintains. "We now have a decade of additional research that confirms lead affects on IQ and behavioral development at much lower levels than the ones they were talking about."

But even this point of view is contested by some. Sanford L. Weiner, a political scientist based in Boston who works for the Milbank Memorial Fund, says he, like Ernhart and Scarr, believes policy actions have outpaced the science. Weiner says it is hard to find a study that clearly demonstrates adverse health effects from lead levels below 25 micrograms per deciliter of blood, the point that the CDC currently uses as its cutoff for lead poisoning. Indeed, agrees Marjorie Smith, a psychologist at the Institute for Child Health at the University of London who directed a lead study in England, 10 micrograms per deciliter—the level to which EPA wants to reduce the lead standard is "unrealistically low" and would "cause unnecessary anxieties for parents."

Binder says that it is extremely hard to find people who don't have strong opinions about lead in the environment "Either they're working on this because they consider it to be an incredible problem and it's worth devoting their life to, or they think everybody else is an idiot, and they have to prove that everyone else is wrong."

So will the day come when both sides can reach a consensus? Not likely, says Binder: "They will all go to their graves thinking the other side is made up of total idiots."

Joseph Palca

APPENDIX III

The MULTTEST Procedure

This appendix lists syntax, data handling methods, and computational details of PROC MULTTEST as it exists at the time of this book was prepared. As with any computer software, there is a good chance that the procedure will be revised over time. Nevertheless, this appendix gives a flavor of the range of application, the ease of use, and the flexibility of the SAS software. We are grateful to SAS Institute Inc. for allowing us to use this material.

AIII.1 PROC MULTTEST SYNTAX

PROC MULTTEST < *options* >;
  BY *variables*;
  CLASS *variable*;
  CONTRAST *'label' values*;
  FREQ *variable*;
  STRATA *variable*;
  TEST *name (variables* < */ options* > *)*;

Items within < > are optional, and statements following the PROC MULTTEST statement can appear in any order. The CLASS and TEST statements are required. The syntax of each statement is described below in alphabetical order after the description of the PROC MULTTEST statement.

PROC MULTTEST Statement

PROC MULTTEST < *options* >;

You can specify the following options in the PROC MULTTEST statement:

BONFERRONI
BON
specifies that the Bonferroni adjustments, (number of tests) × p-value, be computed for each test. These adjustments can be extremely conservative and should be viewed with caution.

BOOTSTRAP
BOOT
specifies that the p-values be adjusted using the bootstrap to resample vectors (with replacement). Resampling is performed independently within levels of the STRATA variable. Continuous variables are mean-centered by default prior to resampling.

If the PERMUTATION= option is used with the CA test, the exact permutation distribution is recomputed for each bootstrap sample. BOOTSTRAP resampling is not allowed with the PETO test for theoretical reasons.

CENTER
requests that continuous variables be mean-centered prior to resampling. The default action is to mean-center for bootstrap resampling and not to mean-center for permutation resampling.

DATA = SASdataset
names the SAS data set to be used by PROC MULTTEST. The default is to use the most recently created data set.

NOCENTER
requests that continuous variables not be mean-centered prior to resampling. The default action is to mean-center for bootstrap resampling and not to mean-center for permutation resampling.

NOPRINT
requests that all printed output be suppressed.

NOTABLES
requests that the MULTTEST TABLES section of the printed output be suppressed.

NSAMPLE = number
N = number
specifies the number of resamples for use with the BOOTSTRAP and PERMUTATION options; it is assumed to be 100 by default. Large values of *number* (e.g., 10,000) are desirable for accuracy, but they can result in long execution times.

ORDER = DATA
ORDER = FORMATTED
ORDER = FREQ
ORDER = INTERNAL
specifies the sorting order for the levels of the classification variable (specified in the CLASS statement). This ordering determines which parameters in the model correspond to each level in the data, so the ORDER= option may be useful when you use the CONTRAST statements. The table below shows how PROC MULTTEST interprets values of the ORDER= option.

Value of ORDER =	Levels Sorted By
DATA	order of appearance in the input data set
FORMATTED	external formatted value
FREQ	descending frequency count; levels with the most observations come first in the order
INTERNAL	internal machine representation

By default, ORDER = FORMATTED. For FORMATTED and INTERNAL, the sort order is machine-dependent.

OUT = SASdataset
names the SAS data set containing variable names, contrast names, and all associated *p*-values.

OUTPERM = SASdataset
names the SAS data set containing entire permutation distributions (upper-tail probabilities) for all tests when the PERMUTATION= option is used. *Caution: this data set can be very large.*

OUTSAMP = SASdataset
names the SAS data set containing information from the resampled data sets when resampling is performed. *Caution: this data set can be very large.*

PERMUTATION
PERM
gives adjusted *p*-values in identical fashion as the BOOTSTRAP option, with the exception that PROC MULTTEST resamples without replacement rather than with replacement. Resampling is performed independently within levels of the STRATA variable. Continuous variables are not mean-centered prior to resampling. PERMUTATION resampling is not allowed with the PETO test for theoretical reasons.

PVALS
requests that a summary table of raw and adjusted *p*-values be included.

SEED = number
S = number
specifies the initial seed for the random number generator used for resampling. *Number* must be a positive integer; the computer clock time is the default. For more details of seed values, see *SAS Language: Reference, Version 6.*

SIDAK
SID
specifies that the Šidák adjustments be computed for each test. These adjustments take the form

$$1 - (1 - p)^n$$

where p is the raw p-value and n is the number of tests. These are slightly less conservative than the Bonferroni adjustments, but still should be viewed with caution.

STEPDOWN
requests that any adjusted p-value be computed using the stepdown techniques.

BY Statement

BY *variables*;

A BY statement can be used with PROC MULTTEST to obtain separate analyses on observations in groups defined by the BY variables. When a BY statement appears, the procedure expects the input data set to be sorted in order of the BY variables. The *variables* are one or more variables in the input data set.

If your input data set is not sorted in ascending order, use one of the following alternatives:

- Use the SORT procedure with a similar BY statement to sort the data.
- Use the options NOTSORTED or DESCENDING in the BY statement for the MULTTEST procedure. As a cautionary note, the NOTSORTED option does not mean that the data are unsorted, but rather means that the data are arranged in groups (according to values of the BY variables) and that these groups are not necessarily in alphabetical or increasing numeric order.
- Use the DATASETS procedure (in Base SAS software) to create an index on the BY variables.

Since sorting the data changes the order that MULTTEST reads observations, this can affect the sorting order for the levels of the CLASS variable if you have specified ORDER = DATA in the PROC MULTTEST statement. This, in turn, will affect specifications in the CONTRAST statements.

For more information on the BY statement, see the discussion in *SAS Language: Reference, Version 6*. For more information on the DATASETS procedure, see the discussion in *SAS Procedures Guide, Version 6.*

CLASS Statement

CLASS *variable*;

The CLASS statement is required. It declares a single variable (character or numeric) used to identify the groups for the analysis. For example, if the variable TRT defines different levels of a treatment, then the statement is

```
class trt;
```

The order of the CLASS levels used by PROC MULTTEST correspond to their formatted values; this order can be changed with the ORDER= option on the PROC MULTTEST statement. You need to be aware of this feature when using the CONTRAST statement. You should check the printed table MULTTEST COEFFICIENTS to verify that the appropriate order is used.

CONTRAST Statement

CONTRAST *'label' values*;

This statement is used to identify tests between the levels of the CLASS variable. The *label* is a string naming the contrast; it contains a maximum of 21 characters. The *values* are scoring coefficients across the CLASS variable levels.

You can supply multiple CONTRAST statements, thereby specifying multiple contrasts for each variable. Multiplicity adjustments are given for all contrasts and all variables simultaneously. The coefficients are applied in the order of the CLASS variables; this order can be changed with the ORDER= option on the PROC MULTTEST statement. For example, consider a four-group experiment with CLASS variable levels A1, A2, B1, and B2 denoting two levels of two treatments. The statements

```
contrast 'a vs b'     -1  -1   1  1;
contrast 'a linear'   -1   1   0  0;
contrast 'b linear'    0   0  -1  1;
```

produce three linear trend tests for each variable identified in the TEST statement. PROC MULTTEST computes the multiplicity adjustments over the entire collection of tests, which would be three times the number of variables.

As another example, consider an animal carcinogenicity experiment with dose levels 0, 4, 8, 16, and 50. One might choose to consider trend tests defined using the following statement:

```
contrast 'arithmetic trend' 0 4 8 16 50;
```

This statement produces a trend test using the indicated scoring coefficients. Multiplicity-adjusted p-values are then computed over the collection of variables identified in the TEST statement.

See Lagakos and Louis (1985) for guidelines on the selection of contrast scoring values.

When a FISHER test is specified on the TEST statement, the CONTRAST coefficients are used to group the CLASS variable's levels. Groups with a -1 scoring coefficient are combined and compared with groups with a 1 scoring coefficient for each test, and groups with a 0 coefficient are not included in the contrast. For example,

```
contrast 'c vs all'  -1  1  1  1;
contrast 'c vs t1'   -1  1  0  0;
contrast 'c vs t3'   -1  0  0  1;
```

computes Fisher exact tests for (a) control vs. the combined treatment groups, (b) control vs. the first treatment group, and (c) control vs. the third treatment group. Multiplicity adjustments are then computed over the entire collection of tests and variables. If FISHER is specified and no CONTRAST statement is given, then all contrasts of control vs. treatment are automatically generated, with the first level of the CLASS variable deemed to be the control. Only -1, 1, and 0 are acceptable CONTRAST scoring coefficients when the FISHER test is specified; PROC MULTTEST ignores the CONTRAST statement if any other coefficients appear.

CONTRAST values are $0, 1, 2, \ldots$ by default, except for the Fisher test described above.

For continuous data (and for the FT tests), the contrast coefficients are centered to have mean zero. The resulting centered scoring coefficients are then applied to the sample means (or to the double-arcsine-transformed proportions in the case of the FT tests).

FREQ Statement

FREQ *variable*;

The FREQ statement names a variable that provides frequencies for each observation in the DATA= data set. Specifically, if n is the value of the FREQ variable for a given observation, then that observation is used n times.

If the value of the FREQ variable is missing or is less than 1, the observation is not used in the analysis. If the value is not an integer, only the integer portion is used.

STRATA Statement

STRATA *variable*;

The STRATA statement identifies a single variable to use as a stratification variable in the analysis. This yields tests similar to those discussed in Mantel and Haenszel (1959) and Hoel and Walburg (1972) for binary data and pooled means tests for continuous data. For example, when testing for prevalence in a carcinogenicity study, it is common to stratify on intervals of the time at death: the first level of the stratification variable may represent weeks 0–52, the second weeks 53–80, etc. In multicenter clinical studies, each level of the stratification variable may represent a particular center.

TEST Statement

TEST *name (variables< / options >);*

The TEST statement is required. It identifies statistical tests to be performed and the discrete and/or continuous variables to be tested. The following tests are permitted as *name* in the TEST statement:

CA
requests the Cochran-Armitage linear trend tests for group comparisons. The test variables should take the value 0 for a failure and 1 for a success. The PERMUTATION= option can be used to request an exact permutation test; otherwise, a Z-score approximation is used. The CONTINUITY= option can be used to specify a continuity correction for the Z-score approximation.

FISHER
requests Fisher exact tests for comparing two treatment groups. The test variables should take the value 0 for a failure and 1 for a success.

FT
requests Z-score CA tests based upon the Freeman-Tukey double arcsine transformation of the frequencies. The test variables should take the value 0 for a failure and 1 for a success.

MEAN
requests the t-test for the mean. The test variables can take on any numeric values.

PETO
requests the Peto mortality-prevalence test. The test variables should take the value 0 for a failure, 1 for a success, and 2 for a fatality. The TIME= option should be used with the PETO test to specify a variable giving the age at death. The CONTINUITY= option can be used to specify a continuity correction for the test.

If the value of a TEST variable is invalid, the observation is not used in the analysis.

You can specify two tests only if one of them is MEAN. For example,

```
test ca(d1-d2) mean(c1-c2);
```

is valid but

```
test ca(d1-d2) ft(d1-d2);
```

is invalid.

The following options can be used in the TEST statement (some only apply to one test):

CONTINUITY = number
C = number
specifies *number* as a particular continuity correction for the Z-score approximation in the CA test or for the Peto test. The default is 0.

LOWERTAILED
LOWER
is used to make all tests lower-tailed. All tests are two-tailed by default.

PERMUTATION = number
PERM = number
specifies that p-values for the CA tests be computed using exact permutation distributions when marginal success or failure totals within a stratum are *number* or less. For values greater than *number* (or when PERMUTATION= is omitted), PROC MULTTEST uses standard normal approximations with a continuity correction chosen to approximate the permutation distribution. PROC MULTTEST computes the appropriate convolution distributions when you use the STRATA statement along with the PERMUTATION= option.

TIME = variable
identifies the PETO test variable containing the age at death. If TIME= is omitted, all ages are assumed to equal 1.

UPPERTAILED
UPPER
is used to make all tests upper-tailed. All tests are two-tailed by default.

AIII.2 PROC MULTTEST DETAILS

Statistical Tests

The following section discusses the statistical tests performed in the MULTTEST procedure. For continuous data, a t-test for the mean is available. For discrete variables, available tests are the Cochran-Armitage (CA) linear trend test, the Freeman-Tukey (FT) double arcsine test, the Peto mortality-prevalence test, and the Fisher exact test.

Throughout this section, the discrete and continuous variables are denoted by S_{vgsr} and X_{vgsr} respectively, where v is the variable, g is the treatment group, s is the stratum, and r is the replication. A + subscript denotes summation over an index.

Cochran-Armitage Linear Trend Test

The Cochran-Armitage linear trend test (Cochran, 1954; Armitage, 1955; Agresti, 1990) is implemented using a Z-score approximation, an exact permutation distribution, or a combination of both.

Z-Score Approximation

Let m_{vgs} denote the sample size for a binary variable v within group g and stratum s. The pooled probability estimate for variable v and stratum s is

$$p_{vs} = S_{v+s+}/m_{v+s}$$

The expected value (under constant within-stratum treatment probabilities) for variable v, group g, and stratum s is

$$E_{vgs} = m_{vgs}p_{vs}$$

The test statistic for variable v has numerator

$$N_v = \sum_s \sum_g t_g(S_{vgs+} - E_{vgs})$$

where t_g denotes a trend coefficient. The variance estimate for this statistic is

$$V_v = \sum_s p_{vs}(1 - p_{vs}) \sum_g m_{vgs}(t_g - \bar{t}_{vs})^2$$

where

$$\bar{t}_{vs} = \sum_g m_{vgs}t_g/m_{v+s}.$$

PROC MULTTEST computes the Z-score statistic

$$Z_v = N_v/V_v^{1/2}$$

and its p-value comes from the standard normal distribution. Whenever a zero is computed for the denominator, the p-value is set to 1. This p-value approximates the probability obtained from the exact permutation distribution, discussed below.

The Z-score statistic can be continuity-corrected to better approximate the permutation distribution. With continuity correction c, the upper-tailed p-value is computed from

$$Z_v = (N_v - c)/V_v^{1/2}$$

For two-tailed, non–continuity-corrected tests, PROC MULTTEST reports the p-value as $2 \min(p, 1 - p)$, where p is the upper-tailed p-value. The same formula holds for the continuity corrected test, with the exception that when the non-continuity-corrected Z and the continuity-corrected Z have opposite signs, the two-tailed p-value is 1.

When PERMUTATION= is specified and no STRATA variable is given, PROC MULTTEST uses a continuity correction selected to optimally approximate the upper-tail probability of permutation distributions with smaller marginal totals (Westfall and Lin, 1988). This optimal continuity correction is computed for each CONTRAST from the first variable specified in the CA test. Otherwise the continuity correction is specified using the CONTINUITY= option of the TEST statement.

The Z-score statistic is virtually identical to the Hoel-Walburg (Mantel-Haenszel) statistic reported in Dinse (1985). The only difference is that the formula above uses the binomial variance estimate (Bickis and Krewski, 1986) rather than the multivariate hypergeometric variance. The rationale for this default is that PROC MULTTEST also computes exact permutation analyses from the multivariate hypergeometric distribution.

Exact Permutation Test

When you use the PERMUTATION= option for CA on the TEST statement, PROC MULTTEST computes the exact permutation distribution of the trend score

$$T_v = \sum_s \sum_g t_g S_{vgs+}$$

then compares the observed value of this trend with the permutation distribution to obtain the p-value

$$p_v = \Pr(X \geqslant \text{observed } T_v)$$

where X is a random variable from the permutation distribution. This probability may be viewed as a binomial probability, where the within-stratum probabilities are constant, and where the probability is conditional with respect to the marginal totals S_{v+s+}. It also can be considered a rerandomization probability.

Because the computations can be quite time-consuming with large data sets, specifying the PERMUTATION = *number* option on the TEST statement limits the situations where PROC MULTTEST computes the exact permutation distribution. When marginal total success and total failure frequencies exceed *number* for a particular stratum, the permutation distribution is approximated using a continuity-corrected normal distribution. You should be cautious in using the PERMUTATION= option in conjunction with bootstrap resampling because the permutation distribution is recomputed for each bootstrap sample. This recomputation is not necessary with permutation resampling.

The permutation distribution is computed in two steps:

1. compute the permutation distributions of the trend scores within each stratum
2. convolve the distributions to obtain the distribution of the total trend

As long as the total success or failure frequency does not exceed *number* for any stratum, the computed distributions are exact. In other words, if $S_{v+s+} \leqslant$ *number* or $m_{v+s} - S_{v+s+} \leqslant$ *number* for all s, then the permutation trend distribution for variable v is computed exactly.

In step 1 above, the distribution of the within-stratum trend

$$\sum_g t_g S_{vgs+}$$

is computed using the multivariate hypergeometric distribution of the S_{vgs+}, provided the threshold *number* is not exceeded. This distribution can be written as

$$\Pr(S_{v1s+}, S_{v2s+}, \ldots, S_{vGs+}) = \prod_{g=1}^{G} \binom{m_{vgs+}}{S_{vgs+}} \Big/ \binom{m_{v+s+}}{S_{v+s+}}$$

The distribution of the within-stratum trend is then computed by summing these probabilities over appropriate configurations. For further information on this technique, see Bickis and Krewski (1986) and Westfall and Lin (1988). In step 2, the exact convolution distribution is obtained for the trend statistic summed over all strata having totals that meet the threshold criterion. This distribution is obtained by applying the fast Fourier transform to the exact within-stratum distributions. A description of this general method can be found in Pagano and Tritchler (1983) and Good (1987).

The convolution distribution of the overall trend is then computed by convolving the exact distribution with the distribution of the continuity-corrected standard normal approximation. To be more specific, let S_1 denote the subset of stratum indices that satisfy the threshold criterion, and let S_2 denote the subset of indices that do not satisfy the criterion. Let T_{v1} denote the combined trend statistic from the set S_1, whose exact distribution was obtained using Fourier analysis as outlined above, and let T_{v2} denote the combined trend statistic from the set S_2. Then the distribution of the overall trend $T_v = T_{v1} + T_{v2}$ is obtained by convolving the analytic distribution of T_{v1} with the continuity-corrected normal approximation for T_{v2}. Using the notation from the previous subsection, this convolution can be written as

$$\begin{aligned} \Pr(T_{v1} + T_{v2} \geqslant u) &= \sum_{u1} \Pr(T_{v1} + T_{v2} \geqslant u \mid T_{v1} = u1) \Pr(T_{v1} = u1) \\ &= \sum_{u1} \Pr(Z \geqslant z) \Pr(T_{v1} = u1) \end{aligned}$$

where Z is a standard normal random variable, and

$$z = \left(u - u1 - \sum_{S_2} p_{vs} \sum_g t_g m_{vgs} - c\right) \Big/ V_v^{1/2}$$

In this expression, the summation of s in V_v is over S_2 and c is the continuity correction discussed under the Z-score approximation.

When a two-tailed test is requested, the "expected" trend

$$E_v = \sum_s \sum_g t_g E_{vgs}$$

is computed, and the two-tailed p-value is reported as the permutation tail probability for the observed trend T_v plus the permutation tail probability for $2E_v - T_v$, the "reflected" trend.

Freeman-Tukey Double Arcsine Test

For this test, the trend scores $t_1, \ldots, t_G$ centered to the values $c_1, \ldots, c_G$, where $c_g = t_g - \bar{t}, \bar{t} = \sum_g t_g/G$, and G is the number of groups. The numerator of this test statistic is

$$N_v = \sum_s \sum_g c_g f(S_{vgs+}, m_{vgs})$$

where the function $f(r, n)$ is the double arcsine transformation:

$$f(r, n) = \arcsin\left[(r/(n+1))^{1/2}\right] + \arcsin\left[((r+1)/(n+1))^{1/2}\right]$$

The variance estimate is

$$V_v = \sum_s \sum_g c_g^2/(m_{vgs} + .5)$$

and the test statistic is

$$Z_v = N_v/V_v^{1/2}$$

The Freeman-Tukey transformation and its variance are given in Miller (1978). Note that its variance is not weighted by the pooled probabilities as is the Cochran-Armitage test.

Peto Mortality-Prevalence Test

The Peto test is a modified Cochran-Armitage procedure incorporating mortality and prevalence information. It represents a special case in PROC MULTTEST because the data structure requirements are different, and the resampling methods used for adjusting p-values are not valid. The TIME= variable is required to specify death times. In addition, the test variables must be given as one of three designations: 0 = no occurrence, 1 = incidental occurrence, and 2 = fatal occurrence. Use the TIME= variable to define the mortality strata, and use the STRATA variable to define the prevalence strata.

The Peto test is computed like two CA Z-score approximations, one for prevalence and one for mortality.

In the following notation, the subscript v represents the variable, g the treatment group, s the stratum, and t the time. Recall that a + in a subscript location denotes summation over that subscript.

Let S^P_{vgst} be the number of incidental occurrences, and let m^P_{vgst} be the number of nonoccurrences before time t.

Let S^F_{vgst} be the number of fatal occurrences, and let m^F_{vgst} be the number of occurrences besides those included in S^F_{vgst}.

The pooled probability estimates are

$$p^P_{vs} = S^P_{v+s+}/m^P_{v+s+}$$
$$p^F_{vt} = S^F_{v++t}/m^F_{v++t}$$

The expected values are

$$E^P_{vgs} = m^P_{vgs+}p^P_{vs}$$
$$E^F_{vgt} = m^F_{vg+t}p^F_{vt}$$

Define the numerator terms:

$$N^P_v = \sum_s \sum_g t_g(S^P_{vgs+} - E^P_{vgs})$$
$$N^F_v = \sum_t \sum_g t_g(S^F_{vg+t} - E^F_{vgt})$$

where t_g denotes a trend coefficient. Define the denominator terms:

$$V^P_v = \sum_s p^P_{vs}(1 - p^P_{vs})\left[\left(\sum_g m^P_{vgs+}t_g^2\right) - \left(\sum_g m^P_{vgs+}t_g\right)^2 \Big/ m^P_{v+s+}\right]$$
$$V^F_v = \sum_t p^F_{vt}(1 - p^F_{vt})\left[\left(\sum_g m^F_{vg+t}t_g^2\right) - \left(\sum_g m^F_{vg+t}t_g\right)^2 \Big/ m^F_{v++t}\right]$$

Note that PROC MULTTEST uses the binomial, not hypergeometric, variance in the above calculations. The Peto statistic is computed as

$$Z_v = (N^P_v + N^F_v - c)/(V^P_v + V^F_v)^{1/2}$$

where c is a continuity correction. The p-value is determined from the standard normal distribution.

For further details about the Peto test, see Lagakos and Louis (1985) and Soper and Westfall (1990).

Fisher Exact Test

The CONTRAST statement in PROC MULTTEST allows you to compute Fisher exact tests for two-group comparisons. No stratification variable is allowed for this test. Note, however, that the FISHER exact test is a special case of the exact permutation tests performed by PROC MULTTEST, and that these permutation tests allow a stratification variable. Recall that contrast coefficients can be -1, 0, or 1 for the Fisher test. The frequencies and sample sizes of the groups scored as -1 are combined, as are the frequencies and sample sizes of the groups scored as 1. Groups scored as 0 are excluded. The "-1" group is then compared with the "1" group using the Fisher exact test.

Letting x and m denote the frequency and sample size of the "1" group, and y and n denote those of the "-1" group, the p-value is calculated as

$$\Pr(X \geqslant x \mid X + Y = x + y) = \sum_{i=x}^{m} \binom{m}{i}\binom{n}{x+y-i} \Big/ \binom{m+n}{x+y}$$

where X and Y are independent binomially distributed random variables with sample sizes m and n and common probability parameter. The hypergeometric distribution is used to determine the stated probability; Yates (1984) is a reference for this technique. PROC MULTTEST computes the two-tailed p-values by adding probabilities from both tails of the hypergeometric distribution. The first tail is from the observed x and y, and the other tail is chosen so that the resulting probability is as large as possible without exceeding the probability from the first tail.

t-Test for the Mean

For continuous variables, PROC MULTTEST automatically centers the trend coefficients, as in the Freeman-Tukey test. These centered coefficients c_g are then used to form a t-statistic contrasting the within-group means. Let n_{vgs} denote the sample size within group g and stratum s; it depends on variable v only when there are missing values. Define

$$\bar{X}_{vgs+} = \frac{\sum_r X_{vgsr}}{n_{vgs}},$$

as the sample mean within a group-and-stratum combination, and

$$s_v^2 = \frac{\sum_s \sum_g \sum_r (X_{vgsr} - \bar{X}_{vgs+})^2}{\sum_s \sum_g (n_{vgs} - 1)},$$

the pooled sample variance. Assume constant variance for all group and stratum combinations. Then the t-statistic for the mean is

$$M_v = \frac{\sum_s \sum_g c_g \bar{X}_{vgs+}}{\left[s_v^2(\sum_s \sum_g c_g^2/n_{vgs})\right]^{1/2}}$$

Let μ_{vgs} denote the treatment means. Then under the null hypothesis that

$$\sum_s \sum_g c_g \mu_{vgs} = 0$$

and assuming normality, independence, and homoscedasticity, M_v follows a t-distribution with $\sum_s \sum_g (n_{vgs} - 1)$ degrees of freedom.

PROC MULTTEST computes M_v internally and only reports the p-value. Whenever a denominator of zero is computed, the p-value is set to 1. When missing data force $n_{vgs} = 0$, then the contribution to the denominator of the pooled variance is 0 and not -1. This is also true for degrees of freedom.

p-Value Adjustments

PROC MULTTEST offers the following p-value adjustments: Bonferroni, Sidak, Bootstrap, and Permutation. The first two methods are simple functions of the raw p-values, while the second two are resampling adjustments. A primary feature of PROC MULTTEST is its ability to perform adjustments for the multiplicity of tests using statistical resampling. Because the resampling techniques incorporate distributional and correlational structures, they are often much less conservative than the Bonferroni and Sidak techniques. A Stepdown procedure is available for all of the methods.

Bonferroni

Suppose PROC MULTTEST performs R statistical tests, yielding p-values $p_1, \ldots, p_R$. Then the Bonferroni p-value for test r is simply Rp_r. If the adjusted p-value exceeds 1, it is set to 1.

Sidak

A technique slightly less conservative than Bonferroni is the Sidak p-value (Šidák, 1967), which is $1 - (1 - p_r)^R$. It can be applied in the case of positive orthant dependence among test statistics (Holland and Copenhaver, 1987).

Bootstrap

The bootstrap method creates pseudo-data sets by sampling observations with replacement from each within-stratum pool of observations. An entire data set is thus created, and p-values for all tests are computed on this pseudo-data set. A counter is set up for each individual test to record whether the minimum p-value from the pseudo-data set is less than or equal to the actual p-value for each base test. This process is repeated a large number of times, and the proportions of occurrences of the minimum pseudo-p-values being less than or equal to each of the actual p-values are the adjusted p-values reported by PROC MULTTEST. This algorithm is described by Westfall and Young (1989b) for the case of binary data.

In the case of continuous data, the pooling of the groups is not likely to re-create the shape of the null hypothesis distribution, since the pooled data is likely to be multimodal. For this reason, PROC MULTTEST automatically mean-centers all continuous variables prior to resampling. Such mean-centering is akin to resampling residuals in a regression analysis, as has been discussed by Freedman (1981). The NOCENTER option can be used if one wishes not to center the data. (In most situations it does not seem to make much difference whether or not one centers the data.)

Permutation

The permutation-style adjusted p-values are computed in identical fashion as the bootstrap adjusted p-values with the exception that the within-stratum resampling is performed without replacement instead of with replacement. This produces a rerandomization analysis. In the spirit of rerandomization analyses, the continuous variables are not centered prior to resampling. This default can be overridden by using the CENTER option.

Stepdown

When the Bonferroni and Šidák methods are used in stepdown fashion, a more powerful procedure results. This method was pioneered by Holm (1979), and further developed by Shaffer (1986), Holland and Copenhaver (1987), and Hochberg and Tamhane (1987). Suppose the base test p-values are ordered as $p_1 < \cdots < p_R$. Then the Bonferroni stepdown p-values are $s_1, \ldots, s_R$ from the following algorithm:

$$\begin{aligned} s_1 &= Rp_1 \\ s_2 &= \max(s_1, (R-1)p_2) \\ s_3 &= \max(s_2, (R-2)p_3) \end{aligned}$$

and so on. As always, if any adjusted p-value exceeds 1, it is set to 1. The Šidák step-down p-values are determined similarly as follows:

$$\begin{aligned} s_1 &= 1-(1-p_1)^R \\ s_2 &= \max(s_1, 1-(1-p_2)^{R-1}) \\ s_3 &= \max(s_2, 1-(1-p_3)^{R-2}) \end{aligned}$$

and so on.

The resampling-style stepdown method is analogous to the above stepdown methods; the most extreme p-value is adjusted according to all R tests, the second-most extreme is adjusted according to $(R - 1)$ tests, and so on. The difference with the resampling-style stepdown adjustments is that the adjustments are made specifically with respect to a particular set of $(R - r + 1)$ tests. In addition, all correlational and distributional characteristics are incorporated when one uses resampling methods. More specifically, assuming the ordering of p-values as above, the resampling-style stepdown adjusted p-value for test r is the probability that the minimum pseudo-p-value of tests $r, \ldots, R$ is less than or equal to p_r.

This probability is evaluated using Monte Carlo, as are the previously described resampling-style adjusted p-values. In fact, the computations for stepdown adjusted p-values are essentially no more time-consuming than the computations for the non-stepdown adjusted p-values. After Monte Carlo, the stepdown adjusted p-values are corrected to ensure monotonicity; this correction leaves the first adjusted p-values alone, then corrects the remaining ones as needed.

Missing Values

If a CLASS or STRATA variable has a missing value, then PROC MULTTEST removes that observation from the analysis.

When there are missing values for test variables, then the within group-and-stratum sample sizes may differ from variable to variable. In most cases this is not a problem; however, it is possible for all data to be missing for a particular group within a particular stratum. For continuous variables and Freeman-Tukey tests, PROC MULTTEST recenters the trend scores within strata where all data for a particular group are missing. The Cochran-Armitage test is unaffected by this situation.

If more than one variable has missing values, then PROC MULTTEST does not perform an exact permutation analysis. One exception is the case where entire groups are missing within a stratum, provided that this pattern holds for every variable.

PROC MULTTEST uses missing values for resampling if they exist in the original data set. If all variables have missing values for any observation, then PROC MULTTEST removes it prior to resampling. Otherwise, PROC MULTTEST treats all missing values as ordinary observations in the resampling. This means that different resampled data sets can have different group sizes. In some cases it means that a resampled data set can have all missing values for a particular variable in a particular group/stratum combination, even when values exist for that combination in the original data. For this reason, PROC MULTTEST recomputes all quantities within each pseudo-data set, including such items as centered scoring coefficients and degrees of freedom for p-values.

While PROC MULTTEST does provide analyses in missing-value cases, you should not feel it solves missing-value problems in an optimal manner. If you are concerned about the adverse effects of missing data on a particular analysis, you should consider imputation and sensitivity analyses to assess the effects of the missing data.

Computational Resources

PROC MULTTEST keeps all of the data in memory to expedite resampling. A large portion of the memory requirement is thus 8*NOBS*NVAR bytes where NOBS is the number of observations in the data set and NVAR is the number of variables analyzed, including CLASS, FREQ, and STRATA variables.

If you specify PERMUTATION = *number* (for exact permutation distributions) then PROC MULTTEST requires additional memory. This requirement is approximately 4 ∗ NTEST ∗ NSTRATA ∗ CMAX ∗ *number* ∗ (*number* + 1) bytes, where NTEST is the number of contrasts, NSTRATA is the number of STRATA levels, and CMAX is the maximum contrast coefficient.

The execution time is linear in the number of resamples.

Output Data Sets

The OUT= data set contains contrast names (_TEST_), variable names (_NAME_), and all associated *p*-values (RAW_P, ADJ_P, BON_P, SID_P).

OUTPERM= Data Set

The OUTPERM= data set contains contrast names (_TEST_), variable names (_NAME_), and the associated permutation distributions (_TREND_ and _TAILP_). PROC MULTTEST computes the permutation distributions when you use the PERMUTATION= option with the CA test. _TREND_ represents the support of the distributions and _TAILP_ their cumulative upper-tail probabilities. The size of this data set depends on the number of variables and the support of their permutation distributions. For information on how this distribution is computed, see the "Exact Permutation Test" section.

OUTSAMP= Data Set

The OUTSAMP= data set contains sampling groups (_CLASS_), sample/observation indicators (_R_), and test variables for the sampled data sets when resampling is performed. Each new data set is randomly drawn from the original data set, either with (bootstrap) or without (permutation) replacement. The size of this data set is thus the number of observations in the original data set times the number of samples.

The _CLASS_ variable indicates the resampling group, that is, the level of the CLASS variable assigned to the new observation. The _R_ variable values are character values of the form "SAMPLE-OBS", where SAMPLE is the sample indicator and OBS is the number of the observation in the original data set. For continuous variables, the mean-centered values are printed instead of the original values.

Printed Output

The output produced by PROC MULTTEST is divided into several sections.

The MULTTEST PROCEDURE section provides a list of the options and settings used for that particular invocation of the procedure. Included in this list are the following items:

- statistical test(s)
- support of the exact permutation distribution for CA test
- continuity correction(s) used for the CA test
- test tails
- strata adjustment
- p-value adjustment(s)
- centering of continuous variables
- number of samples and seed

The MULTTEST COEFFICIENTS section lists the coefficients used in constructing the statistical tests. These coefficients are either specified in CONTRAST statements or generated by default. The coefficients apply to the levels of the CLASS variable.

The MULTTEST TABLES section provides summary statistics for each variable listed in the TEST statement. Included for discrete variables are the count, sample size, and percentage of occurrences. For continuous variables, the mean, sample standard deviation, and sample size are printed. All of the above statistics are computed for distinct combinations of the CLASS and STRATA variables. Also included in this section are the p-values from the tests on each variable.

The MULTTEST P-VALUES section is a collection of the raw and adjusted p-values from this run of PROC MULTTEST. The p-values are listed by variable and test, with each group containing up to four p-values denoted by Raw_p (original), Adj_p (bootstrap or permutation), Bon_p (Bonferroni), and Sid_p (Šidák).

APPENDIX IV

Using Software for the Analysis

In this appendix we show the code used to perform the analyses for selected examples. We have used SAS PROC MULTTEST for many analyses. When PROC MULTTEST does not apply, we have written code for the specific analyses using SAS PROC IML, GAUSS, and RESAMPLING STATS. Examples of code for each of these four software packages are given.

AIV.1 SAS PROC MULTTEST EXAMPLES

PROC MULTTEST Code for Baseball Analysis, Example 7.9 in Section 7.2

```
proc multtest data=batnl pvals bootstrap bon sidak
        nresample=10000 seed=112089;
  class pos;
test mean (h_ab hr_ab rbi_ab r_ab /uppertailed);
   contrast 'Edge vs Center'     1  -1   1  -1  -1   1   1  -1;
   contrast 'In vs Outfield'    -3  -3  -3  -3   5   5   5  -3;
   contrast 'Right vs Left'      1   0  -1   0   0  -1   1   0;
   contrast '1st vs the rest'    7  -1  -1  -1  -1  -1  -1  -1;
```

For the preceding code, the output of PROC MULTTEST notes the following specifications:

```
Test for CONTIN variables:          T test of mean
Test tails:                         Upper-tailed
Block adjustment?                   No
P-value adjustments:                Bonferroni Sidak Bootstrap
Center CONTIN variables?            Yes
Number of resamples:                10000
Seed:                               112089
```

PROC MULTTEST Code for the Clinical Trial Analysis, Example 7.17 of Section 7.4

```
data kochres;
  infile 'kochresraw.dat';
  input center 1 id $ 5-6 sex $ 10 age 14-15 trt $ 19 v0-v4;
  y_o=(age ge 32);
  trt_age=compress(trt|| '-'|| y_o,' ');
  spt=(-4*v0+(v1+v2+v3+v4))/4;
  stl=(-3*v1 -v2 +v3 + 3*v4)/10;
  bpt=(-4*(v0 ge 3)+(v1 ge 3)+(v2 ge 3)+(v3 ge 3)+(v4 ge 3))/4;
  btl=(-3*(v1 ge 3)-(v2 ge 3)+(v3 ge 3)+3*(v4 ge 3))/10;

proc sort data=kochres;
  by center trt_age;

proc multtest data=kochres pvals bootstrap bon sidak
        nresample=10000 seed=953789;
  strata center;
  class trt_age;
  test mean(spt stl bpt btl);

contrast 'Act vs Plac'         -1   -1    1    1;
contrast 'Act vs Plac, Yng'    -1    0    1    0;
contrast 'Act vs Plac, Old'     0   -1    0    1;

 adjustp
title 'Koch_res BLOCKED data :  Continuous';
run;
```

For the preceding code, the output of PROC MULTTEST notes the following specifications:

```
Test for CONTIN variables:        T test of mean
Test tails:                       Two-tailed
Block adjustment?                 Yes
P-value adjustments:              Bonferroni Sidak Bootstrap
Center CONTIN variables?          Yes
Number of resamples:              10000
Seed:                             953789
```

AIV.2 SELECTED PROGRAMS USING OTHER SOFTWARE

For more complex models, resampling-based programs must be programmed individually. In this section we give examples using PROC IML of SAS, GAUSS, and RESAMPLING STATS.

RESAMPLING STATS Code, Example 4.4 of Section 4.4.1

Example 4.4 is a repeated-measures application. The following code computes single-step bootstrap multiplicity adjustments, as well as unadjusted bootstrap p-values, for the hypotheses

$$H_1\colon\ \mu_1 + \mu_2 - \mu_3 - \mu_4 = 0$$
$$H_2\colon\ \mu_1 - \mu_2 + \mu_3 - \mu_4 = 0$$
$$H_3\colon\ \mu_1 - \mu_2 - \mu_3 + \mu_4 = 0$$

For simplicity, the pairwise hypotheses included in Tables 4.6 and 4.7 are not included.

Code	Comment
`READ file "jw219a.dat" x1 x2 x3 x4`	Input data
`  set 3 1 one`	a vector of 1's
`  set 3 0 adjp`	a vector of 0's
`  set 3 0 rawp`	
`  add x1 x2 t1`	
`  add x3 x4 t2`	
`  subtract t1 t2 d1`	Contrast #1
`  add x1 x3 t1`	
`  add x2 x4 t2`	
`  subtract t1 t2 d2`	Contrast #2
`  add x1 x4 t1`	
`  add x2 x3 t2`	
`  subtract t1 t2 d3`	Contrast #3
`  mean d1 m1`	Mean of Contrast
`  mean d2 m2`	
`  mean d3 m3`	
`  stdev d1 s1`	Std Dev of Contrast
`  stdev d2 s2`	
`  stdev d3 s3`	
`  divide m1 s1 t1`	Standardized Contrast
`  divide m2 s2 t2`	
`  divide m3 s3 t3`	
`  concat   t1 t2 t3 t`	
`  square t t`	Squared test statistics used for two-tailed tests
`  subtract d1 m1 d1`	Center the data
`  subtract d2 m2 d2`	
`  subtract d3 m3 d3`	

```
REPEAT 10000
  GENERATE 19 1,19 B                   Generate bootstrap indices
  TAKE d1 B d1$                        Generate bootstrap data
  TAKE d2 B d2$
  TAKE d3 B d3$

  mean d1$ m1
  mean d2$ m2
  mean d3$ m3

  stdev d1$ s1
  stdev d2$ s2
  stdev d3$ s3

  divide m1 s1 t1$
  divide m2 s2 t2$
  divide m3 s3 t3$

  concat t1$ t2$ t3$ t$
  square t$ t$
  max t$ m                             Max test statistic
  multiply m one m$

  subtract t$ t m1$                    Compare bootstrap t with original t
  subtract m$ t m$                     Compare maximum bootstrap t
                                         with original t
  recode m$ <= 0 0 m$                  Construct counting variables
  recode m$ > 0   1 m$
  recode m1$ <= 0 0 m1$
  recode m1$ > 0   1 m1$
  add m$ adjp adjp                     Accumulate sums
  add m1$ rawp rawp
end
  divide adjp 10000 adjp
  divide rawp 10000 rawp
print adjp
print rawp
```

As output, RESAMPLING STATS produced

adjp = 0.0001 0.0351 0.8879

rawp = 0.0001 0.0029 0.5163

Note that the t-statistics are computed incorrectly: they use s rather than $s/\sqrt{n}$ to standardize. This is permissible as long as the same standardization is used for the bootstrap data sets.

PROC IML Code, Example 4.3 of Section 4.3

Table 4.7 displays the results of bootstrap step-down testing of a collection of regression coefficients. The following code shows the use of PROC IML of SAS to perform analysis. The data is from Neter, Wasserman, and Kutner (1989, p. 439–440), and is presented in Table 4.6 of Chapter 4.

Code	Comment
`Data one;`	Read input data
`  INFILE nwk439;`	
`  INPUT x1-x4 y1;`	
`PROC STANDARD out=new mean=0;`	Standardize variables
`  var x1-x4;`	
`DATA two;`	
`   set new;`	
`  y = log(y1);`	
`  x12 = x1*x2;`	
`  x13 = x1*x3;`	
`  x14 = x1*x4;`	
`  x23 = x2*x3;`	
`  x24 = x2*x4;`	
`  x34 = x3*x4;`	
`  x11 = x1*x1;`	
`  x22 = x2*x2;`	
`  x33 = x3*x3;`	
`  x44 = x4*x4;`	
`  one = 1;`	
`PROC IML;`	
`  USE two;`	
`  READ ALL var {one x1 x2 x3 x4`	
`     x12 x13 x14 x23 x24 x34`	
`     x11 x22 x33 x44} into xmat;`	
`  READ ALL var {y} into ymat;`	
`n = nrow(ymat);`	
`p = ncol(xmat);`	
`xpxi = inv(xmat'*xmat);`	$(X'X)^{-1}$ matrix
`b = xpxi*xmat'*ymat;`	OLS estimates
`e = ymat - xmat*b;`	residuals
`mse = e'*e/(n-p);`	
`dxpxi = vecdiag(xpxi);`	
`se = sqrt(mse * dxpxi);`	standard errors
`t = b/se;`	*t*-statistics
`pvals=2*(j(p,1,1)-probt(abs(t),n-p));`	two-tailed *p*-values

```
p0 = pvals(|2:p,1|);                          excludes intercept
print pvals;
t0 = abs(t(|2:p,1|);                          excludes intercept

r = rank(t0);
ind=r;
ind(|r|) = 1:nrow(t0);                        anti-ranks of
                                                absolute t-statistics

k = p-1;
sidp = j(k,1,0);

  do j = 1 to k;
   in = ind(|j,1|);
   sidp(|in,1|) = 1 - (1 - p0(|in,1|))**j;   Šidák step-down p-values,
  end;                                          before enforcing
                                                monotonicity
 k1 = k-1;

 do jind = 1 to k1;
   j1 = k-jind+1;
   j = j1-1;
   in1 = ind(|j1,1|);
   in = ind(|j,1|);
   if sidp(|in,1|) < sidp(|in1,1|)
      then sidp(|in,1|) = sidp(|in1,1|);      Šidák step-down p-values,
   end;                                         after enforcing
                                                monotonicity
print sidp;

 nboot = 10000;
  sum = j(k,1,0);

 do i = 1 to nboot;
 rin=int(n*ranuni(j(n,1,101891))+j(n,1,1)); bootstrap indices
 y = e(|rin,|);                               bootstrap data
b = xpxi*xmat'*y;
   mse = (y -xmat*b)'*(y -xmat*b)/(n-p);
   se = sqrt(mse * dxpxi);
    t = b/se;
    tb0 = abs(t(|2:p,1|));                   bootstrap test statistics

      do j1 = 2 to k;
      in = ind(|j1,1|);
      j = j1-1;
      in1 = ind(|j,1|);
```

```
     if tb0(|in,1|) < tb0(|in1,1|)          define successive
        then tb0(|in,1|) = tb0(|in1,1|);      maxima
     end;

   sum = sum + (tb0 > t0);
  end;

adjp = sum/nboot;                            bootstrap step-down p-values,
                                               before enforcing
  do jind = 1 to k1;                           monotonicity
   j1 = k-jind+1;
   j = j1-1;
   in1 = ind(|j1,1|);
   in = ind(|j,1|);
   if adjp(|in,1|) < adjp(|in1,1|)
       then adjp(|in,1|) = adjp(|in1,1|);
  end;                                       bootstrap step-down,
                                               p-values after
print adjp;                                    enforcing monotonicity
```

The output from this program is given in Table 4.7.

GAUSS Code, Example 3.1 of Section 3.1.1

Table 3.3 displays the results of various multiple testing methods on ANOVA weight loss data. The following GAUSS code was used to compute the *Free Step-Down Heteroscedastic Bootstrap* entries.

```
                    Code                               Comment
-------------------------------------------------------------------
output file = ott.out on;
nboot = 10000;
seed = 20613;
one = ones(10,1);
sumb = zeros(10,1);
d1 =                                        weight loss data
{ 12.4 10.7 11.9 11.0 12.4
  12.3 13.0 12.5 11.2 13.1 ,
9.1 11.5 11.3 9.7 13.2
  10.7 10.6 11.3 11.1 11.7 ,
8.5 11.6 10.2 10.9 9.0
  9.6 9.9 11.3 10.5 11.2 ,
8.7 9.3 8.2 8.3 9.0
  9.4 9.2 12.2 8.5 9.9 ,
12.7 13.2 11.8 11.9 12.2
  11.2 13.7 11.8 11.5 11.7 };
```

```
dmat = d1';
m = meanc(dmat);                                  mean vector
s = stdc(dmat);                                   standard deviation vector
s2 = s^2/10;
tab = (m[1,1] - m[2,1])/                          two-sample t-statistics
    sqrt(s2[1,1] + s2[2,1]);
tac = (m[1,1] - m[3,1])/
    sqrt(s2[1,1] + s2[3,1]);
tad = (m[1,1] - m[4,1])/
    sqrt(s2[1,1] + s2[4,1]);
```

(similar calculations for remaining `t`)

```
t = tde|tad|tce|tac|tbd                           vector of t-statistics,
    |tbe|tab|tcd|tbc|tae;                           ordered largest to smallest
dfab = 9*(s2[1,1] + s2[2,1])^2/                   Satterthwaite's df
    (s2[1,1]^2 + s2[2,1]^2);
dfac = 9*(s2[1,1] + s2[3,1])^2/
    (s2[1,1]^2 + s2[3,1]^2);
dfad = 9*(s2[1,1] + s2[4,1])^2/
    (s2[1,1]^2 + s2[4,1]^2);
```

(similar calculations for remaining `df`)

```
df = dfde|dfad|dfce|dfac|dfbd                     df vector
    |dfbe|dfab|dfcd|dfbc|dfae;
pval = 2*cdftc(abs(t),df);                        p-value vector
tdf = t df;
print tdf;
print pval;
dcent = dmat - one*m';                            centered data
iboot=1;
do while iboot <= nboot;
indvec1 = rndus(10,1,seed);
indvec2 = rndus(10,1,seed);
indvec3 = rndus(10,1,seed);
indvec4 = rndus(10,1,seed);
indvec5 = rndus(10,1,seed);
ind1 = trunc(10*indvec1 + ones(10,1));
ind2 = trunc(10*indvec2 + ones(10,1));
ind3 = trunc(10*indvec3 + ones(10,1));
ind4 = trunc(10*indvec4 + ones(10,1));
ind5 = trunc(10*indvec5 + ones(10,1));
y1 = dcent[ind1,1];                               heteroscedastic bootstrap
y2 = dcent[ind2,2];                                 samples
```

```
y3 = dcent[ind3,3];
y4 = dcent[ind4,4];
y5 = dcent[ind5,5];
y = y1 y2 y3 y4 y5;
mb = meanc(y);
sb = stdc(y);
s2b = sb^2/10;
tbab = (mb[1,1] - mb[2,1])/                   bootstrap t-statistic
    sqrt(s2b[1,1] + s2b[2,1]);
tbac = (mb[1,1] - mb[3,1])/
    sqrt(s2b[1,1] + s2b[3,1]);
tbad = (mb[1,1] - mb[4,1])/
    sqrt(s2b[1,1] + s2b[4,1]);
```

(similar calculations for remaining `tb`)

```
tb = tbde|tbad|tbce|tbac|tbbd
    |tbbe|tbab|tbcd|tbbc|tbae;
dfbab = 9*(s2b[1,1] + s2b[2,1])^2/            bootstrap df
    (s2b[1,1]^2 + s2b[2,1]^2);
dfbac = 9*(s2b[1,1] + s2b[3,1])^2/
    (s2b[1,1]^2 + s2b[3,1]^2);
dfbad = 9*(s2b[1,1] + s2b[4,1])^2/
    (s2b[1,1]^2 + s2b[4,1]^2);
```

(similar calculations for remaining `dfb`)

```
dfb = dfbde|dfbad|dfbce|dfbac|dfbbd
    |dfbbe|dfbab|dfbcd|dfbbc|dfbae;
pvalb = 2*cdftc(abs(tb),dfb);                 bootstrap p-values
j1 = 2;
do while j1 <= 10;
j = 10 -j1 + 1;
if pvalb[j,1] > pvalb[j+1,1];
pvalb[j,1] = pvalb[j+1,1];                    successive minima
endif;
j1 = j1 + 1;
endo;
compr = pvalb .< pval;                        compare bootstrap, actual
sumb = sumb + compr;
iboot = iboot+1;
endo;
adjp = sumb/nboot;
j = 2;
do while j <=10 ;
```

```
if adjp[j,1] < adjp[j-1,1];
adjp[j,1] = adjp[j-1,1];                enforce monotonicity
endif;
j = j+1;
endo;
print adjp;
```

References

Abramovitch, L. and Singh, K. (1985). "Edgeworth Corrected Pivotal Statistics and the Bootstrap," *Annals of Statistics* **13** 116–132.

Agresti, A. (1990). *Categorical Data Analysis*, Wiley, New York.

Aitken, M. A. (1969). "Some Tests for Correlation Matrices," *Biometrika* **56** 443–446.

Aitken, M. A. (1971). "Correction to 'Some Tests for Correlation Matrices'," *Biometrika* **58** 245.

Alexander, W. P. (1989). "MATRIX: An Interpreted Matrix Language," P.O. Box 9266, Schenectady, NY 12309.

Ames, B. N. and Gold, L. S. (1990). "Too Many Rodent Carcinogens: Mitogenesis Increases Mutagenesis," *Science* **249** 970–971.

Amiel, J. F. (1967). "Study of Leucocyte Phenotypes in Hodgkin's Disease," in *Histocompatibility Testing* 1967, E. S. Curtoni, P. L. Mattiuz, and R. M. Tosi (Eds.), 79–81. Munksgaard, Copenhagen.

Anderson, T. W. (1984). *An Introduction to Multivariate Statistical Analysis*, 2nd Edition, Wiley, New York.

Angus, J. E. and Schafer, R. E. (1984). "Improved Confidence Statements for the Binomial Parameter," *The American Statistician* **38** 189–191.

Aptech Systems, Inc. (1992). *GAUSS*, Aptech Systems, Maple Valley, Washington.

Armitage, P. (1955). "Tests for Linear Trend in Proportions and Frequencies," *Biometrics* **11** 375–386.

Babu, G. J. and Singh, K. (1983). "Inference on Means Using the Bootstrap," *Annals of Statistics* **11** 999–1003.

Bailar, J. C., III (1991). "Scientific Inferences and Environmental Health Problems," *Chance* **4**(2) 27–38.

Barnhart, E. R. (1991). *Physicians' Desk Reference®*, Medical Economics Company, Inc., Oradell, New Jersey.

Begun, J. and Gabriel, K. R. (1981). "Closure of the Newman-Keuls Multiple Comparisons Procedure," *Journal of the American Statistical Association* **76** 241–245.

Beran, R. (1982). "Estimated Sampling Distributions: The Bootstrap and Competitors," *Annals of Statistics* **9** 1196–1217.

Beran, R. (1988a). "Balanced Simultaneous Confidence Sets," *Journal of the American Statistical Association* **83** 679–686.

Beran, R. (1988b). "Prepivoting Test Statistics: A Bootstrap View of Asymptotic Refinements," *Journal of the American Statistical Association* **83** 687–697.

Berger, J. O. and Sellke, T. (1987), "Testing a Point Null Hypothesis: The Irreconcilability of *P* Values and Evidence" (with discussion), *Journal of the American Statistical Association* **82** 112–122.

Berger, J. O. and Wolpert, R. L. (1984). "The Likelihood Principle," IMS Lecture Notes–Monograph Series, Vol. 6, S. S. Gupta, Series Editor.

Berry, D. A. (1987). "Logarithmic Transformations in ANOVA," *Biometrics* **43** 439–456.

Berry, D. A. (1988). "Multiple Comparisons, Multiple Tests, and Data Dredging: A Bayesian Perspective" (with discussion), *Bayesian Statistics*, Volume 3, pp. 79–94. J. M. Bernardo, M. H. DeGroot, D. V. Lindley, and A. F. M. Smith (Eds.), Oxford University Press, London.

Besag, J. and Clifford, P. (1991). "Sequential Monte Carlo *P*-Values," *Biometrika* **78** 301–304.

Bickel, P. J. and Freedman, D. A. (1981). "Some Asymptotic Theory for the Bootstrap," *Annals of Statistics* **9** 1196–1217.

Bickis, M. and Krewski, D. (1986). "Statistical Issues in the Analysis of the Long Term Carcinogenicity Bioassay in Small Rodents: An Empirical Evaluation of Statistical Decision Rules," *Environmental Health Directorate*, Health Protection Branch, Health and Welfare Canada, Ottawa, Ontario, Canada K1A OL2.

Bishop, Y. M. M., Fienberg, S. E. and Holland, P. W. (1975). *Discrete Multivariate Analysis: Theory and Practice*, MIT Press, Cambridge, Massachusetts.

Blazer, D., George, L. K., Landerman, R., Pennybacker, M., Melville, M. L., Woodbury, M., Manton, K. G., Jordan, K., and Locke, B. (1985). "Psychiatric Disorders: A Rural/Urban Comparison," *Archives of General Psychiatry* **42** 651–656.

Box, G. E. P. (1950). "Problems in the Analysis of Growth and Wear Curves," *Biometrics* **6** 262–289.

Bradu, D. and Gabriel, K. R. (1974). "Simultaneous Statistical Inference on Interactions in Two-Way Analysis of Variance," *Journal of the American Statistical Association* **69** 428–436.

Breslow, N. (1990). "Biostatistics and Bayes" (with comments), *Statistical Science* **5** 269–284.

Brown, C. C. and Fears, T. R. (1981). "Exact Significance Levels for Multiple Binomial Testing with Application to Carcinogenicity Screens," *Biometrics* **37** 763–774.

Brown, J. D. and Beerstecher, E. (1951). "Metabolic Patterns of Underweight and Overweight Individuals," University of Texas Publication, Biochemical Institute Studies IV, no. 5109.

Casella, G. and Berger, R. L. (1987). "Reconciling Bayesian and Frequentist Evidence in the One-Sided Testing Problem" (with discussion), *Journal of the American Statistical Association* **82** 106–111.

Clopper, C. J. and Pearson, E. S. (1934). "The Use of Confidence or Fiducial Limits Illustrated in the Case of the Binomial," *Biometrika* **26** 404–413.

Cochran, W. (1954). "Some Methods for Strengthening the Common χ^2 Tests," *Biometrics* **10** 417–451.

Conover, W. J. and Salsburg, D. S. (1988). "Locally Most Powerful Tests for Detecting Treatment Effects When Only a Subset of Patients Can Be Expected to 'Respond' to Treatment," *Biometrics* **44** 189–196.

Cox, D. R., and Snell, E. J. (1981). *Applied Statistics: Principles and Examples*, Chapman and Hall, New York.

D'Agostino, R. B., Chase, W., and Belanger, A. (1988). "The Appropriateness of Some Common Procedures for Testing Equality of Two Independent Binomial Proportions," *The American Statistician* **42** 198–202.

D'Amaro, J., Bach, F. H., van Rood, J. J., Rimm, A. A., and Bortin, M. M. (1984). "HLA C Associations with Acute Leukemia," *The Lancet* November 24 1176–1178.

Dare, W. H. (1992). "Divergence of Opinion and the Information Content in Point Spreads," Unpublished Ph.D. Dissertation, Texas Tech University.

Davis, L. J. (1986). "Exact Tests for (2 × 2) Contingency Tables," *The American Statistician* **40** 139–141.

Dessler, A. J. (1991). "The Small-Comet Hypothesis," *Reviews of Geophysics* **29(3)** 355–382.

Dettmer, R. (1989). "Cold Water on Cold Fusion: Harwell Stops Experiments," *IEEE Review* **35** July/August, 256.

Diaconis, P. (1978). "Statistical Problems in ESP Research," *Science* **201** 131–136.

Diaconis, P. (1985). "Theories of Data Analysis: From Magical Thinking Through Classical Statistics," in *Exploring Data Table, Trends, and Shapes* 1–36. D. Hoaglin, F. Mosteller, and J. Tukey (Eds.), Wiley, New York.

Diaconis, P. and Efron, B. (1983). "Computer-Intensive Methods in Statistics," *Scientific American* May, 1983, 116–130.

Dinse, G. E. (1985). "Testing for Trend in Tumor Prevalence Rates: I. Nonlethal Tumors," *Biometrics* **41** 751–770.

Do, K. A. and Hall, P. (1991). "On Importance Sampling for the Bootstrap," *Biometrika* **78** 161–167.

Dunnett, C. W. (1955). "A Multiple Comparison Procedure for Comparing Several Treatments with a Control," *Journal of the American Statistical Association* **50** 1096–1121.

Dunnett, C. W. and Tamhane, A. C. (1992). "A Step-Up Multiple Test Procedure," *Journal of the American Statistical Association* **87** 162–170.

Edgington, E. S. (1987). *Randomization Tests*, 2nd Ed., Marcel Dekker, New York.

Edwards, D. and Berry, J. J. (1987). "The Efficiency of Simulation-Based Multiple Comparisons," *Biometrics* **43** 913–928.

Efron, B. (1979). "Bootstrap Methods: Another Look at the Jackknife," *Annals of Statistics* **7** 1–26.

Efron, B. (1981). "Nonparametric Standard Errors and Confidence Intervals," *Canadian Journal of Statistics* **9** 139–172.

Efron, B. (1986). "Comment on 'Jackknife, Bootstrap and Other Resampling Methods in Regression Analysis'," *Annals of Statistics* **14** 1301–1304.

Efron, B. (1990). "More Efficient Bootstrap Computations," *Journal of the American Statistical Association* **85** 79–89.

Efron, B. and Gong, G. (1983). "A Leisurely Look at the Bootstrap, the Jackknife, and Cross-Validation," *The American Statistician* **37** 36–48.

Efron, B. and Tibshirani, R. (1986). "Bootstrap Methods for Standard Errors, Confidence Intervals, and Other Methods of Statistical Accuracy" (with discussion), *Statistical Science* **1** 54–77.

Einot, I. and Gabriel, K. R. (1975). "A Study of the Powers of Several Methods of Multiple Comparisons," *Journal of the American Statistical Association* **70** 574–583.

Ernhart, C. B., Landa, B. and Schnell, N. B. (1981). "Subclinical Levels of Lead and Development Deficit—A Multivariate Follow-Up Reassessment," *Pediatrics* **67** 911–919.

Fama, E. (1970). "Efficient Capital Markets: A Review of Theory and Empirical Work," *Journal of Finance* **25** 383–417.

Farrar, D. B. and Crump, K. S. (1988). "Exact Tests for Any Carcinogenic Effect in Animal Bioassays," *Fundamental and Applied Toxicology* **11** 652–663.

Feinstein, A. R. (1988). "Scientific Standards in Epidemiological Studies and the Menace of Daily Life," *Science* **242** 1257–1263.

Feller, W. (1969). "Are Life Scientists Overawed by Statistics?," *Scientific Research* February 23 24–29.

Fienberg, S. E. (1980). *The Analysis of Cross-Classified Categorical Data*, Second Ed., The MIT Press, Cambridge, Massachusetts.

Fienberg, S. E., Martin, M. E., and Straf, M. L. (1985). *Sharing Research Data*, National Academy Press.

Fisher, R. A. (1935). *The Design of Experiments*, Oliver and Boyd, Edinburgh.

Fisher, R. A. (1958). *Statistical Methods for Research Workers*, Hafner, New York.

Frank, L. A., Sigwarth, J. B. and Craven, J. D. (1986). "On the Influx of Small Comets into the Earth's Upper Atmosphere, II, Interpretation," *Geophysical Research Letters* **13** 307–310.

Freedman, D. A. (1981). "Bootstrapping Regression Models," *Annals of Statistics* **9** 1218–1228.

Freedman, D. A. and Peters, S. C. (1984). "Bootstrapping a Regression Equation: Some Empirical Results," *Journal of the American Statistical Association* **79** 97–106.

Freeman, M. F. and Tukey, J. W. (1950). "Transformations Related to the Angular and Square Root," *Annals of Mathematical Statistics* **21** 607–611.

Gayen, A. K. (1950). "Significance of Difference Between the Means of Two Non-Normal Samples," *Biometrika* **37** 399–408.

Geary, R. C. (1936). "The Distribution of "Student's" Ratio for Non-normal Samples," *Journal of the Royal Statistical Society Supplement* (superseded by *Series B*) **3** 178–184.

Geary, R. C. (1947). "Testing for Normality," *Biometrika* **34** 209–242.

Good, I. J. (1987). "A Survey of the Use of the Fast Fourier Transform for Computing Distributions," *Journal of Statistical Computation and Simulation* **28** 87–93.

Graham, R. L., Hinkley, D. V., John, P. W. M. and Shi, S. (1990). "Balanced Design of Bootstrap Simulations," *Journal of the Royal Statistical Society B* **52** 185–202.

Grizzle, J. E. and Allen, D. M. (1969). "Analysis of Growth and Dose Response Curves," *Biometrics* **25** 357–381.

Hájek, J. and Šidák, Z. (1967). *Theory of Rank Tests*, Academic Press, New York.

Hall, P. and Wilson, S. R. (1991). "Two Guidelines for Bootstrap Hypothesis Testing," *Biometrics* **47** 757–762.

Hamada, M. and Wu, C. F. J. (1990). "A Critical Look at Accumulation Analysis and Related Methods" (with discussions), *Technometrics* **32** 119–162.

Hamilton, D. P. (1991). "Brain Food?," *Science* **251** 1566.

Hartley, H. O. (1950). "The Maximum *F*-Ratio as a Short-Cut Test for Heterogeneity of Variance," *Biometrika* **37** 308–312.

Haseman, J. K., Winbush, J. S, and O'Donnell, M. W. (1986). "Use of Dual Control Groups to Estimate False Positive Rates in Laboratory Animal Carcinogenicity Studies," *Fundamental and Applied Toxicology* **7** 573–584.

Hayter, A. J. (1985). "A Study of the Tukey Multiple Comparisons Procedure Including a Proof of the Tukey Conjecture for Unequal Sample Sizes," doctoral dissertation, Cornell University, Ithaca, New York.

Hedges, L. V. and Olkin, I. O. (1985). *Statistical Methods for Meta-Analysis*, Academic Press, London.

Heyse, J. F. and Rom, D. (1988). "Adjusting for Multiplicity of Statistical Tests in the Analysis of Carcinogenicity Studies," *Biometrical Journal* **30** 883–896.

Hill, A. B. (1966). *Principles of Medical Statistics*, 8th ed., Oxford University Press, London.

Hills, M. (1969). "On Looking at Large Correlation Matrices," *Biometrika* **56** 249–253.

Hochberg, Y. (1976). "A Modification of the T-Method of Multiple Comparisons for a One-Way Layout with Unequal Variances," *Journal of the American Statistical Association* **71** 200–203.

Hochberg, Y. (1988). "A Sharper Bonferroni Procedure for Multiple Tests of Significance," *Biometrika* **75** 800–802.

Hochberg, Y. and Tamhane, A. C. (1987). *Multiple Comparison Procedures*, Wiley, New York.

Hoel, D. G. and Walburg, H. E. (1972). "Statistical Analysis of Survival Experiments," *Journal of the National Cancer Institute* **49** 361–372.

Holland, B. S. and Copenhaver, M. D. (1987). "An Improved Sequentially Rejective Bonferroni Test Procedure," *Biometrics* **43** 417–424.

Holm, S. (1979). "A Simple Sequentially Rejective Multiple Test Procedure," *Scandinavian Journal of Statistics* **6** 65–70.

Hommel, G. (1988). "A Comparison of Two Modified Bonferroni Procedures," *Biometrika* **75** 383–386.

Hope, A. C. A. (1968). "A Simplified Monte Carlo Significance Test Procedure," *Journal of the Royal Statistical Society* **30** 582–598.

Iyengar, S. (1991). "Much Ado About Meta-Analysis," *Chance* **4(1)** 33–40.

Jogdeo, K. (1977). "Association and Probability Inequalities," *The Annals of Statistics* **5** 495–504.

John, J. A. and Draper, N. R. (1980). "An Alternative Family of Transformations," *Applied Statistics* **29** 190–197.

Johnson, M. E. (1987). *Multivariate Statistical Simulation*, Wiley, New York.

Johnson, R. A. and Wichern, D. W. (1988). *Applied Multivariate Statistical Analysis*, 2nd Edition, Prentice Hall, New Jersey.

Johnsson, T. (1988). "Bootstrap Multiple Test Procedure," *Journal of Applied Statistics* **15** 335–339.

Kaiser, H. F. (1960). "Directional Statistical Decisions," *Psychological Review* **67** 160–167.

Kempthorne, P., Mukhopadhyay, N., Sen, P. K. and Zacks, S. (1991). "Research — How to Do It: A Panel Discussion," *Statistical Science* **6** 149–162.

Kim, W. C., Stefánsson, G. and Hsu, J. C. (1987). "On Confidence Sets in Multiple Comparisons," in *Statistical Decision Theory and Related Topics*, Vol. IV, S. S. Gupta and J. O. Berger (Eds.), Academic Press, New York.

Kimeldorf, G., Sampson, A. R., and Whitaker, L. R. (1992). "Min and Max Scorings for Two-Sample Ordinal Data," *Journal of the American Statistical Association* **87** 241–247.

Koch, G. G., Carr, G. J., Amara, I. A., Stokes, M. E. and Uryniak, T. J. (1990). "Categorical Data Analysis." Ch. 13 in *Statistical Methodology in the Pharmaceutical Sciences*, Donald A. Berry (Ed.), Marcel Dekker, New York.

Koziol, J. A. and Perlman, M. D. (1978). "Combining Independent Chi-Squared Tests," *Journal of the American Statistical Association* **73** 753–763.

Lagakos, S. W. and Louis, T. A. (1985). "The Statistical Analysis of Rodent Tumorigenicity Experiments,"pp. 144–163, in Clayson, D. B., Krewski, D. and Munro, I. (Eds.), *Toxicological Risk Assessment*, Vol. 1, CRC Press, Boca Raton, Florida.

Lancaster, H. O. (1961). "Significance Tests in Discrete Distributions," *Journal of the American Statistical Association* **56** 223–234.

Lindley, D. V. (1990). "The 1988 Wald Memorial Lectures: The Present Position in Bayesian Statistics," *Statistical Science* **5** 44–65.

Little, R. J. A. (1989). "Testing the Equality of Two Independent Binomial Proportions," *The American Statistician* **43** 283–288.

Loh, W. (1987). "Calibrating Confidence Coefficients," *Journal of the American Statistical Association* **82** 155–162.

Louv, W. C. and Littell, R. C. (1986). "Combining One-Sided Binomial Tests," *Journal of the American Statistical Association* **81** 550–554.

Lovell, M. C. (1983). "Data Mining," *The Review of Economics and Statistics* **65** 1–12.

Madsen, M. (1976). "Statistical Analysis of Multiple Contingency Tables," *Scandinavian Journal of Statistics* **3** 97–106.

Mantel, N. and Haenszel, W. (1959). "Statistical Aspects of the Analysis of Data from Retrospective Studies of Disease," *Journal of the National Cancer Institute* **22** 719–748.

Mantel, N., Tukey, J. W., Ciminera, J. L. and Heyse, J. F. (1982). "Tumorigenicity Assays, Including Use of the Jackknife," *Biometrical Journal* **24** 579–596.

Marcus, R., Peritz, E. and Gabriel, K. R. (1976). "On Closed Testing Procedures with Special Reference to Ordered Analysis of Variance," *Biometrika* **63** 655–660.

Martin, G. (1984). "Munchausen's Statistical Grid, Which Makes All Trials Significant," *The Lancet* December 22/29, 1457.

Mehta, C. R. and Patel, N. R. (1983). "A Network Algorithm for Performing Fisher's Exact Test in $r \times c$ Contingency Tables," *Journal of the American Statistical Association* **78** 427–434.

Mehta, C. R., Patel, N. R. and Senchaudhuri, P. (1988). "Importance Sampling for Estimating Exact Probabilities in Permutational Inference," *Journal of the American Statistical Association* **83** 999–1005.

Mehta, C. R., Patel, N. R. and Tsiatis, A. A. (1984). "Exact Significance Testing to Establish Treatment Equivalence for Ordered Categorical Data," *Biometrics* **40** 819–825.

Mehta, C. R., Patel, N. R. and Wei, L. J. (1988). "Computing Exact Permutational Distributions with Restricted Randomization Designs," *Biometrika* **75** 295–302.

Meng, C. Y. K. and Dempster, A. P. (1987). "A Bayesian Approach to the Multiplicity Problem for Significance Testing with Binomial Data," *Biometrics* **43** 301–311.

Miller, J. J. (1978). "The Inverse of the Freeman-Tukey Double Arcsine Transformation," *The American Statistician* **32** 138.

Miller, R. G. (1977). "Developments in Multiple Comparisons 1966–1976," *Journal of the American Statistical Association* **72** 779–788.

Miller, R. G. (1981). *Simultaneous Statistical Inference*, 2nd Ed., Springer-Verlag, New York.

Miller, R. G. (1986). *Beyond ANOVA, Basics of Applied Statistics*, Wiley, New York.

Morgan, M. G. (1989). "Electric and Magnetic Fields from 60 Hz Electric Power: Possible Health Risks?," *Chance* **2(4)** 12–20, 37.

Morgenthaler, S. and Tukey, J. (1991). *Configural Polysampling: A Route to Practical Robustness*, Wiley, New York.

Morrison, D. F. (1990). *Multivariate Statistical Methods*, 3rd Edition, McGraw-Hill, New York.

National Toxicology Program (1982). "Carcinogenesis Bioassay of D&C Red No. 9 in F344 Rats and B6C3F1 Mice," NTP Technical Report 225, U.S. Department of Health and Human Services, Research Triangle Park, NC.

Needleman, H. L., Gunnoe, C., Leviton, A., Reed, R., Peresie, H., Maher, C. and Barrett, P. (1979). "Deficits in Psychologic and Classroom Performance of Children with Elevated Dentine Lead Levels," *The New England Journal of Medicine* **300** 689–695.

Neter, J., Wasserman, W. and Kutner, M. H. (1985). *Applied Linear Statistical Models*, 2nd Edition, Richard Irwin, Homewood, Illinois.

Neter, J., Wasserman, W. and Kutner, M. H. (1989). *Applied Linear Regression Models*, 2nd Edition, Richard Irwin, Homewood, Illinois.

Noreen, E. W. (1989). *Computer-Intensive Methods for Testing Hypotheses*, Wiley-Interscience, New York.

Oden, N. L. (1991). "Allocation of Effort in Monte Carlo Simulation for Power of Permutation Tests," *Journal of the American Statistical Association* **86** 1074–1076.

Ott, L. (1988). *An Introduction to Statistical Methods and Data Analysis*, 3rd Edition, PWS-Kent, Boston.

Pagano, M. and Tritchler, D. (1983). "On Obtaining Permutation Distributions in Polynomial Time," *Journal of the American Statistical Association* **78** 435–440.

Palca, J. (1991). "Get-the-Lead-Out Guru Challenged," *Science* **253** 842–844.

Pankov, L. D. (1968). "Market Efficiency and Football Betting," *Journal of Business* **41** 203–214.

Peter, J. B. and Hawkins, B. R. (1981) "HLA Antigens and Disease: What the Associations Signify for Diagnosis" *Diagnostic Medicine*, January/February 1981, 54–65.

Peto, R., Pike, M., Day, N., Gray, R., Lee, P., Parish, S., Peto, J., Richards, S. and Wahrendorf, J. (1980). "Guidelines for Simple, Sensitive Significance Tests for Carcinogenic Effects in Long-Term Animal Experiments," in *Long-Term and Short-Term Screening Assays for Carcinogens: A Critical Appraisal*, IARC Monographs, Annex to Supplement 2, 311–426. International Agency for Research on Cancer, Lyon.

Petrondas, D. A. and Gabriel, K. R. (1983). "Multiple Comparisons by Randomization Tests," *Journal of the American Statistical Association* **78** 949–957.

Pitman, E. J. G. (1938). "Significance Tests Which May Be Applied to Samples from Any Population III. The Analysis of Variance Test," *Biometrika* **29** 322–335.

Pocock, S. J., Geller, N. L. and Tsiatis, A. A. (1987). "The Analysis of Multiple Endpoints in Clinical Trials," *Biometrics* **43** 487–498.

Rao, C. R. (1973). *Linear Statistical Inference and Its Applications*, 2nd Edition, Wiley, New York.

Rice, W. R. (1988). "A New Probability Model for Determining Exact *P* Values for 2×2 Contingency Tables" (with discussion), *Biometrics* **44** 1–22.

Ries, P. N. and Smith, H. (1963). "The Use of Chi-Square for Preference Testing in Multidimensional Problems," *Chemical Engineering Progress* **59** 39–43.

Rom, D. M. (1990). "A Sequentially Rejective Test Procedure Based on a Modified Bonferroni Inequality," *Biometrika* **77** 663–665.

Romano, J. P. (1989). "Bootstrap and Randomization Tests of Some Nonparametric Hypotheses," *Annals of Statistics* **17** 141–159.

Rosenthal, R. (1979). "The 'File Drawer' Problem and Tolerance for Null Results," *Psychological Bulletin* **86** 638–641.

Rosenthal, R. (1990). "An Evaluation of Procedures and Results," Chapter 11 in *The Future of Meta-Analysis*, K. Wachter and M. Straf (Eds.), Russell Sage Foundation, New York.

Rubinstein, R. Y. (1981). *Simulation and the Monte Carlo Method*, Wiley, New York.

Russo, B., Gandar, J. M. and Zuber, R. A. (1989). "Market Rationality Tests Based on Cross-Equation Restrictions," *Journal of Monetary Economics* **24** 455–470.

Ryan, T. A. (1960). "Significance Tests for Multiple Comparison of Proportions, Variances, and Other Statistics," *Psychological Bulletin* **57** 318–328.

SAS Institute, Inc. (1990). *SAS® Language: Reference, Version 6, First Edition*, SAS Institute Inc., Cary, North Carolina.

Satterthwaite, F. E. (1946). "An Approximate Distribution of Estimates of Variance Components," *Biometrics Bulletin* **2** 110–114.

Saville, D. J. (1990). "Multiple Comparison Procedures: The Practical Solution," *The American Statistician* **44** 174–180.

Schweder, T. and Spjøtvoll, E. (1982). "Plots of P-Values to Evaluate Many Tests Simultaneously," *Biometrika* **69** 493–502.

Shafer, G. and Olkin, I. (1983). "Adjusting P Values to Account for Selection Over Dichotomies," *Journal of the American Statistical Association* **78** 674–678.

Shaffer, J. P. (1986). "Modified Sequentially Rejective Multiple Test Procedures," *Journal of the American Statistical Association* **81** 826–831.

Shorack, G. R. (1982). "Bootstrapping Robust Regression," *Communications in Statistics—Theory and Methods* **11** 961–972.

Šidák, Z. (1967). "Rectangular Confidence Regions for the Means of Multivariate Normal Distributions," *Journal of the American Statistical Association* **62** 626–633.

Siddiqui, M. M. (1967). "A Bivariate t Distribution," *Annals of Mathematical Statistics* **38** 162–166.

Simes, R. J. (1986). "An Improved Bonferroni Procedure for Multiple Tests of Significance," *Biometrika* **73** 751–754.

Simon, J. L. and Bruce, P. (1991a). "RESAMPLING STATS: Probability and Statistics a Radically Different Way," 612 N. Jackson St., Arlington, VA 22201.

Simon, J. L. and Bruce, P. (1991b). "Resampling: A Tool for Everyday Statistical Work," *Chance* **4(1)** 22–32.

Singh, K. (1981). "On the Asymptotic Accuracy of Efron's Bootstrap," *Annals of Statistics* **9** 1187–1195.

Smith, H., Gnanadesikan, R. and Hughes, J. B. (1962). "Multivariate Analysis of Variance (MANOVA)," *Biometrics* **18** 22–41.

Smouse, P. E. and Williams, R. C. (1982). "Multivariate Analysis of HLA–Disease Association," *Biometrics* **38** 757–768.

Snedecor, G. W. and Cochran, W. G. (1989). *Statistical Methods*, 8th Edition, Iowa State University Press, Ames, Iowa.

Soper, K. A. and Westfall, P. H. (1990). "Monte Carlo Estimation of Significance Levels for Carcinogenicity Tests Using Univariate and Multivariate Models," *Journal of Statistical Computation and Simulation* **37** 189–209.

Spurrier, J. D. and Isham, S. P. (1985). "Exact Simultaneous Confidence Intervals for Pairwise Comparisons of Three Normal Means," *Journal of the American Statistical Association* **80** 438–442.

Staudte, R. G. and Sheather, S. J. (1990). *Robust Estimation and Testing*, Wiley-Interscience, New York.

Steel, R. G. D. and Torrie, J. H. (1980). *Principles and Procedures of Statistics: A Biometrical Approach*, 2nd Edition, McGraw-Hill, New York.

Stoline, M. R. (1983). "The Hunter Method of Simultaneous Inference and its Recommended Use in Applications Having Large Known Correlation Structures," *Journal of the American Statistical Association* **78** 366–370.

Strawderman, R. and Mehta, C. (1990). "On the Validation of Exact Tests in Software Packages," unpublished manuscript.

Svejgaard, A., Plats, P., Ryder, L. P., Nielsen, L. S., and Thomsen, M. (1975). "HL-*A* and Disease Associations–a Survey," *Transplantation Reviews* **22** 3–43.

Taguchi, G., and Wu, Y. (1980). *Introduction to Off-Line Quality Control*, Central Japan Quality Association, Nagoya, Japan.

Tarone, R. E. (1990). "A Modified Bonferroni Method for Discrete Data," *Biometrics* **46** 515–522.

Tukey, J. W. (1953). "The Problem of Multiple Comparisons," unpublished manuscript.

Tukey, J. W. (1977). "Some Thoughts on Clinical Trials, Especially Multiplicity," *Science* **198** 679–684.

Tukey, J. W. (1988). "Randomization and Rerandomization: The Wave of the Past in the Future," unpublished manuscript.

Tukey, J. W. (1991). "The Philosophy of Multiple Comparisons," *Statistical Science* **6** 100–116.

Upton, G. J. G. (1982). "A Comparison of Alternative Tests for the 2 × 2 Comparative Trial," *Journal of the Royal Statistical Society, Ser. A* **148** 86–105.

Utts, J. (1991). "Replication and Meta-Analysis in Parapsychology," *Statistical Science* **6** 363–378.

Weinberg, A. M. and Storer, J. B. (1985). "Ambiguous Carcinogens and Their Regulation," *Risk Analysis* **5** 151–156.

Welsch, R. E. (1977). "Stepwise Multiple Comparison Procedures," *Journal of the American Statistical Association* **72** 566–575.

Westfall, P. H. (1985). "Simultaneous Small-Sample Multivariate Bernoulli Confidence Intervals," *Biometrics* **41** 1001–1013.

Westfall, P. H. and Lin, Y. (1988). "Estimating Optimal Continuity Corrections in Run Time," *Proceedings of the Statistical Computing Section*, American Statistical Association, New Orleans, 297–298.

Westfall, P. H., Lin, Y. and Young, S. S. (1990). "Resampling-Based Multiple Testing," *SAS Users Group International Conference Proceedings* **15** 1359–1364.

Westfall, P. H. and LuValle, M. J. (1985a). "Estimating the Simultaneous Confidence Level of Multivariate Normal Intervals," *American Statistical Association Proceedings of the Statistical Computing Section* 271–274.

Westfall, P. H. and LuValle, M. J. (1985b). "Using Estimated Correlations in the Construction and Evaluation of Rectangular Confidence Regions," Unpublished Manuscript, Westfall: Contact Author.

Westfall, P. H. and Rom, D. (1990). "Bootstrap Step-Down Testing with Multivariate Location Shift Data," unpublished manuscript, Westfall: Contact Author.

Westfall, P. H. and Young, S. S. (1989a). "*P*-Value Adjustments for Multiple Tests in Multivariate Binomial Models," *Journal of the American Statistical Association* **84** 780–786.

Westfall, P. H. and Young, S. S. (1989b). "*P*-Value Adjustments for Multiple Tests in Multivariate Binomial Models," *SAS Users Group International Conference Proceedings* **14** 1381–1384.

Wilcox, R. R. (1987). *New Statistical Procedures for the Social Sciences: Modern Solutions to Basic Problems*, Lawrence Erlbaum Associates, New Jersey.

Wilkinson, L. (1988). *SYSTAT: The System for Statistics*, SYSTAT, Inc., Evanston, Illinois.

Wishart, J. (1938). "Growth-Rate Determination in Nutrition Studies with the Bacon Pig, and their Analysis," *Biometrika* **30** 16–28.

Wu, C. F. J. (1986). "Jackknife, Bootstrap and Other Resampling Methods in Regression Analysis" (with discussions by several authors), *Annals of Statistics*, **14** 1261–1350.

Yates, F. (1984). "Tests of Significance for 2 × 2 Contingency Tables" (with discussion), *Journal of the Royal Statistical Society*, Series A, **147** 426–463.

Young, S. S. (1989). "Are There Location/Cage/Systematic Nontreatment Effects in Long-Term Rodent Studies? A Question Revisited," *Fundamental and Applied Toxicology* **13** 183–188.

Young, S. S. and Gries, C. L. (1984). "Exploration of the Negative Correlation Between Proliferative Hepatocellular Lesions and Lymphoma in Rats and Mice—Establishment and Implications," *Fundamental and Applied Toxicology* **4** 632–640.

List of Algorithms

List of Examples

Example **Page**

Example Index

E = Example, § = Section.

Author Index

Subject Index

WILEY SERIES IN PROBABILITY
AND MATHEMATICAL STATISTICS

ESTABLISHED BY WALTER A. SHEWHART AND SAMUEL S. WILKS

Probability and Mathematical Statistics

ANDERSON · An Introduction to Multivariate Statistical Analysis, *Second Edition*
ARNOLD, BALAKRISHNAN, AND NAGARAJA · A First Course in Order Statistics
BARNETT · Comparative Statistical Inference, *Second Edition*
BERNARDO and SMITH · Bayesian Statistical Concepts and Theory
BHATTACHARYYA and JOHNSON · Statistical Concepts and Methods
BILLINGSLEY · Probability and Measure, *Second Edition*
BOROVKOV · Asymptotic Methods in Queuing Theory
BRANDT, FRANKEN, and LISEK · Stationary Stochastic Models
CAINES · Linear Stochastic Systems
CHEN · Recursive Estimation and Control for Stochastic Systems
CONSTANTINE · Combinatorial Theory and Statistical Design
COVER and THOMAS · Elements of Information Theory
*DOOB · Stochastic Processes
DUDEWICZ and MISHRA · Modern Mathematical Statistics
ETHIER and KURTZ · Markov Processes: Characterization and Convergence
FELLER · An Introduction to Probability Theory and Its Applications, Volume I, *Third Edition*, Revised; Volume II, *Second Edition*
FULLER · Introduction to Statistical Time Series
FULLER · Measurement Error Models
GIFI · Nonlinear Multivariate Analysis
GUTTORP · Statistical Inference for Branching Processes
HALD · A History of Probability and Statistics and Their Applications before 1750
HALL · Introduction to the Theory of Coverage Processes
HANNAN and DEISTLER · The Statistical Theory of Linear Systems
HEDAYAT and SINHA · Design and Inference in Finite Population Sampling
HOEL · Introduction to Mathematical Statistics, *Fifth Edition*
HUBER · Robust Statistics
IMAN and CONOVER · A Modern Approach to Statistics
KAUFMAN and ROUSSEEUW · Finding Groups in Data: An Introduction to Cluster Analysis
LARSON · Introduction to Probability Theory and Statistical Inference, *Third Edition*
LESSLER and KALSBEEK · Nonsampling Error in Surveys
MORGENTHALER and TUKEY · Configural Polysampling: A Route to Practical Robustness
MUIRHEAD · Aspects of Multivariate Statistical Theory
OLIVER and SMITH · Influence Diagrams, Belief Nets and Decision Analysis
*PARZEN · Modern Probability Theory and Its Applications
PILZ · Bayesian Estimation and Experimental Design in Linear Regression Models
PRESS · Bayesian Statistics: Principles, Models, and Applications
PUKELSHEIM · Optimal Design of Experiments
PURI and SEN · Nonparametric Methods in General Linear Models
PURI, VILAPLANA, and WERTZ · New Perspectives in Theoretical and Applied Statistics
RAO · Asymptotic Theory of Statistical Inference
RAO · Linear Statistical Inference and Its Applications, *Second Edition*
ROBERTSON, WRIGHT, and DYKSTRA · Order Restricted Statistical Inference

*Now available in a lower priced paperback edition in the Wiley Classics Library.

Probability and Mathematical Statistics (Continued)

ROGERS and WILLIAMS · Diffusions, Markov Processes, and Martingales, Volume II: Îto Calculus
ROHATGI · A Introduction to Probability Theory and Mathematical Statistics
ROSS · Stochastic Processes
RUBINSTEIN · Simulation and the Monte Carlo Method
RUZSA and SZEKELY · Algebraic Probability Theory
SCHEFFE · The Analysis of Variance
SEBER · Linear Regression Analysis
SEBER · Multivariate Observations
SEBER and WILD · Nonlinear Regression
SERFLING · Approximation Theorems of Mathematical Statistics
SHORACK and WELLNER · Empirical Processes with Applications to Statistics
STAUDTE and SHEATHER · Robust Estimation and Testing
STOYANOV · Counterexamples in Probability
STYAN · The Collected Papers of T. W. Anderson: 1943–1985
WHITTAKER · Graphical Models in Applied Multivariate Statistics
YANG · The Construction Theory of Denumerable Markov Processes

Applied Probability and Statistics

ABRAHAM and LEDOLTER · Statistical Methods for Forecasting
AGRESTI · Analysis of Ordinal Categorical Data
AGRESTI · Categorical Data Analysis
ANDERSON and LOYNES · The Teaching of Practical Statistics
ANDERSON, AUQUIER, HAUCK, OAKES, VANDAELE, and WEISBERG · Statistical Methods for Comparative Studies
ASMUSSEN · Applied Probability and Queues
*BAILEY · The Elements of Stochastic Processes with Applications to the Natural Sciences
BARNETT · Interpreting Multivariate Data
BARNETT and LEWIS · Outliers in Statistical Data, *Second Edition*
BARTHOLOMEW, FORBES, and McLEAN · Statistical Techniques for Manpower Planning, *Second Edition*
BATES and WATTS · Nonlinear Regression Analysis and Its Applications
BELSLEY · Conditioning Diagnostics: Collinearity and Weak Data in Regression
BELSLEY, KUH, and WELSCH · Regression Diagnostics: Identifying Influential Data and Sources of Collinearity
BHAT · Elements of Applied Stochastic Processes, *Second Edition*
BHATTACHARYA and WAYMIRE · Stochastic Processes with Applications
BIEMER, GROVES, LYBERG, MATHIOWETZ, and SUDMAN · Measurement Errors in Surveys
BLOOMFIELD · Fourier Analysis of Time Series: An Introduction
BOLLEN · Structural Equations with Latent Variables
BOX · R. A. Fisher, the Life of a Scientist
BOX and DRAPER · Empirical Model-Building and Response Surfaces
BOX and DRAPER · Evolutionary Operation: A Statistical Method for Process Improvement
BOX, HUNTER, and HUNTER · Statistics for Experimenters: An Introduction to Design, Data Analysis, and Model Building
BROWN and HOLLANDER · Statistics: A Biomedical Introduction
BUCKLEW · Large Deviation Techniques in Decision, Simulation, and Estimation
BUNKE and BUNKE · Nonlinear Regression, Functional Relations and Robust Methods: Statistical Methods of Model Building
CHATTERJEE and HADI · Sensitivity Analysis in Linear Regression
CHATTERJEE and PRICE · Regression Analysis by Example, *Second Edition*
CLARKE and DISNEY · Probability and Random Processes: A First Course with Applications, *Second Edition*

Continued on back end papers

*Now available in a lower priced paperback edition in the Wiley Classics Library.

Applied Probability and Statistics (Continued)

COCHRAN · Sampling Techniques, *Third Edition*
*COCHRAN and COX · Experimental Designs, *Second Edition*
CONOVER · Practical Nonparametric Statistics, *Second Edition*
CONOVER and IMAN · Introduction to Modern Business Statistics
CORNELL · Experiments with Mixtures, Designs, Models, and the Analysis of Mixture Data, *Second Edition*
COX · A Handbook of Introductory Statistical Methods
*COX · Planning of Experiments
CRESSIE · Statistics for Spatial Data
DANIEL · Applications of Statistics to Industrial Experimentation
DANIEL · Biostatistics: A Foundation for Analysis in the Health Sciences, *Fifth Edition*
DAVID · Order Statistics, *Second Edition*
DEGROOT, FIENBERG, and KADANE · Statistics and the Law
*DEMING · Sample Design in Business Research
DILLON and GOLDSTEIN · Multivariate Analysis: Methods and Applications
DODGE and ROMIG · Sampling Inspection Tables, *Second Edition*
DOWDY and WEARDEN · Statistics for Research, *Second Edition*
DRAPER and SMITH · Applied Regression Analysis, *Second Edition*
DUNN · Basic Statistics: A Primer for the Biomedical Sciences, *Second Edition*
DUNN and CLARK · Applied Statistics: Analysis of Variance and Regression, *Second Edition*
ELANDT-JOHNSON and JOHNSON · Survival Models and Data Analysis
FLEISS · The Design and Analysis of Clinical Experiments
FLEISS · Statistical Methods for Rates and Proportions, *Second Edition*
FLEMING and HARRINGTON · Counting Processes and Survival Analysis
FLURY · Common Principal Components and Related Multivariate Models
GALLANT · Nonlinear Statistical Models
GROSS and HARRIS · Fundamentals of Queueing Theory, *Second Edition*
GROVES · Survey Errors and Survey Costs
GROVES, BIEMER, LYBERG, MASSEY, NICHOLLS, and WAKSBERG · Telephone Survey Methodology
HAHN and MEEKER · Statistical Intervals: A Guide for Practitioners
HAND · Discrimination and Classification
HEIBERGER · Computation for the Analysis of Designed Experiments
HELLER · MACSYMA for Statisticians
HOAGLIN, MOSTELLER, and TUKEY · Exploratory Approach to Analysis of Variance
HOAGLIN, MOSTELLER, and TUKEY · Exploring Data Tables, Trends and Shapes
HOAGLIN, MOSTELLER, and TUKEY · Understanding Robust and Exploratory Data Analysis
HOCHBERG and TAMHANE · Multiple Comparison Procedures
HOEL · Elementary Statistics, *Fifth Edition*
HOGG and KLUGMAN · Loss Distributions
HOLLANDER and WOLFE · Nonparametric Statistical Methods
HOSMER and LEMESHOW · Applied Logistic Regression
IMAN and CONOVER · Modern Business Statistics
JACKSON · A User's Guide to Principle Components
JOHN · Statistical Methods in Engineering and Quality Assurance
JOHNSON · Multivariate Statistical Simulation
JOHNSON and KOTZ · Distributions in Statistics
Continuous Univariate Distributions—1
Continuous Univariate Distributions—2
Continuous Multivariate Distributions
JOHNSON, KOTZ, and KEMP · Univariate Discrete Distributions, *Second Edition*
JUDGE, GRIFFITHS, HILL, LÜTKEPOHL, and LEE · The Theory and Practice of Econometrics, *Second Edition*

*Now available in a lower priced paperback edition in the Wiley Classics Library.

Applied Probability and Statistics (Continued)

JUDGE, HILL, GRIFFITHS, LÜTKEPOHL, and LEE · Introduction to the Theory and Practice of Econometrics, *Second Edition*

KALBFLEISCH and PRENTICE · The Statistical Analysis of Failure Time Data

KASPRZYK, DUNCAN, KALTON, and SINGH · Panel Surveys

KISH · Statistical Design for Research

KISH · Survey Sampling

LAWLESS · Statistical Models and Methods for Lifetime Data

LEBART, MORINEAU, and WARWICK · Multivariate Descriptive Statistical Analysis: Correspondence Analysis and Related Techniques for Large Matrices

LEE · Statistical Methods for Survival Data Analysis, *Second Edition*

LePAGE and BILLARD · Exploring the Limits of Bootstrap

LEVY and LEMESHOW · Sampling of Populations: Methods and Applications

LINHART and ZUCCHINI · Model Selection

LITTLE and RUBIN · Statistical Analysis with Missing Data

MAGNUS and NEUDECKER · Matrix Differential Calculus with Applications in Statistics and Econometrics

MAINDONALD · Statistical Computation

MALLOWS · Design, Data, and Analysis by Some Friends of Cuthbert Daniel

MANN, SCHAFER, and SINGPURWALLA · Methods for Statistical Analysis of Reliability and Life Data

MASON, GUNST, and HESS · Statistical Design and Analysis of Experiments with Applications to Engineering and Science

McLACHLAN · Discriminant Analysis and Statistical Pattern Recognition

MILLER · Survival Analysis

MONTGOMERY and PECK · Introduction to Linear Regression Analysis, *Second Edition*

NELSON · Accelerated Testing, Statistical Models, Test Plans, and Data Analyses

NELSON · Applied Life Data Analysis

OCHI · Applied Probability and Stochastic Processes in Engineering and Physical Sciences

OSBORNE · Finite Algorithms in Optimization and Data Analysis

PANKRATZ · Forecasting with Dynamic Regression Models

PANKRATZ · Forecasting with Univariate Box-Jenkins Models: Concepts and Cases

RACHEV · Probability Metrics and the Stability of Stochastic Models

RÉNYI · A Diary on Information Theory

RIPLEY · Spatial Statistics

RIPLEY · Stochastic Simulation

ROSS · Introduction to Probability and Statistics for Engineers and Scientists

ROUSSEEUW and LEROY · Robust Regression and Outlier Detection

RUBIN · Multiple Imputation for Nonresponse in Surveys

RYAN · Statistical Methods for Quality Improvement

SCHUSS · Theory and Applications of Stochastic Differential Equations

SCOTT · Multivariate Density Estimation: Theory, Practice, and Visualization

SEARLE · Linear Models

SEARLE · Linear Models for Unbalanced Data

SEARLE · Matrix Algebra Useful for Statistics

SEARLE, CASELLA, and McCULLOCH · Variance Components

SKINNER, HOLT, and SMITH · Analysis of Complex Surveys

STOYAN · Comparison Methods for Queues and Other Stochastic Models

STOYAN, KENDALL, and MECKE · Stochastic Geometry and Its Applications

THOMPSON · Empirical Model Building

THOMPSON · Sampling

TIERNEY · LISP-STAT: An Object-Oriented Environment for Statistical Computing and Dynamic Graphics

TIJMS · Stochastic Modeling and Analysis: A Computational Approach

TITTERINGTON, SMITH, and MAKOV · Statistical Analysis of Finite Mixture Distributions

UPTON and FINGLETON · Spatial Data Analysis by Example, Volume I: Point Pattern and Quantitative Data

Applied Probability and Statistics (Continued)

UPTON and FINGLETON · Spatial Data Analysis by Example, Volume II: Categorical and Directional Data

VAN RIJCKEVORSEL and DE LEEUW · Component and Correspondence Analysis

WEISBERG · Applied Linear Regression, *Second Edition*

WESTFALL and YOUNG · Resampling-Based Multiple Testing: Examples and Methods for *p*-Value Adjustment

WHITTLE · Optimization Over Time: Dynamic Programming and Stochastic Control, Volume I and Volume II

WHITTLE · Systems in Stochastic Equilibrium

WONNACOTT and WONNACOTT · Econometrics, *Second Edition*

WONNACOTT and WONNACOTT · Introductory Statistics, *Fifth Edition*

WONNACOTT and WONNACOTT · Introductory Statistics for Business and Economics, *Fourth Edition*

WOOLSON · Statistical Methods for the Analysis of Biomedical Data

Tracts on Probability and Statistics

BILLINGSLEY · Convergence of Probability Measures

TOUTENBURG · Prior Information in Linear Models

Printed in the United States
52150LVS00001B/85-87

9 780471 557616